Engineering Thermodynamics of Thermal Radiation

For Solar Power Utilization

Ryszard Petela

New York   Chicago   San Francisco
Lisbon   London   Madrid   Mexico City
Milan   New Delhi   San Juan
Seoul   Singapore   Sydney   Toronto

The *McGraw·Hill* Companies

Cataloging-in-Publication Data is on file with the Library of Congress

1 2 3 4 5 6 7 8 9 0   DOC/DOC   1 6 5 4 3 2 1 0

ISBN 978-0-07-163962-0
MHID 0-07-163962-4

Sponsoring Editor	**Copy Editor**
Taisuke Soda	Lisa Maxwell Meyer
Editing Supervisor	**Proofreader**
Stephen M. Smith	Devdutt Sharma
Production Supervisor	**Art Director, Cover**
Pamela A. Pelton	Jeff Weeks
Acquisitions Coordinator	**Composition**
Michael Mulcahy	Aptara, Inc.
Project Manager	
Deepa Krishnan, Aptara, Inc.	

Printed and bound by RR Donnelley.

McGraw-Hill books are available at special quantity discounts to use as premiums and sales promotions, or for use in corporate training programs. To contact a representative, please e-mail us at bulksales@mcgraw-hill.com.

This book is printed on acid-free paper.

To my Grażyna

Contents

Preface

From the very beginning, humans have lived together with the cardinal laws of thermodynamics. Some of the laws—e.g., about temperature, conservation of energy, and the irreversibility of processes—were sensed intuitively since the days of ancient civilizations. However, according to our obtainable knowledge, only in the nineteenth and the beginning of the twentieth centuries were these three laws inspiringly articulated together with the other two laws added later in the last century. Today, the cardinal laws of thermodynamics are applied to more and more problems, on both the micro and macro scales of specific objects, and they often are formulated, sometimes unnecessarily, in a complex and sophisticated mathematical way—multidimensional, differential, vectorial, matrix, statistical, etc. Through inventing ways the science mainly develops applications of these old laws to explore newly arising problems or objects.

The reader should not expect to find any new cardinal discoveries described in the present book, but what will be found here is only application of some old laws for exploration of one of the most admirable natural phenomena—thermal radiation. In the continual quest for new energy sources, solar radiation, or other radiation, grows in significance and becomes more and more attractive because its utilization does not pollute the environment. However, we should not forget that besides solar radiation there are also other sources of thermal radiation, e.g., hot walls radiating at a temperature not as high as that of the sun but still significant enough to be considered in various processes, mostly industrial.

Therefore, the aim of this book is to study radiation from any arbitrary source. One specific case is cold radiation, well disclosed by exergy, which comes from remote cosmic space and is represented by the lower sky temperature which slightly differs from the temperature of bodies surrounding daily human existence on earth. However, although such a source exists potentially, it is still not practically considered because of the relatively small power available at such a small temperature difference.

It was the end of the 1950s when I tried for the first time to approach radiation from an exergy viewpoint. In those days exergy analysis was already relatively well advanced but applied only to the thermodynamics of substance. The exchange of information between researchers was not as good as it is today, and generally the communication gap between thermodynamic physicists and engineering thermodynamicists was visible.

For example, the works on entropy by Planck were not popular in engineering circles, and some distinguished scientists in engineering did not recognize the practical benefit of implementing entropy in the analysis of engineering processes related to radiation. Also, the Second Law of Thermodynamics was not commonly applied to radiation.

My own doctoral thesis in 1960, in which I derived the formulae for the exergy of radiation, was met with astonishment mixed with skepticism and only formally proceeded thanks to a few supportive individuals (e.g., S. Ochęduszko and J. Szargut). In 1964, I published in ASME a brief overview of the thesis, but it did not awaken any significant interest until years later in the late 1970s. Gradually, but still very slowly, interest grew to the relatively large focus that is noticed today due to the growth of the solar energy role.

My theory of radiation exergy seemed to me very simple and basic; therefore, from the beginning I tried to incorporate it into textbooks on either thermodynamics or heat transfer. But it was usually rejected as not fitting, neither to substance thermodynamics nor to engineering calculation of heat transferred by radiation. Time flew by, and only recently I came to the conclusion that radiation exergy could be the pivotal target in a new book defined around the area of the engineering thermodynamics of thermal radiation. Thus, the present book is proposed as an introduction to such an area.

Writing my book was also inspired by the solar power chapter in Bejan's outstanding book, *Advanced Engineering Thermodynamics*, which introduced thermodynamics in many new areas. The present book, however, is focused mainly only on radiation, which is an important part of overall thermodynamics. I assume that the reader is familiar with the fundamentals of engineering thermodynamics and radiative heat transfer, and only a brief outline of these areas is discussed here, mostly for comparison of the substance and photon gas. The book is addressed to the designers, users, and researchers of different devices or installations in which radiation—in particular, solar radiation—plays a role in generating heat, power, or green plants.

I will be grateful to readers for any comments and suggestions that could lead to improvement of the present book.

Ryszard Petela

CHAPTER 1

Introduction

1.1 Objective and Scope of This Book

Heat transfer books consider mainly the heat rate during conduction, convection, and radiation. However, radiation is distinguished from convection and conduction by the fact that it is not a phenomenon but a kind of matter and it has properties similar to a substance matter. In contrast, thermodynamics books consider mainly processes with substances, neglecting the presence of radiation.

The present book is not about radiant heat transfer, although it is discussed, but about the thermodynamics of a nonsubstantial medium—the radiation. Thus, the objective of the book is to fill the gap between most heat transfer and thermodynamics books and explores the thermodynamics of radiation matter, which recently has become an attractive source of energy. All the laws and thermodynamic aspects of substances are reconsidered with including of radiation. The working fluid in the considered thermodynamic systems can be either the substance or a photon gas.

However, because the radiation and substance processes occur mostly together, some brief background on the thermodynamics of substances has also been included here. Some processes in which radiation plays an important role, such as solar heating, a solar chimney power plant, photosynthesis, and the photovoltaic effect are analyzed as examples. Thermodynamic analyses of these processes are developed from both the energy and exergy viewpoints. A new element is introduced by including the exergetic influence of the terrestrial gravity field, which contributes to the buoyancy driven by solar heating.

An introduction to the present discussion of the thermodynamics of radiation will be similar to any general introduction to thermodynamics. Thermodynamics is a part of physics and addresses the energetic phenomena occurring in a collection of sufficiently large amounts of matter.

1

The three categories of thermodynamics are the following:

- *theoretical*—as a part of theoretical physics;
- *chemical* (founded by J. W. Gibbs)—concerning the interrelations of heat and work with chemical reactions; and
- *engineering* (considered in the present book)—concerning particularly the energy conversion processes occurring in numerous thermal installations, usually on an industrial scale.

In engineering thermodynamics single atoms, molecules, electrons, or photons are excluded from consideration. Engineering thermodynamics intentionally gives up complete precision in order to make the considerations simpler and more comprehensible.

The two methods for investigating energetic processes are called *phenomenological* and *statistical*. Phenomenological thermodynamics neglects the microstructure of matter, and the mechanisms are analyzed only on the basis of macroscopic results of experiments. Statistical thermodynamics, assuming the particle structure of matter, applies the methods of statistical mechanics and probability. Statistical thermodynamics, based on the determined microstructure, allows for the explanation and calculation of many thermodynamic macroconcepts such as the pressure or specific heat of a substance. However, the separate considerations of phenomenological and statistical thermodynamics are applied only in theoretical thermodynamics.

Energetic processes occur according to Nature's rules, which are known as the laws of thermodynamics. These laws are not, and cannot be, derived, but they have been articulated based on experiments and many years of observing Nature; never event discordant with the thermodynamics law was noticed. The following laws of thermodynamics can be articulated:

The *zeroth law* defines the concept of temperature.

The *first law* determines the conservation of energy, which is a feature of matter, whereas heat and work are examples of energy transfer.

The *second law* is quantitatively defined by entropy (R. Clausius's concept), which can be applied to determining the possible direction of any phenomenon. However, the application of the second law is limited to the phenomena occurring in a very large amount of matter.

The *third* law (Nernst's theorem), relating to chemical reactions, was formulated based on the second law and has no application to radiation. Planck's interpretation of the third law is used only in Paragraph 5.9 for comparison of a substance and radiation depending on the temperatures near absolute zero.

The rule often called the *fourth* law of thermodynamics concerns nonequilibrium thermodynamics and can be explained in further detail. Classical thermodynamics considers processes occurring through successive states of equilibrium. Instead of the name classical thermodynamics, the name *thermostatics* was proposed (but it was not accepted). In principle, application of the classical thermodynamic laws to nonequilibrium phenomena is possible only if thermodynamic equilibrium exists in the initial and final states. The increments of the thermodynamic functions occurring during the phenomenon can be then calculated because they depend only on these extreme states and do not depend on the transition path. For nonequilibrium phenomena, classical thermodynamics allows us to formulate only inequality relations instead of respective equations.

Generally, in Nature and in engineering processes the states of thermodynamic equilibrium do not occur and, moreover, the nonequilibrium phenomena most often do not occur alone. Usually any phenomenon, driven by any impulse, is reciprocated by other phenomena. For example, the temperature gradient in a gas mixture generates not only the heat conduction but also diffusion of gas components (due to the density gradient), or, in another example, the gradient of electrical potential causes not only the electric current but also the generation of heat due to electrical resistance.

Nonequilibrium phenomena are irreversible and are accompanied by energy dissipation that manifests itself by the growth of the overall entropy of the participating bodies. However, a reciprocated phenomenon restrains the entropy growth. If this phenomenon were to occur separately, then it would contradict the second law of thermodynamics. For example, spontaneous generation of the concentration gradient within a single-phase mixture of gases is not possible from the viewpoint of the second law of thermodynamics. However, the reciprocated phenomenon cannot occur separately but only in the presence of the other phenomenon. The joint phenomenon, comprised of all reciprocated phenomena, is irreversible.

Thus, in contrast to classical equilibrium thermodynamics, which is concerned with matter states in equilibrium, *nonequilibrium thermodynamics* is concerned with the thermodynamic systems of irreversible transformation processes, when the systems are time-dependent, usually not isolated, and continuously sharing energy with other systems. Consideration of such systems becomes difficult because, due to the possibility of fluctuations, the concepts in equilibrium thermodynamics such as entropy production, equipartition of energy, the definition of temperature, or predicting the heat transfer cannot be applied. The system behaves as a collection of component processes that are mutually dependent of one another according to the fourth law of thermodynamics described in the Onsager reciprocal relations.

One of the simplest examples of reciprocated processes is green plant vegetation, which reciprocates with the phenomenon of irreversible solar energy transferred from sun to earth. The energy dissipation is reciprocated by the accumulation of energy in the plants. The overall effect of these two phenomena is irreversible because the accumulation of energy does not fully compensate the entropy growth due to the irreversible solar energy transfer. However, the energy accumulation in the plants would be impossible without the reciprocation with the related phenomena of solar radiation.

In practice, recognition of the thermodynamic processes as being in equilibrium is usually sufficient, and nonequilibrium thermodynamics may not need to be considered. More information on nonequilibrium thermodynamics can be found, e.g., in books by de Groot and Mazur (1962, 1984) and Kuiken (1994).

The full scope of thermodynamics also embraces the law of conservation of substance (which is not valid for radiation and is applied only in nonnuclear processes), the equation of state and thermodynamic concepts such as parameters, functions, and so on.

Analysis of radiation problems, aside from knowledge of the thermodynamics laws, also requires specific assumptions and principles for radiation. Thus, the subject of this book is the set of additional rules applicable to typical processes involving both substantial and nonsubstantial working fluids (the radiation product).

Optics is not considered in this book.

The ability to emit radiation is a feature of substance. Only thermal radiation is considered in this book. Other types of radiation such as emission of light by certain materials when they are relatively cool— e.g., the phenomena of luminescence or chemiluminescence created as a low-temperature emission of light by a chemical or physiological process—are excluded from our analysis.

These types of radiation are in contrast to the radiation emitted from the sun or by incandescent bodies such as burning wood or coal, molten iron, or wire heated by an electric current. Luminescence may be seen, e.g., in neon and fluorescent lamps, television, radar, and X-ray fluoroscope screens. Luminescence can also be generated by organic substances such as luminol or the luciferins in fireflies and glowworms, or by natural electrical phenomena such as lightning or auroras. The practical value of luminescent materials lies in their natural ability to transform invisible forms of energy into visible light. In all these phenomena, light emission comes from the material at about room temperature, and so luminescence is often called "cold light." However, the so-called cold radiation coming at a low temperature (e.g., thermal radiation from the sky) belongs to the category of thermal radiation considered in this book. Thermal radiation appears when a substance has a temperature greater than absolute zero with the exception of the model substance, which, regardless of its temperature, is passive; it neither radiates nor absorbs radiation (a perfect mirror).

The structure of this book is similar to traditional books on engineering thermodynamics. After some description of the basic definitions, properties of substance, and radiation matter, the conservation of mass, thermodynamics laws, and efficiency definitions are discussed.

Based on the discussed properties of the photon gas, Chapter 6 presents the key derivation of the fundamental concept, which is the exergy of emission. The reference state for the calculation of radiation exergy is determined as the temperature of the environment, and we discuss the effects of variation in this temperature.

The exergy of any thermal radiation is the pivotal problem in radiation thermodynamics. Chapter 7 presents the derivation of exergy of arbitrary radiation characterized by an arbitrarily irregular spectrum.

Based on the discussion in earlier chapters, the reader is prepared for Chapter 9 in which we discuss the existing literature on the exergy of radiation and develop the critical analyses of theories given by other authors.

Chapters 10–13 present examples of analyses of real thermodynamic processes in which the substance and radiation take place jointly. The examples show the possibility of drawing conclusions both about the processes of solar radiation harvested by heating or power in a solar chimney power plant and also about the simplified interpretations of the photosynthesis and photovoltaic processes. Thermodynamic analyses are approached from the viewpoints of both energy and exergy.

The considerations are illustrated by quantitative calculation examples in which only the SI (metric) system of units is used. Some general data related to considered problems are presented in the Appendix. Nomenclature is given separately for each chapter at its end. The book concludes with an index of names and subjects.

1.2 General Thermodynamic Definitions

Various situations can be the focus of thermodynamic considerations, which, to be carried out, require the determination of the *system* representing the problem. The system is the part of space separated by an imagined shell called the *system boundary*. The system boundary is sometimes identified the other way around, e.g., a control volume or reference frame. The size and shape of the system space is arbitrary. The system can change in size, shape, and location.

The precise establishment of the system boundary is necessary for the correct balance of matter, energy, and exergy, as well as for determination of the overall entropy growth for the system. All determinations of the matter's parameters and fluxes entering and exiting the system should be determined at the place where they pass through the system boundary.

A system is said to be *closed* if no matter flows through the system boundary. The system is said to be *open* if matter flows through the system boundary. The rules of conservation of matter are discussed in Chapter 4. The *secluded* system is the case when there is no exchange of energy and matter through the system boundary.

The state of a system is determined by the values of the state parameters. These are the values of macroscopic magnitudes related to the system, which can be determined based on measurement—with no need, however, to know the history or future of the system. Examples of state parameters are temperature, pressure, volume, amount of substance, component velocity, coordinates of location in a field of external forces, and so on. To make sure that a magnitude is really the state parameter, one has to judge if an increase in the considered magnitude depends only on the initial and final state of this transformative increase. If this increase depends on the mode of system transformation, i.e., it depends on system history, then the considered magnitude is not a state parameter (e.g., heat and work).

The state parameters can be intensive or extensive. The *intensive parameters* (e.g., temperature, pressure, and specific volume of a substance) do not depend on the size of a system and do not change value after division of the uniform system into parts. The *extensive parameters* (e.g., volume, energy, entropy and exergy) depend on the system size and have an additive quality, i.e., the extensive parameter of the whole system is the sum of the extensive parameters of the system's parts (subsystems). *Extensive parameters* related to the unit amount of a substance are usually identified by the adjective "specific," e.g., specific enthalpy (J/kg), specific internal energy (J/kmol), and specific heat (J/kg K), although this latter, beside kg, is also related to the temperature, K.

Not all the state parameters can vary independently from one another. However, there can always be an established set of independent parameters from which, if they are known, it is sufficient to determine all other state parameters. The independent parameters can be selected arbitrarily; however, their number is limited. The state parameter that does not belong to the independent parameters is called the *state function*, e.g., the not directly measurable parameter of the internal energy, entropy or exergy.

To analyze a system, only essential independent parameters should be included in the consideration. The essential parameters for thermodynamic systems can be the *thermal parameters* such as temperature, pressure, or volume. Only in a special system should the independent *mechanical parameters* be considered, e.g., for determination of component velocities or system coordinates. A *uniform system* has the same value for intensive parameters at every point of the system.

Thermodynamic equilibrium is achieved in a secluded system spontaneously after a sufficiently long time. In an equilibrium state the state parameters are established as being constant. In general, thermodynamic equilibrium requires three equilibriums to be fulfilled—*mechanical* (equilibrium of forces), *thermal* (equality of temperature), and *chemical* (constant composition). In the state of thermodynamic equilibrium the number of independent state parameters is the smallest. If the system is not in thermodynamic equilibrium, then the univocal determination of some state parameters may be impossible. For example, in a system with intensive chemical reactions the temperature in the system cannot be univocally determined. As explained in Paragraph 1.1, the engineering thermodynamics considers only the states of equilibrium (thermostatics) and the phenomena of transformations from one to another equilibrium state.

CHAPTER 2

Definitions and Laws of Substance

2.1 Equation of State

Everything that has mass is called *matter*. *Mass* is a property of matter that determines momentum and gravitational interactions of bodies. Matter appears in substantial and nonsubstantial forms. *Substance* is matter for which the rest mass is not zero. Thus, substance is the macroscopic body composed of elemental particles (i.e., atoms, molecules). Matter for which the rest mass equals zero (e.g., a radiation photon) appears in the form of different fields; e.g., the field of electromagnetic waves (radiation), the gravity field, the surface tension field, and so on.

The significance of substance in engineering thermodynamics is that the substance amount expresses the number of particles participating in thermodynamic processes. Units of substance amount can be kg, kmol, or a standard cubic meter (defined by values of standard temperature and pressure). Substance can also be the object of a conservation equation. Nonsubstantial matter (sometimes called *field matter*) can also be considered as a component in processes of energy conversion; however, it does not fulfill the matter conservation equation.

The thermal parameters of substance can be determined in numerical value and units. The numerical value depends on the selected units. To obtain an easily imaginable number, the units multiplied by 10^n can be used, where n is an integer larger or smaller than 1. The name of a multiplied unit is created with use of the proper prefix (see Section A.1).

In practice, a gas that appears in nature consists of a large number of particles that are in continual motion. The particles translate (linear replacement), rotate, and can oscillate (the vibrations of atoms in the molecule). The particles have a volume and they interact with mutual attraction forces. Because the thermodynamic properties of real gases are complex, imagined models of gases are introduced to simplify the calculation of the thermal parameters of the gas.

One model of an *ideal gas* is a hypothetical gas in which molecules do not interact mutually, have a molecule volume equal to zero (i.e., the atoms are considered as material points), and are rigid (i.e., there is no oscillation within the molecules). A more general and less restricted model of an ideal gas (sometimes called a *semi-ideal gas*) allows for oscillation within the molecules. Thus, the atoms in the molecules of semi-ideal gas are mutually bonded elastically.

The relatively simple dependence between gas parameters such as pressure, temperature and volume can be derived for a semi-ideal gas, which obviously is also valid for an ideal gas. For a gas that is not at excessive density, the following experimentally established laws exist:

- Boyle's law, according to which, for a given mass of gas maintained at a constant absolute temperature T, the pressure p is inversely proportional to the volume V;

- Charles' and Gay-Lussac's law which states that for a given mass of gas held at constant pressure, the volume is directly proportional to the temperature.

These two experimental results conclude in the form of the relation $p \times V/T = $ constant, for a fixed mass of gas. It can be interpreted that the volume occupied by a gas, at a given pressure and temperature, is proportional to its mass. Thus, the constant $p \times V/T$ has also to be proportional to the mass of gas. Expressing mass μ in kmol, the *universal gas constant* R can be experimentally determined and the equation of state of an ideal gas or of semi-ideal gas at a not-excessive density is established as follows:

$$pV = \mu RT \tag{2.1}$$

The constant R has the same value for all gases, $R = 8314.3\,\text{J}/(\text{kmol K})$. If mass in equation (2.1) is expressed in kilograms, then R becomes the *individual gas constant* which has individual values for different gases. If the molar gas density $\rho_\mu = \mu/V$ is introduced to equation (2.1), then the following formula is obtained:

$$p = \rho_\mu RT \tag{2.1a}$$

It clearly results from equation (2.1a) that the thermodynamic state of a gas is determined completely by the arbitrarily chosen pair of possibly three parameters, either p and ρ_μ, p and T, or T and ρ_μ. The presented considerations are later compared to the considerations of the thermodynamic state of a photon gas.

2.2 State Parameters of Substance

2.2.1 Pressure

Pressure is defined as the force exerted by a fluid (e.g., liquid, gas, or radiation product) on an enclosed surface, divided by the area of the surface. The total pressure is the sum of both static pressure and dynamic pressure. Static pressure is measured by an apparatus that is motionless relative to a flowing fluid. Static pressure can be the absolute pressure when it is measured relative to a vacuum, or the relative pressure when it is measured from any pressure reference level (e.g., measured as a surplus above the atmospheric pressure). The dynamic pressure is equal to the kinetic energy of a fluid. As a principle, only absolute static pressure is used in thermodynamic equations.

Pressure can be calculated based on the *kinetic theory of an ideal gas*. Consider a gas in a cubical vessel (i.e., each edge has length L), the walls of which are perfectly elastic. Consider an ith molecule of velocity w_i and of mass m_i, which collides with the wall and rebounds with the same velocity w_i. Thus, the change ΔP in the particle's momentum is:

$$\Delta P = m_i w_i - (-m_i\ w_i) = 2\ m_i w_i \qquad (2.2)$$

Assume that the particle reaches the wall without striking any other particle on the way. The time required to cross the cube is L/w_i and the time required for the round trip is $2 \times L/w_i$. The number of collisions of the particle with the wall per unit time is $w_i/(2 \times L)$, and the rate at which the particle transfers momentum to the wall is:

$$2\ m_i w_i \frac{w_i}{2L} = \frac{m_i\ w_i^2}{L} \qquad (2.3)$$

The particles in the cube are moving entirely at random. There is no preference among the particles for motion along any one of the three coordinate directions. The classical statistical mechanics involves the equipartition theorem, which is a general formula of equal distribution and in relation to the gas parameters it concerns the different components of gas energy. According to the theorem, in a thermal equilibrium, energy is shared equally among its various forms and orientation directions. The theorem allows for quantitative predictions, and when applied to the molecules it states that the molecules in thermal equilibrium have the same average energy associated with each independent degree of freedom of their motion. To find the pressure p imparted to the wall by all the gas molecules, but traveling only perpendicularly to the wall, the momentum force represented, e.g., by the right-hand side of equation (2.3), has to be divided by three, also divided by the wall area L^2 and summed up for all the particles:

$$p = \frac{1}{3}\frac{m}{L^3} \sum_{i=1}^{i=N} w_i^2 \qquad (2.4)$$

where N is the total number of particles in the vessel. Introducing the number n of particles per unit volume, $n = N/L^3$, one obtains from (2.4):

$$p = \frac{1}{3}m\,n\frac{\sum_{i=1}^{i=N} w_i^2}{N} \qquad (2.5)$$

Taking into account that $m \times n$ is the gas density ρ, $m\times n = \rho$, and that the sum divided by N is the average value w^2 of the square velocities w_i^2, the following formula is received:

$$p = \frac{1}{3}\rho w^2 \qquad (2.6)$$

The result (2.6) was derived by neglecting the collisions between particles; however, it can be assumed that the result is true even when such collisions are considered. Based on the probability viewpoint, it is usually argued that despite the collision between particles during the numerous exchanges of velocities in the entire system, there are always certain molecules that collide with the considered wall, which correspond to certain other molecules exiting the opposite wall with the same momentum. In addition, the time spent during collisions is negligible compared to the time spent between collisions. Therefore, neglecting collisions between particles is only a convenience for mathematical derivation. A vessel of any shape can be selected for derivation. The cubic shape of the vessel above is also assumed to simplify calculations. The pressure exerted only on one wall was calculated; however, following Pascal's law for fluids, the same pressure is exerted on all walls and everywhere inside the vessel interior.

2.2.2   Temperature

Temperature is a state parameter that determines ability for heat transfer. The temperature T' of a body is higher than the temperature T'' of another body if after contact between the bodies the first one transfers heat to the second one. However, if the heat transfer does not appear between these bodies when separated from their surroundings, then between these bodies there is a thermal equilibrium and the bodies have the same temperature ($T' = T''$).

Maxwell formulated the following law regarding temperature, known as the *zeroth law of thermodynamics*. If three systems A, B, and C are in a state of respective internal thermal equilibrium, and systems A and B are in thermal equilibrium with system C, then systems A

and B are in mutual thermal equilibrium, i.e., they have the same temperature. This law is the basis for using thermometers for the measurement of temperature. Thus, thermometers allow for different systems for measuring temperature. As a principle, in thermodynamic equations the absolute temperature is given in kelvins (K). Another commonly used scale of temperature is the Celsius scale, where $t = T - 273.15$, where T is the absolute temperature. The value 273.15 is the absolute temperature for the triple point of water, which is the temperature at which the three phases (solid, liquid, and gas) of water can exist in equilibrium.

The temperature of a gas can be also measured based on the kinetic theory of an ideal gas. Each side of equation (2.6) can be multiplied by the volume V, and the product $V \times \rho$ represents the gas mass m:

$$pV = \frac{1}{3}mw^2 \qquad (2.7)$$

Equation (2.7) can be interpreted with the expression for kinetic energy $m \times w^2/2$:

$$pV = \frac{2}{3}\frac{mw^2}{2} \qquad (2.8)$$

which reveals that the right-hand side of equation (2.7) represents two-thirds of the total kinetic energy of the translation of the molecule. A mass m (kg) can also be expressed as the mass μ (kmol) according to the relation $\mu = m/M$, where M is the molecular weight of gas. Thus equation (2.8) changes to the form:

$$pV = \frac{2}{3}\frac{\mu Mw^2}{2} \qquad (2.9)$$

Combining the state equation (2.1) with (2.9):

$$\frac{1}{2}Mw^2 = \frac{3}{2}RT \qquad (2.10)$$

one obtains the result that the total translational kinetic energy per kmol of the molecules of an ideal gas is proportional to the temperature. Equation (2.10) can be considered as the definition of gas temperature on a kinetic theory basis or on a microscopic basis.

Let us divide each side of equation (2.10) by Avogadro's number N_0, which represents the number of molecules per kmol ($N_0 = 6.02283 \times 10^{26}$):

$$\frac{1}{2}\left(\frac{M}{N_0}\right)w^2 = \frac{3}{2}\left(\frac{R}{N_0}\right)T \qquad (2.11)$$

The ratio M/N_0 is the mass of a single molecule and the left-hand side of equation (2.11) represents the average translation kinetic energy of the molecule. On the other hand, the ratio R/N_0 is equal to the Boltzmann constant k:

$$k = \frac{R}{N_0} = 1.38053 \times 10^{-23} J/K \qquad (2.12)$$

which plays the role of the universal gas constant of the molecule.

2.3 Energy of Substance

Generally, *energy* is the ability to perform work and is determined for an arbitrarily chosen reference state, whereas *exergy* is also the ability to perform work, however, when the environment is assumed as the reference state. Work, heat, energy, and exergy are determined with the same units. Energy, or exergy, are functions of the state and do not depend on the history of matter for which energy, or exergy, is considered. Work and heat are phenomena, lasting longer or shorter, during which energy, or exergy, is transferred. Work and heat are not a function of the matter state and depend on the history of such phenomena.

In addition to the macroscopic components of energy, such as kinetic or potential, a substance has its *internal energy*, the components of which are:

- kinetic energy of translations and rotations of molecules
- energy of oscillations of atoms in molecules
- potential energy in the field of mutual attraction of molecules
- inner energy related to the possibility of chemical restructuring of molecules (called *chemical energy*)
- energy of electrons states
- nuclear energy

Internal energy does not depend on the velocity of the body and its location. Most often only changes in internal energy play a role; thus, the reference state for calculation of the internal energy has to be assumed and such reference can be chosen arbitrarily. Usually not all the mentioned components of internal energy vary in thermodynamic processes; thus the nonvarying components can be neglected. In engineering thermodynamics it is usually assumed that the considered components of the internal energy depend only on temperature, pressure, and volume, and only two of these three parameters can vary independently from each other.

Very convenient in engineering thermodynamic considerations is the concept of *enthalpy*, which is the formally incorrect combination of two different kinds of magnitudes—internal energy U and work ($p \times V$). This informality however, does not produce any erroneous consequences. The enthalpy H is defined as:

$$H = U + pV \tag{2.13}$$

The expression $p \times V$ represents "transportation" work required during exchanging substance with the considered system through the determined system boundary. One of the ways of exchanging energy with a system is the stream of substance passing the system boundary. The exchanged energy in such a way is determined by the enthalpy. The enthalpy can be also interpreted as the sum of the internal energy (U) of matter in a vacuum and of the required work ($p \times V$) on the environment to ensure the room for the matter.

In practice, especially important is the consideration of change in internal energy U of a fluid that absorbs heat Q and, at the same time, performs work W. The first law of thermodynamics applied to such process, which starts at parameters with subscripts 1 and ends with parameters with subscripts 2, takes the form:

$$Q_{1-2} = U_2 - U_1 + W_{1-2} \tag{2.14}$$

Equation (2.14) illustrates well that, as mentioned, work and heat are not forms of energy, because energy is a property of matter; thus, energy is a function of the matter state, whereas work and heat are phenomena that disappear. This—what is left after work or heat—is the changed value of the energy ($U_2 - U_1$) of the bodies that participated in the phenomena. The values of work W_{1-2} or heat Q_{1-2} depend not only on the initial and final states of the considered system but also on the path of the transition between states 1 and 2. Work and heat are not state functions, so, e.g., saying that a body contains heat would be incorrect.

The work W_{1-2} in equation (2.14) is *absolute work,* and the heat Q_{1-2} in the case of a real process occurring with friction comprises the friction heat that is absorbed by a fluid. If the internal energy is eliminated from equation (2.14) by using equation (2.13), then:

$$Q_{1-2} = H_2 - H_1 + W_{t,1-2} \tag{2.15}$$

where $W_{t,1-2}$ is the work interpreted as the *technical work.* Equations (2.14) and (2.15) can be presented, respectively, in differential form and using the specific (related to the unit of mass) magnitudes, e.g., as follows:

$$dq = du + dw \qquad (2.16)$$

$$dq = dh + dw_t \qquad (2.17)$$

2.4 Energy Transfer

2.4.1 Work

Work is one of the many ways of exchanging energy.

Mechanical *work* is defined as the scalar product of force and replacement. The force is taken as the projection on the replacement direction. A *force* gives to a mass its acceleration. The force is expressed in N (newton), which gives to 1 kg of mass the 1 m/s^2 acceleration. The force that is perceptible due to gravitational acceleration is called weight. *Power* in W (watt) is the ratio of work and time. Thermal power can be interpreted as a ratio of transferred heat and time.

The formula for the absolute work performed by a fluid is:

$$dW = p\,dV \qquad (2.18)$$

where p is the static absolute pressure of fluid and V is its volume.

The useful work W_u is defined in a case when the part of the absolute work is used for compression of the environment at pressure p_0:

$$W_{u,1-2} = W_{1-2} - p_0(V_2 - V_1) \qquad (2.19)$$

The same performed work can be interpreted also as a technical work:

$$dW_t = -V\,dp \qquad (2.20)$$

However, a certain generalization of the concept of work is required because, beside mechanical work, there also are other forms of work such as the work performed by an electrical current or a magnetic field, etc. In such cases, work has to be determined in a specific way. If work is the only way of interaction between a system and its environment, then the system is called *adiabatic*. Another case of the adiabatic system occurs if there is no heat transfer between a system and its surroundings.

Work performed by an adiabatic system causes a change in the energy function of the system. This change is equal to the difference of the system energy after and before performing work.

Performing work is one of the ways of energy exchange between systems. If the systems are closed then only two ways of energy transfer can occur—work and heat. Heat exchange occurs when the temperatures of the considered systems are different.

2.4.2 Heat

Heat as the process of exchanging energy with a system can be considered from different viewpoints. One viewpoint is the calculation of heat exchanged between different objects and this is the subject of many manuals on heat transfer, e.g., Holman (2009).

Another viewpoint, discussed shortly, is the calculation of heat absorbed by a substance. Heat Q_{1-2} absorbed by a body, during a change in its temperature from T_1 to T_2, is generally the sum of the heat delivered from outside and the heat of friction occurring within the body. The elemental amount of heat, dq, absorbed by the unit mass of the body increases appropriately its temperature:

$$dq = c\, dT \tag{2.21}$$

where c is the specific heat of the body. Equation (2.21) is solved either for a given function $c(T)$:

$$q_{1-2} = \int_{T_1}^{T_2} c\, dT \tag{2.22}$$

or for a known mean specific heat used as follows:

$$q_{1-2} = c\,\Big|_{T_1}^{T_2} (T_2 - T_1) \tag{2.23}$$

If during the heating process the substance changes its phase, equations (2.22) and (2.23) cannot be applied directly because the latent heat (at constant temperature) of the phase change has to be included.

The specific heat depends also on the kind of the heating process. Using (2.21) and (2.20) in (2.16):

$$c\, dT = du + p\, dv \tag{2.24}$$

The total differential du of the function $u(T, v)$, expressed as follows,

$$du = \left(\frac{\partial u}{\partial T}\right)_v dT + \left(\frac{\partial u}{\partial v}\right)_T dv \tag{2.25}$$

can be introduced to (2.24) and after division by dT,

$$c = \left(\frac{\partial u}{\partial T}\right)_v + \left[\left(\frac{\partial u}{\partial v}\right)_T + p\right]\left(\frac{\partial v}{\partial T}\right)_\pi \qquad (2.26)$$

where the partial derivative for a given kind π of the considered process is used. Generally, the specific heat c depends on the property of the body, can vary during varying parameters of the body, and depends also on the kind of process. For example, if the process occurs at constant volume (v = constant) then the second term of equation (2.26) is zero. Thus, the specific heat for constant volume is:

$$c_v = \left(\frac{\partial u}{\partial T}\right)_v \qquad (2.27)$$

For a semi-ideal gas the internal energy u does not depend on volume; thus:

$$\left(\frac{\partial u}{\partial v}\right)_T = 0 \qquad (2.28)$$

Interpreting (2.28) in (2.25):

$$u = c_v dT \qquad (2.29)$$

Further derivations can lead to the formulae:

$$h = c_p dT \qquad (2.30)$$

$$c_p - c_v = R \qquad (2.31)$$

The result is that specific heat at constant volume is used to calculate the increase of internal energy, whereas the specific heat at constant pressure allows for calculation of enthalpy. The difference of these two specific heats is equal to the individual gas constant R discussed earlier.

The ratio of specific heats

$$\frac{c_p}{c_v} = \kappa \qquad (2.32)$$

is equal to the isentropic exponent κ in the equation representing varying parameters during an ideal (no friction) adiabatic process (called the *isentropic process*). For example, parameters p and V in the isentropic process change according to the equation:

$$pV^\kappa = \text{const.} \qquad (2.33)$$

The relations discussed above will be compared to similar relations for radiation.

2.5 Entropy of Substance

Entropy, expressed in J/K, is a measure of thermodynamic probability of the system's disorder. Entropy can be interpreted from a macroscopic viewpoint (classical thermodynamics), a microscopic viewpoint (statistical thermodynamics), and an information viewpoint (information theory). The latter viewpoint differs from the concept of thermodynamic entropy and contributes to the mathematical theory of communication. The statistical definition of entropy is a basic definition from which the other two can be mathematically derived, but not vice versa. All properties of entropy, as well as the second law of thermodynamics, follow from this definition. In statistical thermodynamics the entropy S is defined as the number Ω of microscopic configurations that are possible in the observed macroscopic thermodynamic system:

$$S = k \ \ln \Omega \qquad (2.34)$$

where k is the Boltzmann constant discussed earlier.

In classical thermodynamics the entropy is derived from analysis of the heat engine generating work according to the theoretical model of the Carnot cycle, which is reversible and has the maximum possible efficiency:

$$\eta_C = \frac{T_I - T_{II}}{T_I} \qquad (2.35)$$

where T_I and T_{II} are the temperatures, respectively, of hot and cold heat sources available for the cycle. In Section 4.6 it will be proven that the Carnot efficiency expressed by formula (2.35) is independent of the working fluid; therefore, it can be also applied for a photon gas.

The entropy is a function of the thermodynamic state, as any other state function, with property depending only on the current state of the system and independent of how the state was achieved. Entropy S is defined as follows:

$$dS = \frac{dQ}{T} \qquad (2.36)$$

where T is the temperature at which the elemental amount dQ of heat is exchanged, whereas dS is a total differential of entropy. The integral of the total differential is the difference of function between the initial and final state, whereas, for comparison, the elemental amount is determined by integrating based on knowledge of the history between initial and final state. Entropy, introduced by Clausius, allows for quantitative application of the second law of thermodynamics.

Equation (2.36) can be developed by using, e.g., (2.1), (2.16), and (2.20), for consideration of a gas with specific values of the magnitudes

$$ds = c_p \frac{dT}{T} - R \frac{dp}{p} \qquad (2.37)$$

which after integration is

$$s_2 - s_1 = c_p \ln \frac{T_2}{T_1} - R \ln \frac{p_2}{p_1} \qquad (2.38)$$

Equation (2.36) for entropy can be simply integrated for heat Q taken from the heat source at constant temperature T:

$$S = \frac{Q}{T} \qquad (2.39)$$

2.6 Exergy of Substance

2.6.1 Traditional Exergy

Exergy is one of several thermodynamic functions of state. The functions make consideration easier, allow for interpretation of phenomena, and (most of them) have practical application in thermodynamic calculations. For example, enthalpy is used to determine the energy of exchanged matter with the system considered. Internal energy expresses the energy of the substance remaining within the system at the time of consideration. Entropy determines the thermodynamic probability of a given matter state. Exergy was introduced to express the practical energetic value of matter existing in the environment given by nature. This practical value is determined by the ability of matter to perform mechanical work. The work was selected as the measure not only due to the human inclination toward laziness but also because work represents the energy exchange at the unlimited level.

However, full utilization of the energetic value of matter to perform work within the determined environment could not occur without cooperation from the environment. For example, to utilize the energetic value of natural gas from its combustion, a certain amount of oxygen contained in the environment air has to be taken. To fully utilize compressed air at the temperature of an environment, depressurizing of the air has to occur at a constant temperature, and to keep this temperature steady, heat from the environment has to be taken. The full definition of exergy was given by Szargut as follows: *Exergy*

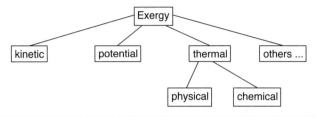

FIGURE 2.1 Traditional exergy components.

of matter is the maximum work the matter could perform in a reversible process in which the environment is used as the source of worthless heat and worthless substances, if at the end of the process all the forms of participating matter reach the state of thermodynamic equilibrium with the common components of the environment. The environment is the natural reference state in nature, which consists of an arbitrary amount of the "worthless" components. The components in the environment that have apparent energetic value, in limited amounts, are the exception, and are recognized as natural resources, e.g. natural fuels. The matter considered in the definition of exergy can be a substance or any field matter, e.g., radiation.

The total exergy of a substance is composed of components as shown schematically in Figure 2.1. Usually, only such components are used, which vary during the consideration. Most often used is the thermal exergy, which is the sum of physical and chemical exergies. The physical exergy results from the different temperature and pressure of the considered substance in comparison to its temperature and pressure in equilibrium with the environment (i.e., dead state). The chemical exergy results from the different chemical composition of the considered substance in comparison to the common substance components of the environment.

If the considered substance has significant velocity, then the kinetic exergy can be recognized as being equal to the kinetic energy calculated for the velocity relative to the environment. Potential exergy is equal to the potential energy if it is calculated for the reference level, which is the surface of the earth. The other possible components, e.g., nuclear or interfacial tension, are rarely used and are excluded from the present discussion. The sum of most important components in engineering thermodynamics, physical exergy B_{ph} and chemical exergy B_{ch}, is called thermal exergy B:

$$B = B_{ph} + B_{ch} \qquad (2.40)$$

The model shown in Figure 2.2 is used for derivation of the formula defining the drop of exergy $(-\Delta B)$ of a substance medium. The thermodynamic medium at enthalpy H_1 and entropy S_1 enters the machine,

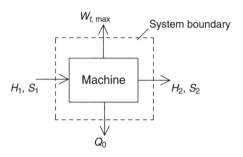

FIGURE 2.2 The model for calculation of the exergy drop.

which operates continuously, reversibly, and at a steady mode. The medium at enthalpy H_2 and entropy S_2 leaves the machine. The machine utilizes the environment as a source of heat Q_0 which is worthless because it is released at the temperature T_0 of environment. According to the definition of exergy the maximum technical work $W_{t,max}$ performed by the machine is equal to the exergy drop; $W_{t,max} = -\Delta B$. The exergy balance equation for the system boundary, shown in Figure 2.2, is:

$$-\Delta B = H_1 - H_2 - Q_0 \qquad (2.41)$$

The maximum work can occur only in the reversible process and according to the second law of thermodynamics when the overall entropy growth for the considered system is zero:

$$S_2 - S_1 + \frac{Q_0}{T_0} = 0 \qquad (2.42)$$

From equations (2.41) and (2.42) it results:

$$-\Delta B = H_1 - H_2 - T_0 (S_1 - S_2) \qquad (2.43)$$

If we interpret state 2 as the state of thermodynamic equilibrium with the environment ($H_2 = H_0$, and $S_2 = S_0$), state 1, as representing any state of the considered medium ($H_1 = H$, and $S_1 = S$), then, from (2.43), the general formula for the exergy of substance, ($B_1 - B_2 = B - 0 = B$) is obtained as:

$$B = H - H_0 - T_0 (S - S_0) \qquad (2.45)$$

where B is the thermal exergy of thermodynamic medium at enthalpy H and entropy S, and H_0 and S_0 are the enthalpy and entropy, respectively, of the medium in an eventual state of thermodynamic equilibrium within the parameters of the environment. The thermal exergy B expressed by equation (2.45) is for a substance passing through the system boundary. The exergy B of the substance can be positive or negative (e.g., for each medium flowing through the pipeline one can select the sufficiently low pressure at which thermal exergy is smaller than zero). However, the exergy B_s of any part of the system remaining within the system boundary is always positive. It is possible to derive that the exergy B_s of the substance remaining within the system is calculated as:

$$B_s = B - V(p - p_0) \tag{2.46}$$

From (2.46) there results a particular case when the space of volume V is empty, thus $p = 0$, as well as $B = 0$ because there is no substance. Then the exergy of the empty space is:

$$(B)_{P=0} = Vp_0 \tag{2.47}$$

A similar effect of the finite exergy of the empty vessel, as shown later by equation (6.19), is observed also for a photon gas.

2.6.2 Gravitational Interpretation of Exergy

The purpose of the concept of exergy is to develop a particular interpretation. However any interpretation is always characterized by a certain freedom; thus some modification of exergy can be justified. For example, consider the potential component of exergy. Among possible potentials that act on a substance, the potential exergy, shown in Figure 2.1, takes into account only the effect of the gravitational field. However, to fully reflect the effect of the Earth's gravity field, Petela (2008) proposed application of a new component, "mechanical exergy," which replaces traditional physical and potential components of exergy. Mechanical exergy can be called *eZergy*.

The mechanical exergy concept b_m is derived from the difference between the density ρ of the considered substance and the density ρ_0 of the environment. Regardless of the temperature T and the pressure p of the substance under consideration, the substance instability and, thus, ability to work in the environment at respective parameters T_0 and p_0 is sensed if either an anchor ($\rho < \rho_0$) or a supporting basis ($\rho > \rho_0$) is removed. In the first case, the substance moves upwards; in the other case, the substance sinks.

The altitude of the considered substance is measured from an actual level $x = 0$. In both cases the substance tends to achieve an equilibrium altitude ($x = H$), at which point the density of the local environment $\rho_{0,x}$ is equal to the density of the considered substance, $\rho = \rho_{0,x}$. The substance motion (at constant T and p) to reach the equilibrium altitude would generate work called the *buoyant exergy*, b_b. At level H the substance would be allowed to generate additional work, denoted by b_H, which would occur during the reversible process of equalization of parameters T and p with the respective local environment parameters, $T_{0,H}$ and $p_{0,H}$.

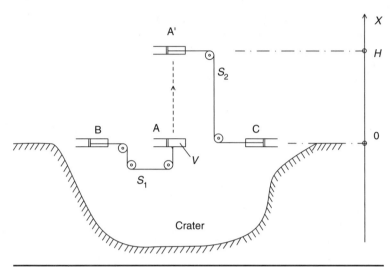

Figure 2.3 An ideal arrangement for discussing the exergy in a gravitational field (from Petela, 2008).

For example, Figure 2.3 represents a theoretical model for determining the exergy of a gas in a gravitational field. The gas of a volume V, a density ρ smaller than the density $\rho_{S,0}$ of the surrounding air ($\rho < \rho_{S,0}$), a temperature T, and a pressure p exists within an imagined ideal cylinder–piston system (system A), at an environmental altitude ($x = 0$), i.e., at the locally common level of the earth's surface.

The cylinder and piston walls at this stage provide perfect heat insulation (i.e., they do not allow any heat exchange), are hermetic (i.e., they do not allow exchange of substance), are rigid (i.e., they do not allow for changing the volume V), and are perfectly weightless (i.e., the mass of the cylinder and the piston is zero). System A is connected by a system S_1 of fixed pulleys and nonmaterial thread to the system B, containing also a cylinder and piston, and remaining always at the level $x = 0$. Due to buoyancy the considered gas lifts the whole system A upward to position A', at height H, at which the density ρ, remaining unchanged, is now equal to the density $\rho_{S,H}$ of the surroundings at the level H. During the lifting, work on system B is performed.

After taking position A' the system is now able to perform additional work during an ideal process of equalization of pressure p and temperature T with the respective parameters of surrounding air at the level H. Such equalization of parameters is possible because the restricting assumptions at location A' are now relieved and the considered substance can exchange worthless heat (at the surroundings temperature level $T_{S,0}$), and the piston can move within the cylinder to change appropriately the initial volume V.

Thanks to the system S_2 of the ideal pulleys-and-thread, the work is fully transferred to the cylinder–piston system C located at the environment level ($x = 0$). The above two portions of maximum work (in systems B and C) performed by the considered gas of initial volume V represent the full exergy of the gas as a result of the terrestrial gravitational field and relative to the environment level ($x = 0$).

By an appropriate rearrangement of systems S_1 and S_2 used in Figure 2.3, one can present also a scheme for consideration when the density ρ of the considered substance is larger than the density $\rho_{S,0}$ of the outside air ($\rho > \rho_{S,0}$) and the volume V moves downward to the imagined local crater. A liquid or solid substance usually has high density ($\rho \gg \rho_{S,0}$) and thus can represent the remarkable ability to work when falling down into an imagined crater or sinking into a sea. For example, theoretically, leveling a 3000-m high conical mountain with a 45° slope, located by the sea, can take over 10 years, releasing the power of about 1500 MW, with the additional benefit of acquiring new land.

The buoyant exergy b_b does not depend on the kind of substance. During repositioning of the substance from actual altitude $x = 0$ to $x = H$, the gravitational acceleration g_x is changing, e.g., decreasing with growing altitude x; thus:

$$b_b = \int\limits_{x=0}^{x=H} g_x \left(\frac{\rho_{0,x}}{\rho} - 1 \right) dx \tag{2.48}$$

Equation (2.48) is significant in the procedure of calculating the eZergy.

Example 2.1 In practice, the integral in formula (2.48) can be determined as follows. The formula is valid for any kind of considered substance determined by density ρ, whereas $\rho_{S,x}$ and g_x have to be determined for the atmospheric air. Petela (2008) proposed using the literature data for calculation of any required parameter (y) from the general linear approximation:

$$y = a + b H \tag{a}$$

where H is the altitude in meters, and the coefficients values a and b are shown in Table 2.1.

Thus, the height H can be calculated with appropriate substitution $x = H$ and $\rho_{S,x} = \rho$ in the approximation for the density with the respective coefficients a and b:

$$H = 9.973 \times 10^5 (1.217 - \rho) \tag{b}$$

Integrating of equation (2.48) leads to the solution composed of the two integrals, I_1 and I_2:

$$b_b = \frac{I_1 - I_2}{\rho} \tag{c}$$

Dependent variable y	Coefficients	
	a	b
$T_{S,x}$ (K)	288.16	−0.0065
$p_{S,x}$ (Pa)	100339.5	−9.699
$\rho_{S,x}$ (kg/m^3)	1.217	−9.973 × 10^{-5}
g_x (m/s^2)	9.7807	−3.086 × 10^{-6}

TABLE 2.1 Data for Approximation Formula

where:

$$I_1 \equiv \int_{x=0}^{x=H} g_x \rho_{S,x} dx = \frac{a_1 a_3}{a_4}(\rho - a_3) + \frac{a_1 a_4 + a_3 a_2}{2\,a_4^2}(\rho - a_3)^2 + \frac{a_2}{3\,a_4^2}(\rho - a_3)^3 \quad \text{(d)}$$

$$I_2 \equiv \int_{x=0}^{x=H} g_x \rho\, dx = \frac{a_1}{a_4}\rho\,(\rho - a_3) + \frac{a_2}{2\,a_4^2}\rho(\rho - a_3)^2 \quad \text{(e)}$$

and where the coefficients (according to Table 2.1) are:

$a_1 = 9.7807$ m/s^2 the gravitational acceleration assumed at the level $x = 0$,

$a_2 = -3.086 \cdot 10^{-6}$ 1/s^2 the rate of growth of gravitational acceleration with the growing level x,

$a_3 = 1.217$ kg/m^3 the density of environmental air assumed at the level $x = 0$,

$a_4 = -9.973 \cdot 10^{-5}$ kg/m^4 the rate of growth of air density with growing x.

Introducing (d) and (e) to (c):

$$b_b = -\frac{1}{a_4 \rho}\left[\frac{a_2}{6\,a_4}(\rho - a_3)^3 + \frac{a_1}{2}(\rho - a_3)^2\right] \quad J/kg \quad \text{(f)}$$

The calculated b_b from formula (f) is in J/kg. The values of b_b, in the more convenient unit kJ/kg, can be approximated by the third-order polynomial:

$$b_b = 164.186 - 357.258\,\rho + 253.398\,\rho^2 - 58.096\,\rho^3\, kJ/kg \quad \text{(g)}$$

Figure 2.4 shows the comparison of the values from formula (f) (points), to the values from formula (g) (solid line), and the characteristic equilibrium altitude H (dashed line) which is the linear function (b) of density ρ. With growing altitude H the buoyant exergy b_b decreases for heavy substances ($\rho > \rho_{S,0}$) and grows for light substances ($\rho < \rho_{S,0}$). The approximation (g) is acceptable for practical calculations although it is inconveniently imprecise in the vicinity of the density $\rho_{S,0}$ in which it can produce small negative values of b_b (not truly representing the real values which are always nonnegative). For example, for $\rho = \rho_{S,0}$ exact value from (f) is $b_b = 0$ but from formula (g) the value $b_b = -0.00918$ kJ/kg is obtained.

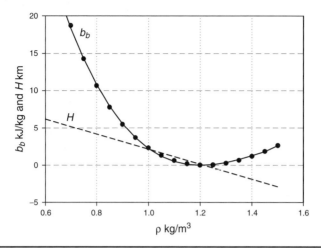

FIGURE 2.4 Specific exergy b_b and altitude H as function of density ρ (from Petela, 2008).

In case of a gas, during equalizing of the gas parameters T and p with the parameters $T_{0,H}$ and $p_{0,H}$ at the altitude H, the following work (exergy b_H) can be done:

$$b_H - c_p\,(T - T_{0,H}) - T_{0,H}\left(c_p \ln \frac{T}{T_{0,H}} - R \ln \frac{p}{p_{0,H}}\right) \qquad (2.49)$$

where c_p and R are specific heat at constant pressure and individual gas constant, respectively.

On the other hand, the gas at the actual altitude ($x = 0$) has the traditional physical exergy b, equal to the work that can be done by the gas during equalizing its parameters, T and p, with respective environment parameters T_0 and p_0:

$$b = c_p\,(T - T_0) - T_0\left(c_p \ln \frac{T}{T_0} - R \ln \frac{p}{p_0}\right) \qquad (2.50)$$

The definition of exergy postulates it to be the maximum possible work. Therefore, the larger work of the two, $b_b + b_H$ or b, is the true exergy, called the *mechanical exergy*; $b_m = \max[(b_b + b_H), b]$.

For some considerations Petela (2008) introduces also the term of gravitational exergy b_g of substance:

$$b_g = b_b + (b_H - b) \qquad (2.51)$$

which is the sum of the buoyant exergy and the difference between physical exergies for altitude $x = H$ and $x = 0$. Therefore, the

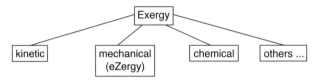

FIGURE 2.5   Exergy components including the mechanical exergy.

mechanical exergy can be expressed also as

$$b_m = \max[(b_g + b), b] \qquad (2.52)$$

or in the form of analytical equation:

$$b_m = \frac{\left(b_g + \sqrt{b_g^2}\right)(b_g + b) - \left(b_g - \sqrt{b_g^2}\right)b}{2\,b_g} \qquad (2.53)$$

where $b_g \neq 0$. If $b_g \leq 0$, then $b_m = b$.

Analogically to mechanical exergy, Petela (2008) proposed also the correspondent concept of mechanical energy e_m. Including the mechanical exergy (eZergy) into consideration, the scheme of the all components of the exergy of substance can be shown in Figure 2.5.

The application of mechanical exergy in the classic exergy balance equation has some implications that will be discussed in Section 4.5.4. It is worth emphasizing that both exergy and eZergy are only interpretative concepts. In contrast to exergy, eZergy allows introduction of the additional factor of gravity into the considered process. Therefore, disclosure of the gravity input requires application of the eZergy concept, i.e., the mechanical exergy B_m. eZergy is applied only for the substance (not for heat or radiation) and replaces the two traditional exergy components—physical (B_{ph}) and potential (B_p). Thus, $B_m \equiv Z = f(B_{ph}, B_p)$. To better distinguish exergy of substance from eZergy of substance, different symbols will be used: B for exergy and Z for eZergy (as shown in Chapter 11).

2.6.3   Exergy Annihilation Law

In realty, there is no exergy conservation equation. Exergy can be conserved only in ideal processes (e.g., model processes), which are reversible because they occur without friction at infinitely small differences of concentration and temperature. All real processes occur irreversibly; the energy is dissipated and thus the processes are accompanied by unrecoverable exergy loss.

The exergy loss caused by irreversibility of the process can be determined by comparison of operation of real and ideal installations for which the initial and final states of the driving medium are

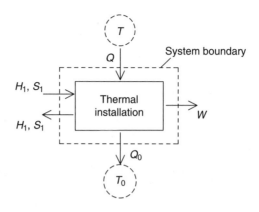

FIGURE 2.6 Thermal installation performing work.

respectively the same. Figure 2.6 shows the scheme of thermal installation, the purpose of which is performing work W. The type of analyzed processes does not affect the results. The installation receives a thermodynamic medium of enthalpy H_1 and entropy S_2, which leaves the installation at the enthalpy H_2 and entropy S_2. The installation absorbs valuable heat Q from the source at temperature T, which differs from environment temperature T_0. At the same time the installation extracts worthless heat Q_0 to the environment.

If the installation is considered to be real, then the energy conservation equation can be written as:

$$W = Q + H_1 - H_2 - Q_0 \qquad (2.54)$$

To have the ideal installation comparable to the real one, the same heat Q and enthalpies H_1 and H_2 should be considered. Work performed by ideal installation is maximum, W_{max}, and it requires changing the real heat Q_0 to a value $Q_{0,i}$ for an ideal process. Thus, the energy conservation equation for ideal installation can be used in the form:

$$W_{max} = Q + H_1 - H_2 - Q_{0,i} \qquad (2.55)$$

The exergy loss ΔB caused by irreversibility of the real installation is equal to the loss of work ($W_{max} - W$), and such loss can be determined from equations (2.54) and (2.55) as follows:

$$\Delta B = W_{max} - W = Q_0 - Q_{0,i} \qquad (2.56)$$

According to the second law of thermodynamics, the overall entropy growth Π for all bodies participating in the considered process is larger than zero for real installation:

$$\Pi = -\frac{Q}{T} + S_2 - S_1 + \frac{Q_0}{T_0} \qquad (2.57)$$

and is zero ($\Pi = 0$) for the compared ideal process:

$$0 = -\frac{Q}{T} + S_2 - S_1 + \frac{Q_{0,i}}{T_0} \qquad (2.58)$$

From equations (2.57) and (2.58) is:

$$\Pi \, T_0 = Q_0 - Q_{0,i} \qquad (2.59)$$

Using equation (2.59) in (2.56) one obtains the formula:

$$\Delta B = \Pi \, T_0 \qquad (2.60)$$

which is known as the Gouy–Stodola law expressing the law of exergy annihilation due to irreversibility. Although the sum Π depends on temperature T_0 the minimum of Π corresponds to minimum ΔB.

The exergy loss expressed by formula (2.60) is called the *internal exergy loss*, because it occurs within the considered system. This loss is totally nonrecoverable. Internal loss of exergy for multicomponent system is calculated by summing up the internal loss of exergy occurring in the particular system components.

Each exergy loss contributes to an increase of the consumption of the energy carrier that sustains the process or to the reduction of the useful effects of the process. One of an engineer's principal tasks is operating the process in such a way that the exergy loss is kept at the minimum. However, most often, the reduction of the exergy loss is possible only by increasing capital costs of the process. For example, the reduction of exergy loss in a heat exchanger is reachable by increasing the surface area of the heat exchange. Therefore, the economy of such a reduction of exergy loss can be verified by economic calculations. The exergy analysis explains the possibilities of improvement of the thermal process; however, only economic analysis can finally motivate an improvement.

Usually, the thermal process releases one or more waste thermodynamic mediums (e.g. combustion products), of which the parameters are different from the respective parameters of such medium being in equilibrium with environment. The waste medium represents certain exergy unused in the process. Such exergy, if released into the environment, is destroyed due to irreversible equalization of the parameters of the waste medium with the parameters of the environmental components. The exergy loss of the system, caused in such a way, is called *external exergy loss* and its numerical value is equal to the exergy of the waste medium released by the system. External exergy loss is recoverable, at least in part, e.g., by utilization in another system.

2.6.4 Exergy Transfer During Heat and Work

By releasing heat Q the exergy of the heat source diminishes. The exergy decrease B_Q of the source at temperature T is measured by the mechanical work that can be performed with the heat Q in the reversible engine according to the Carnot cycle for which, as the other required heat source, the environment at temperature T_0 would be utilized. Therefore:

$$B_Q = \left| Q \frac{T - T_0}{T} \right| \tag{2.61}$$

where Q is the heat exchanged between the source and the considered system. The value B_Q can be recognized either as a positive input to the system at $T > T_0$ or a positive output from the system if $T < T_0$.

For convenience, the acceptable jargon expression "exergy of heat" can be used while having in mind the exact expression "change of exergy of a heat source" (exergy is a function of matter state, whereas heat is only a phenomenon; however, a heat source is recognized as matter).

Obviously, exergy of any work is directly equal to the work, because exergy is measured by performed work.

2.7 Chemical Exergy of Substance

In this section we outline the significance of the concept of *devaluation reaction* and the resulting concept of *devaluation enthalpy* used for calculation of chemical energy. Based on this concept, as discussed, e.g., by Szargut et al. (1988), the following quantities can be calculated: enthalpy devaluation of substance (appearing in the energy conservation equation), standard entropy of devaluation reaction (in the entropy considerations), and, consequently, chemical exergy of substance (appearing in the exergy balance equation).

In a chemical process, in contrast to a physical process, substances change and only the chemical elements remain unchanged. Therefore, to calculate the chemical energy of substances the reference substances have to be assumed.

Existing methods for determination of the chemical energy of substances differ mainly by the definition of reference substances. For example, in the *enthalpy formation method*, the reference substances are the chemical elements at standard temperature and pressure. In the *devaluation reaction method* the number of reference substances is the same; however, they are not the chemical elements but the devaluated substances (compounds or chemical elements most commonly appearing in the environment). For example, the reference substance of C is gaseous CO_2, for H it is gaseous H_2O, and for O it is just O_2. In any particular case, when a substance is composed only of C, O, H, N, and S, the devaluation enthalpy of the substance is equal to its calorific value.

Contrary to the devaluation enthalpies, the values of the enthalpy of formation are not practical. For example, the enthalpy of formation for C is zero and the enthalpy of formation of CO_2 is significantly different than zero (–394 MJ/kmol). However, the devaluation enthalpy of C is equal to the calorific value ~394 MJ/kmol, whereas the calorific value of CO_2 is zero (as it is the reference substance for C).

The reference substances for the devaluation enthalpy and chemical exergy are the same. Also, the reference temperature and pressure are the same. Thus, only the devaluation enthalpy method, contrary to the formation enthalpy method, allows for fair comparison of the values of chemical energy and chemical exergy.

For comparison, the chemical exergy of C is ~413 MJ/kmol and devaluation enthalpy (calorific value) of C is only ~394 MJ/kmol. Only the devaluation enthalpy method should be used in thermodynamic analysis, which simultaneously includes the energy and exergy aspects.

Devaluation enthalpy is determined based on the stoichiometric devaluation reaction for a substance. The devaluation reaction is a combination only of the considered substance and the various reference substances. A good example of a devaluation reaction is reaction of photosynthesis:

$$6\ H_2O + 6\ CO_2 \rightarrow C_6H_{12}O_6 + 6\ O_2 \qquad (2.62)$$

in which, beside the considered substance of sugar ($C_6H_{12}O_6$), only the reference substances appear: CO_2, H_2O and O_2.

The devaluation enthalpy, d_n, is calculated from the energy conservation equation for the chemical process in which substrates are supplied, and products are extracted, all at standard temperature and pressure.

The physical exergy b_{ph} of a substance, at the state determined by enthalpy H and entropy S, is calculated based on definition (2.45) in which H_0 and S_0 are the enthalpy and entropy of this substance at environment parameters T_0 and p_0 respectively.

However, calculation of the chemical exergy b_{ch} of a substance is more complex, depending on its composition and based on the devaluation reaction. The calculation procedure is discussed by Szargut and Petela (1965b) and Szargut et al. (1988), and the calculated standard values of the devaluation enthalpy and chemical exergy are tabulated. If the environment temperature T_0 differs from the standard environment temperature T_n, then, when using the standard data on d_n and b_n, the formula for the chemical exergy of condensed substances (solid or liquid) should be corrected as shown, e.g., for the specific chemical exergy $b_{ch,su}$ of sugar:

$$b_{ch,su} = b_{n,su} + \frac{T_n - T_0}{T_n}(d_{n,su} - b_{n,su}) \tag{2.63}$$

where the chemical exergy of sugar is determined based on the standard tabulated value $b_{n,su} = 2{,}942{,}570$ kJ/kmol.

If a substance has a temperature different from the surrounding environment, then a physical component of energy or exergy has to be included as shown, e.g., again for the physical exergy $b_{ph,su}$ of the sugar:

$$b_{ph,su} = c_{su}(T - T_0) - T_0\,c_{su}\,\ln\frac{T}{T_0} \tag{2.64}$$

Note as well that based on the devaluation reaction, the so-called standard entropy σ_n of the devaluation reaction can be determined. For example, again for the photosynthesis reaction, based on equation (2.62), the standard entropy of the devaluation reaction, $\sigma_{n,su}$, is:

$$\sigma_{n,su} = 6\left(s_{H_2O} + s_{CO_2}\right)_n - \left(s_{O_2}\right)_n - s_{n,su} \tag{2.65}$$

where s_{H_2O}, s_{CO_2}, and s_{O_2} are the absolute standard entropies of the respective gases. The stoichiometric factor of six results from equation (2.62). The above formulae presented for sugar are utilized in the consideration of the photosynthesis in Chapter 12.

Nomenclature for Chapter 2

A, B, C	different cylinder–piston systems
a	coefficient in Table 2.1
a_1, a_2, a_3, a_4	coefficients for calculation of b_b
B	exergy, J
b	specific exergy, J/kg
b	coefficient in Table 2.1
c	specific heat of substance, J/(kg K)

d_n	devaluation enthalpy, J/kg
e	specific energy, J/kg
g	gravitational acceleration, m/s^2
H	level height, m
H	enthalpy, J
h	specific enthalpy, J/kg
k	Boltzmann constant, $k = 1.38053 \times 10^{-23}$ J/K
L	length, m
M	molecular weight, kg/kmol
m	mass, kg
N	number of particles
N_0	Avogadro's number, $N_0 = 6.02283 \times 10^{26}$ molecule/kmol
n	number of molecules per unit volume, molecule/m^3
P	momentum, kg m/s
p	static absolute pressure, Pa
Q	heat, J
q	specific heat, J/kg
R	universal gas constant, J/(kmol K)
R	individual gas constant, J/(kg K)
S	entropy, J/K
s	specific entropy, J/(K kg)
T	absolute temperature, K
T_n	standard environment temperature, K
t	temperature, C
U	internal energy, J
u	specific internal energy, J/kg
V	volume, m^3
v	specific volume, m^3/kg
W	absolute work, J
w	velocity, m/s
w	specific work, J/kg
x	vertical coordinate, m
y	dependent variable
Z	eZergy, J

Greek

Δ	increment
η	efficiency
κ	isentropic exponent
μ	mass of gas, kmol
Π	overall entropy growth, J/K
ρ	mass density, kg/m^3
ρ_μ	molar mass density, kmol/m^3
σ	entropy of the devaluation reaction, J/(K kmol)
Ω	number of microscopic configurations

Subscript

b	buoyant
C	Carnot
ch	chemical
g	gravitational
H	level at $x = H$
i	successive number
i	ideal
m	mechanical
max	maximum
n	standard
p	potential
p	constant pressure
ph	physical
S	system
S	surroundings
SU	sugar
S_1, S_2	different mechanical arrangements (systems)
T	constant temperature
t	technical
u	useful
v	constant volume
x	coordinate
π	any process
0	environment
$1, 2$	denotation
I, II	denotation

CHAPTER 3

Definitions and Laws of Radiation

3.1 Radiation Source

Radiation—caused only by the fact that a radiating body has a temperature higher than absolute zero—can be considered from many different viewpoints. According to its purpose, radiation can be considered to be electromagnetic waves or a collection of radiation energy quanta, i.e., photons, which are the matter particles.

The photons constitute a so-called photon gas. Therefore, analogously to a substance gas, a photon gas can be the subject of statistical (microscopic) or phenomenological (macroscopic) consideration.

Energy supplied to a body, e.g., by heating, sustains oscillations of atoms in molecules that then become like the emitters of electromagnetic waves. At expend of internal energy or enthalpy of the body substance the energy propagates from the body via the waves in a process called *thermal radiation*. The terms *radiation* and *emission* are two homonyms and can be used not only for the process but also for the product of the radiation or emission process, respectively, i.e., the collection of emitted energy quanta or photon gas. The product of radiation is comprised of matter, the rest mass of which, in contrast to a substance, is equal to zero.

According to the Prevost law, a body at a temperature greater than absolute zero radiates energy that can differ depending on different types of body substance, surface smoothness, and temperature. The energy of this radiation does not depend on the parameters, properties, or presence of neighboring bodies.

The different bodies also absorb oncoming radiation in different amount. Thus, energy exchange by radiation depends on the difference in emitted and absorbed radiation. For example, if the energy emitted is greater than the energy absorbed, and the energy of the body is not supplemented, then the temperature of the body decreases.

Phenomenologically, heat exchange by radiation is interpreted as a transformation of internal energy (or enthalpy) into the energy of electromagnetic waves of thermal radiation, which then travels through the surrounding medium to another body, at which point the radiation energy transforms again to internal energy (or enthalpy). Statistically, heat exchange by radiation is defined as the transportation of energy by photons that emit from excited atoms and move until they are absorbed by other atoms.

Radiation energy is composed of electromagnetic waves of length theoretically from 0 to ∞. The length λ of the waves is correlated with the oscillation frequency ν and the speed of propagation c as follows:

$$\lambda \nu = c \tag{3.1}$$

The speed c_0 of the propagation of electromagnetic waves in a vacuum is largest: $c_0 = 2.9979 \times 10^8$ m/s. The ratio n of speed c_0 to the propagation speed c in a given medium

$$n = \frac{c_0}{c} \tag{3.2}$$

is called the *refractive index* and is always larger than 1. For gases, n is close to 1, but, e.g., for glass it is about 1.5.

In experimental investigations it is usually more convenient to measure the wavelength. In theoretical investigations, however, it is usually more convenient to use frequency, which does not change when radiation travels from one medium to another at different speeds.

The shorter are the wavelengths, the more penetrable are the waves. Figure 3.1 shows approximately some characteristic regions of the wavelengths. As the wavelength decreases, i.e., the frequency increases, the penetration of the radiation within the matter grows deeper and deeper. For example, X-rays at $\sim 10^{17}$ Hz (Hz $\equiv 1/s$) travel through the human body, finding only slight difficulty in penetrating bones. Gamma rays at $\sim 10^{22}$ Hz have no problem penetrating most substances including metals. Shields used against gamma rays are made of dense metals, e.g., lead. However, natural cosmic waves have far greater penetrating power than manmade gamma radiation and can pass through a thickness even of 2 m of lead. With increasing radiation frequency, the wavelength becomes very short in comparison to even the densest metal lattices. For extremely large frequencies

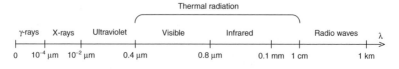

FIGURE 3.1   Scheme of the characteristic wavelength regions.

even the heaviest metals lose their shielding ability and are not able to reflect the radiation. With diminishing wavelengths the radiation energy decreases significantly. The shortest possible wavelength limit is equal to the so-called Planck's length, which corresponds to a frequency of 7.4×10^{42} Hz.

All the regions shown in Figure 3.1 overlap and, e.g., radiation of wavelength 10^{-3} m, can be produced either by microwave techniques (microwave oscillators) or by infrared techniques (incandescent sources). All these waves are electromagnetic and propagate with the same speed c_0 in a vacuum. The properties of radiation depend on their wavelengths. From the viewpoint of heat transfer, most essential are the rays that, when absorbed by bodies, cause a noticeable increase of energy of these bodies. The rays that indicate such properties at practical temperature levels are called *thermal radiation*.

An electromagnetic wave is said to be *polarized* if its electric field oscillates up and down along a single axis. For example, polarized radiation is comprised of the waves generated by a radio broadcasting with a vertical antenna, which makes the electric field point either up or down, but never sideways. The light from an electric bulb is an example of nonpolarized radiation: the radiating atoms are not organized. Such radiation arriving in the eyes can have, for a while a vertical electric field, but then it rotates around to horizontal, then back to vertical in random fashion. The radiation can be polarized, e.g., with use of a material such as Polaroid that absorbs radiation in one direction while transmitting radiation in the other direction. For example, Polaroid sunglasses can absorb horizontally polarized radiation emitted mostly from reflective surfaces such as glass, water, etc.

3.2 Radiant Properties of Surfaces

The principles of propagation, deflection, and refraction of visible rays are valid for all rays, thus also for all invisible rays.

An energy portion E from any surface, striking the considered body of finite thickness, splits into three parts as schematically shown in Figure 3.2. Generally, part E_ρ is reflected, part E_α is absorbed, and part E_τ can be transmitted through the body. The energy conservation equation for the portion E comes in the following form:

$$E = E_\rho + E_\alpha + E_\tau \qquad (3.3)$$

The parts can be expressed in relation to the portion E. Thus we have the definitions: reflectivity $\rho = E_\rho/E$, absorptivity $\alpha = E_\alpha/E$, and transmissivity $\tau = E_\tau/E$, where:

$$1 = \rho + \alpha + \tau \qquad (3.4)$$

The magnitudes ρ, α, and τ are dimensionless and can vary for different bodies from 0 to 1.

FIGURE 3.2 Split of emission energy E arriving at the considered body.

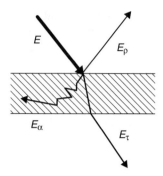

In practice, there are some bodies with different specific properties that make the characteristic magnitudes of equation (3.4) take values very close to 1 or 0. In order to systemize considerations, some idealized body models with extreme values of radiation are introduced.

If a body is able to totally absorb any radiation striking the body, i.e. $\alpha = 1$, and thus from equation (3.4) has the result $\rho = \tau = 0$, then such a perfectly absorbing body is called *perfectly black* (i.e., a *blackbody*).

If a body is able to totally reflect any radiation striking the body, then in such a case $\rho = 1$ and $\alpha = \tau = 0$, and the body is called perfectly *white*. If, due to the perfect smoothness of the surface, the reflection is not dispersed, i.e., the incident and reflection angles are identical (*specular reflection*), then the body is additionally called a *mirror*. However, if reflected radiation is dispersed in many directions (*diffuse reflection*), then the surface is called *dull*.

Monatomic gases (e.g., He, Ar) and diatomic gases (e.g., O_2, N_2) are examples of bodies that practically transmit total radiation. Such bodies can be considered as a model called perfectly *transparent* ($\tau = 1$), and from equation (3.4) we get $\alpha = \rho = 0$.

Some bodies are permeable only for waves of a determined length. For example, a window glass transmits only visible radiation and almost entirely does not transmit other thermal radiation. Quartz glass is also practically nontransmittable for thermal radiation except for visible and ultraviolet radiation.

Solid and liquid bodies, even of very small thickness, practically do not transmit thermal radiation. They can be considered as a model of perfectly *radiopaque* body for which $\tau = 0$ and:

$$\alpha + \rho = 1 \tag{3.5}$$

As the results from equation (3.5) show, the better a body reflects radiation, the worse it absorbs, and vice versa. The reflecting ability of thermal radiation can be significantly larger for smooth and polished surfaces in comparison to rough surfaces.

The body for which the reflecting and absorbing abilities are constant for any wavelength, (i.e., α_λ = const and ρ_λ = const) is called perfectly *gray*, and equation (3.5) is then:

$$\alpha_\lambda + \rho_\lambda = 1 \qquad (3.6)$$

The bodies with $\alpha_\lambda \neq$ const, which are commonly met, are sometimes called *varicolored* bodies.

Radiation occurring only at a certain value of frequency v or wavelength λ, i.e., within a narrow frequency band dv (or $d\lambda$), is called *monochromatic*, and the radiation occurring within some finite frequency (or wavelength) band is called *selective*. In comparison to *monochromatic*, the term *panchromatic*, rarely used, means relevancy to all wavelengths.

In reality there are no bodies that ideally fulfill the assumptions for the discussed models. Even black-looking soot has absorptivity $\alpha = 0.9$– 0.96, and thus is clearly smaller than 1. The perfectly gray surface does not exist in nature. The absorptive ability of real bodies is not constant for all wavelengths and temperatures. For example, the reflectivity of polished metals (which are good electrical conductors) is large and grows with increasing wavelength. However, the reflectivity of other technical materials (which are poor electrical conductors) is large for short waves and small for long waves. In spite of intense shining these materials have a significant ability for absorbing radiation.

3.3 Definitions of the Radiation of Surfaces

Emission E of a surface is the energy radiated at the temperature of the surface and emitted into the front hemisphere. The emission expressed in watts (W), related to the emitting surface area A, is called the *density of emission*:

$$e = \frac{E}{A} \qquad (3.7)$$

and is expressed in W/m^2.

However, generally, the radiation propagating from a considered surface can be composed of both the emission from such a surface and the radiation from other surfaces that are reflected by the considered surface. The particular radiation components can differ depending on their temperature. In energetic consideration of radiation, the temperature of such components is not distinguished and the total radiation (emission and reflected radiation) is called the *radiosity, J*. The radiosity is expressed in the same units as emission and, analogously, the radiosity density j is related also to the surface area:

$$j = \frac{J}{A} \qquad (3.8)$$

For a blackbody, which does not reflect radiation, the radiosity equals the emission ($J = E$). Usually the general term *radiation* can mean either emission or radiosity.

The exchange of radiation energy can occur between surfaces of different size and configuration. In calculations of the exchange between any two surfaces n and m, generally only a part of the radiation from surface n arrives at surface m. Therefore, one can use the view factor φ_{n-m}, which is defined as the ratio of the radiosity J_{n-m}, arriving from surface n at surface m, to the radiosity J_n leaving surface n:

$$\varphi_{n-m} = \frac{J_{n-m}}{J_n} \qquad (3.9)$$

The factor value can be within the range from 0 to 1. If each of the considered surfaces is uniform in terms of temperature and radiative properties, i.e., the density of radiosity is constant at every point of the respective surfaces, then the factor depends only on the location of both the surfaces in space and is sometimes called the *view factor*. However, if j is not the same at any point of the considered surface area A, then the radiosity density has to be considered locally ($j = dJ/dA$) as will be discussed later.

The density of emission e consists of the energy emitted at the wavelength λ from zero to infinity. The very small part de of the emission corresponds to the wavelength range $d\lambda$. Therefore, for the given wavelength the monochromatic density e_λ of emission is defined as follows:

$$e_\lambda = \frac{de}{d\lambda} \qquad (3.10)$$

The monochromatic emission density e_λ, W/m^3, depends on the wavelength, temperature, and radiative properties of the emitting surface. However, the model of a black surface has determined radiative properties and the monochromatic density $e_{b,\lambda}$ of emission of the black surface

$$e_{b,\lambda} = \frac{de_b}{d\lambda} \qquad (3.11)$$

is only a function of temperature and wavelength.

As shown in Figure 3.3, the total, i.e. panchromatic (for all wavelengths), emission density e_b of the black surface is represented by the

FIGURE **3.3** The area representing the elemental energy of emission.

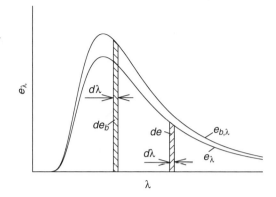

area under the $e_{b,\lambda}$ spectrum, whereas the total panchromatic emission of the gray surface corresponds to the smaller area, under the e_λ spectrum. The quantity e_b can be determined based on equation (3.11) by its integration over the whole range of wavelengths from 0 to ∞.

3.4 Planck's Law

Figure 3.4 shows the theoretical model of a blackbody, called the *cavity radiator*, which has played an important role in the study of radiation. The analysis of the nascent radiation in the model led to the birth of modern quantum physics.

The virtual model of the black surface (Figure 3.4) appears as a small hole in the wall embracing a certain space. Any radiation portion P entering the space through the hole is the subject of successive multiple deflections. Each deflection attenuates the portion P, especially when the interior is lined up with material with high absorptivity. It can be assumed that the portion P is entirely absorbed by the

FIGURE **3.4** Cavity radiator.

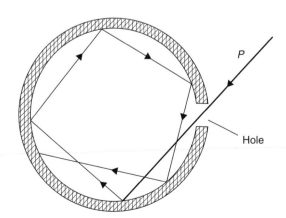

hole; therefore the hole behaves like a perfect blackbody ($\alpha = 1$). The radiosity of the hole does not contain any reflected radiation, but it represents the density of the emission e_b of a perfectly black surface. Thus, the density of black radiosity j_b of the hole is equal to the density of emission e_b of the black surface, $j_b = e_b$. The cavity space does not contain any substance; the refractive index $n = 1$. The emission density e_b expresses radiation energy emitted from the hole into the front hemisphere, i.e., within the solid angle 2π sr.

In 1900, Planck announced his hypothesis with a detailed model of the atomic processes taking place at the wall of the cavity radiator. The atoms that make up the cavity wall behave like tiny electromagnetic oscillators. Each oscillator emits electromagnetic energy into the cavity and absorbs electromagnetic energy from the cavity. The oscillators do not exchange energy continuously, but only in jumps called quanta $h\nu$, where ν is the oscillator frequency and h is Planck's constant, $h = 6.625 \times 10^{-34}$ J s.

Thus, in radiation processes there arise discrete quanta for which, if the principle of quantum-statistical thermodynamics is applied, the following expression can be derived for the energy density u_λ, J/m^4, of radiation per unit volume and per unit wavelength:

$$u_\lambda = \frac{8\pi h c_0}{\lambda \left(e^{\frac{hc}{\lambda k T}} - 1 \right)} \tag{3.12}$$

where $k = 1.3805 \times 10^{-23}$ J/K is the Boltzmann constant.

In order to obtain the radiation energy flux, i.e., the energy emission $e_{b,\lambda}$, instead of the radiation energy remaining within a certain volume, the energy density u_λ should be multiplied by the factor $c_0/4$ resulting from the geometrical considerations discussed, e.g., by Guggenheim (1957). Thus, based on the quantum theory, initially empirically and later proven theoretically, the Planck's formula for the black monochromatic emission density $e_{b,\lambda}$, can be established as follows:

$$e_{b\,\lambda} = \frac{c_1}{\lambda^5 \left(e^{\frac{c_2}{\lambda T}} - 1 \right)} \tag{3.13}$$

where

$$c_1 = 2\pi h c_0^2 = 3.74 \times 10^{-16}\,\text{Wm}^2 \quad \text{and}$$

$$c_2 = \frac{h c_0}{k} = 1.4388 \times 10^{-2}\,\text{m K}$$

are the first and the second, respectively, Planck's constants and T is the absolute temperature of black radiation. Figure 3.5 presents

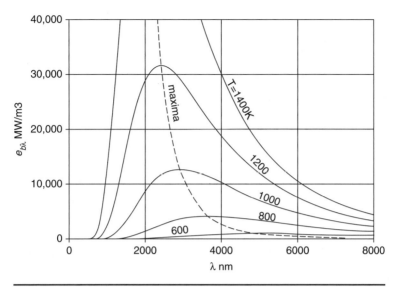

Figure 3.5 Monochromatic density of emission as a function of temperature and wavelength.

the curves of the black monochromatic density of emission $e_{b,\lambda}$ as a function of wavelength λ and for some different temperatures T. The higher is the temperature T, the larger is the area between the λ-axis and the respective curve. The dashed line in Figure 3.5 represents points of the maximum values of $e_{b,\lambda}$ and it shows that the higher is the temperature T the smaller is the wavelength λ_m corresponding to the maximum.

For the model of a perfectly gray surface it is assumed that the panchromatic emissivity ε, defined later by equation (3.22), is equal to the monochromatic emissivity ε_λ as follows:

$$\varepsilon = \frac{e}{e_b} = \varepsilon_\lambda = \frac{e_\lambda}{e_{b,\lambda}} \tag{3.14}$$

For comparison, Figure 3.6 presents four examples of the different surface spectra e_λ for the same temperature. The largest and always the maximum values of the spectrum appear for the black surface (dashed–dotted line). The real surfaces (solid line) have the smaller values of the monochromatic emission e_λ, (always $e_\lambda \leq e_{b,\lambda}$), which can be represented by the regular averaged curve (dashed line) corresponding to the appropriately selected model of a perfectly gray surface with a constant value of emissivity ε_λ. Thus, the spectra for the models of black and gray surfaces reach the maximum for the same wavelength. An entirely different type of spectrum can appear for a gas. The gas spectrum can be irregular (e.g., dotted line) so that application of the gray model is too inexact.

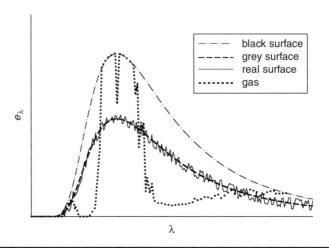

Figure 3.6 Examples of spectra of three surfaces; black, gray (at $\varepsilon = 0.6$), and real, compared to the spectrum of gas (H_2O), at the same temperature.

For some cases the Planck's formula (3.13) can be simplified to the two forms; each giving an error smaller than only 1%. First, if $\lambda \times T < 3000$ μm K, then $c_2/(\lambda T) \gg 1$ and the following formula derived by Wien, is obtained:

$$e_{b,\lambda} = \frac{c_1}{\lambda^5} e^{-\frac{c_2}{\lambda T}} \tag{3.15}$$

Second, if $\lambda \times T \gg c_2$, i.e., if $\lambda \times T > 7.8 \times 10^{-5}$ μm K, then expanding the expression in brackets in the denominator of equation (3.13)

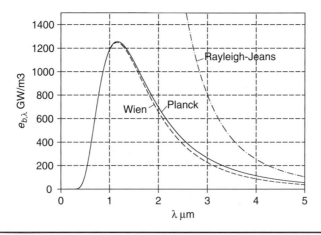

Figure 3.7 Comparison of $e_{b,\lambda}$ values for 2500 K.

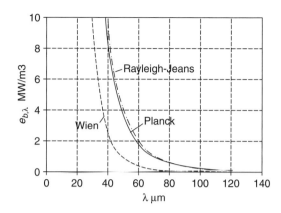

FIGURE 3.8
Comparison of $e_{b,\lambda}$ values for 1000 K.

in series:

$$e^{\frac{c_2}{\lambda T}} - 1 = \frac{c_2}{\lambda T} + \frac{1}{2!}\left(\frac{c_2}{\lambda T}\right)^2 + \cdots$$

and neglecting further terms, the Rayleigh–Jeans formula can be applied:

$$e_{b,\lambda} = \frac{c_1}{c_2}\frac{T}{\lambda^4} \tag{3.16}$$

The precision of the Wien formula (3.15), in comparison to Planck's formula (3.13), is illustrated in Figure 3.7 for $T = 2500$ K. The convergence for this temperature is better the smaller is the wavelength. The Rayleigh-Jeans formula (3.16) for the shown range of wavelength gives significantly inexact values.

The precision of the Rayleigh–Jeans formula (3.16) in comparison to the Planck's formula (3.13) is illustrated in Figure 3.8 for T = 1000 K. The convergence for this temperature is better the larger is the wavelength. The Wien formula (3.15) for the shown range of wavelength gives significantly inexact values.

3.5 Wien's Displacement Law

The wavelength λ_m, for which the spectrum of black emission reaches maximum, can be determined by considering the derivative of equation (3.13) as equal to zero:

$$\frac{de_{b,\lambda}}{d\lambda} = 0 \tag{3.17}$$

Introducing a new variable x as follows:

$$\lambda = \frac{c_2}{Tx}, \qquad d\lambda = -\frac{c_2}{Tx^2}dx$$

it could be written as:

$$\frac{d}{dx}\left(\frac{x^5}{e^x - 1}\right) = 0$$

which leads to the transcendental equation:

$$\frac{xe^x}{e^x - 1} = 5$$

with only one real solution, $x = 4.965$. Thus, the considered maximum value in the spectrum appears for the condition, called the Wien's displacement law:

$$\lambda_m T = c_3 \tag{3.18}$$

where $c_3 = c_2/x = 2.8976 \times 10^{-3}$ m K.

Substituting (3.18) into (3.13), the value of the maximum of the monochromatic intensity of the blackbody emission is:

$$e_{b\lambda m} = c_4 T^5 \tag{3.19}$$

where

$$c_4 = \frac{c_1}{c_3^5 \left(e^{4.965} - 1\right)} = 1.2866 \times 10^{-5} \ \frac{W}{m^3 \ K^5}$$

Equation (3.19) presents the hyperbole with asymptotes that are the axes of the coordination system $(\lambda, e_{b,\lambda})$ as shown in Figure 3.5 (dashed line).

3.6 Stefan–Boltzmann Law

In order to determine the emission density e_b of a black surface, equation (3.11) can be applied in integrated form:

$$e_b = \int_0^\infty e_{b\lambda}d\lambda \tag{3.20}$$

Applying Planck's relation (3.13) into (3.20), with substitution $x \equiv c_2/(\lambda T)$, yields:

$$e_b = c_1 \left(\frac{T}{c_2}\right)^4 \int_{x=0}^{x=\infty} x^3 \frac{1}{e^x - 1}dx \tag{a}$$

The fraction in equation (a) can be represented as the sum of the infinite geometric series:

$$\frac{1}{e^x - 1} = \sum_{m=1}^{m=\infty} e^{-mx} \tag{b}$$

Using (b) in (a):

$$e_b = c_1 \left(\frac{T}{c_c}\right)^4 \sum_{m=1}^{m=\infty} \int_0^\infty x^3 e^{-mx} dx \tag{c}$$

Then, combining consecutively the integration solution

$$\int x^n e^{ax} dx = \frac{1}{a} x^n e^{ax} - \frac{n}{a} \int x^{n-1} e^{ax} dx \qquad (n > 0) \tag{d}$$

given, e.g., by Korn and Korn (1968; integral #452, p. 966, with their symbols n and a), and after substitution for the present considerations: $n = m$ and $a = -n$, integral (3.20) comes finally to the following Stefan–Boltzmann law:

$$e_b = \sigma T^4 = \frac{a c_0}{4} T^4 \tag{3.21}$$

where the Boltzmann constant for black radiation:

$$\sigma = \frac{\pi^4}{15} \frac{c_1}{c_2^4} = 5.6693 \times 10^{-8} \qquad \frac{W}{m^2 \, K^4}$$

and the universal constant:

$$a = 7.564 \times 10^{-16} \quad \frac{J}{m^3 \, K^4}$$

are determined theoretically.

From the assumption for the gray surface model, expressed by relations in equation (3.14), the emission density e_b of the black surface, given by equation (3.21), can be used for determination of the emission density e of the gray surface as follows:

$$e = \varepsilon \sigma T^4 \tag{3.22}$$

For convenience in practical calculations, equation (3.22) is sometimes applied in the form:

$$e = \varepsilon C_b \left(\frac{T}{100}\right)^4 \tag{3.23}$$

in which the radiation constant for a black surface $C_b = 10^8 \times \sigma$. The experimental value is $C_b = 5.729$ W/(m^2 K^4), which is a little larger than $\sigma / 10^8 = 5.6693$.

Surface material	Surface temperature, °C	Emissivity, $(\varepsilon)_{\beta=0}$	Average emissivity, ε
Gold	20	0.02–0.03	—
Silver, polished	20	0.02–0.03	—
Copper, polished	20	0.03	—
Copper, oxidized	130	0.76	0.725
Aluminum	170	0.039	0.049
Steel, polished	20	0.24	—
Steel, red rust	20	0.61	—
Steel, scale	130	0.60	—
Zinc, oxidized	20	0.23–0.28	—
Lead, oxidized	20	0.28	—
Bismuth, shining	80	0.34	0.366
Clay, burnt	70	0.91	0.86
Brick	20	0.93	—
Ceramics	—	—	0.85
Porcelain	20	0.91–0.94	—
Glass	90	0.94	0.876
Ice, liquid water	0	0.966	—
Frost	0	0.985	—
Paper	90	0.92	0.89
Wood	70	0.935	0.91
Soot	—	—	0.96
Asbestos	23	—	0.96

TABLE 3.1   Emissivity Values of Different Materials

In practice, the choice of a proper value of emissivity ε is difficult. Some averaged values of ε for different materials are shown in Table 3.1 and more values can be found in related literature, e.g., Holman (2009).

3.7 Lambert's Cosine Law

The radiosity density j can be considered for a body surface or for any cross section in a space. The radiosity density j determines the total energy radiated in unit time, corresponding to the unit of surface area

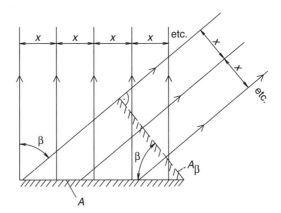

and in all directions into the front hemisphere, i.e., within the solid angle 2π sr:

$$j = \int_{\omega=0}^{\omega=2\pi} i_\beta d\omega \qquad (3.24)$$

where i_β is the *directional radiation intensity*, W/(m² sr), expressing the total radiation propagating within solid angle $d\omega$ and along a direction determined by the flat angle β with the normal to the surface (Figure 3.9).

Usually, the practical observations motivate the assumption that a certain surface A (Figure 3.9), is seen at the same brightness under any angle β. It means that for any direction determined by β the radiation intensity is the same as is schematically represented by equal spacing "x" of the normal rays (at $β = 0$) and for the rays propagating from surface A under arbitrary angle β. Thus, the directional radiation intensity i_β of surface A along angle β can be replaced by the normal radiation intensity i_0 of equivalent surface A_β:

$$A_\beta = A \cos \beta \qquad (3.25)$$

If the surfaces A and A_β have the same temperature and properties, then the energetic equivalence of radiation of both surfaces leads to the statement:

$$A i_\beta = A_\beta i_0 \qquad (3.26)$$

Substituting (3.25) into (3.26) the Lambert's cosine law is obtained which states that for the flat surface the radiation intensity i_β along a direction determined by angle β with the normal to the surface is:

$$i_\beta = i_0 \cos \beta \qquad (3.27)$$

Figure 3.10 Circular diagram of radiation intensity.

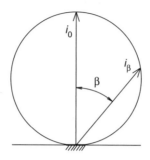

where i_0 is the normal radiation intensity. Equation (3.27) can be illustrated by a circular diagram shown in Figure 3.10. With the growing angle β from 0 to $\pi/2$ deg, the intensity i_β decreases respectively from values i_0 to 0 deg.

Based on Lambert's cosine law the following consideration can be developed. As shown in Figure 3.11, the solid angle $d\omega$, under which a surface dA' is seen from surface dA, is measured as a surface area dA' divided by the square of distance r of this surface from the observation point at surface dA:

$$d\omega = \frac{dA'}{r^2} \tag{a}$$

where

$$dA' = r\,d\beta\,2\pi r\,\sin\beta \tag{b}$$

Substitute (3.27), (a), and (b) into (3.24):

$$j = \pi i_0 \int_{\beta=0}^{\beta=\pi/2} 2\sin\beta\cos\beta\,d\beta = \pi\,i_0 \int_{\beta=0}^{\beta=\pi/2} \sin 2\beta\,d\beta = \pi\,i_0\frac{1}{2}\left(-\cos 2\beta\right)\Big|_0^{\pi/2}$$

Figure 3.11 Radiation of element dA on element dA'.

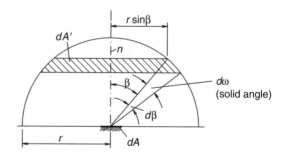

and finally:

$$j = \pi i_0 \qquad (3.28)$$

Based on equations (3.27) and (3.28):

$$i_\beta = \frac{j}{\pi}\cos\beta \qquad (3.29)$$

For given values of the radiosity density j and angle β, formula (3.29) allows for calculation of directional radiation intensity i_β.

The result is that when Lambert's law is fulfilled, the surface emitting radiation has the same radiosity intensity regardless of the direction from which the surface is seen. For example, this is why heavenly bodies make an impression like shining flat walls and not like a lump body.

3.8 Kirchhoff's Law

The relation between the absorptivity α and emissivity ε of the surface can be derived with use of the model of heat exchange shown in Figure 3.12. There are two flat, infinite, and parallel surfaces facing each other; one is perfectly gray (with any constant values of emissivity ε and reflectivity ρ), the other is perfectly black ($\varepsilon_b = 1$ and $\rho_b = 0$). The same and uniform temperature T prevails over both surfaces. Emission $e = \varepsilon \times e_b$ of the gray surface is totally absorbed by the black surface. Emission e_b of the black surface is partly absorbed ($\alpha \times e_b$) and partly reflected ($\rho \times e_b$). The system boundary (the dashed line in Figure 3.12) defines the considered system, which is the very thin layer next to the gray surface.

FIGURE 3.12
Scheme of energy
radiation balance.

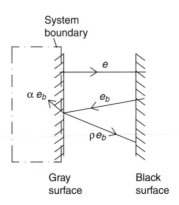

The energy conservation equation, applied for the system, yields:

$$e_b - \rho e_b = e \qquad (3.30)$$

After elimination of ρ and e from equation (3.30) by using, respectively, equations (3.5) and (3.14), one obtains:

$$\alpha = \varepsilon \qquad (3.31)$$

which is *Kirchhoff's law* (also called *Kirchhoff's identity*); the surface emissivity is equal to the surface absorptivity at the same temperature. In practice, equation (3.31) can be applied if $\varepsilon > 0.5$. For smaller values of ε, Kirchhoff's law can be inexact.

Derivation of the obtained result (3.31) did not require assumptions about any parameters, i.e., the result does not depend on the wavelength λ, temperature T, and the angle β; thus, for any wavelength, temperature, or direction, we have also:

$$\alpha_{\lambda T\beta} = \varepsilon_{\lambda T\beta} \qquad (3.32)$$

Emissivities of real materials differ from the values for discussed models, e.g., Lambert's cosine law, especially for polished surfaces. The directional *emissivity* ε_β, in a direction determined by angle β, is the following ratio of the respective directional radiation intensities

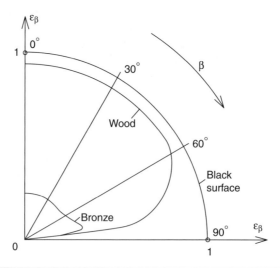

Figure 3.13 Real directional emissivity ε_β of bronze and wood as a function of angle β (from Petela, 1983).

i_β and $i_{b,\beta}$ for gray and black surfaces:

$$\varepsilon_\beta = \frac{i_\beta}{i_{b\beta}} \qquad (3.33)$$

For example, Figure 3.13 shows the comparison of the directional emissivities ε_β for bronze and wood to the emissivity $\varepsilon_{b,\beta}$ for a black surface ($\varepsilon_{b,\beta} = 1$). There are different surfaces, e.g., bronze, for which, in the significant range of angle β, the emissivity ε_β can grow with the increased angle β. However, for all materials, with the angle β approaching 90°, the directional emissivity ε_β rapidly decreases to zero. Table 3.1 presents some illustrative data on emissivity of different surfaces, selected from data given by McAdams (1954) and Schmidt (1963).

Nomenclature for Chapter 3

A	surface area, m^2
a	universal radiation constant, $a = 7.764 \times 10^{-16}$ J/(m^3 K^4)
c	speed of propagation of radiation, m/s
c_0	speed of propagation of radiation in vacuum $c_0 = 2.9979 \times 10^8$ m/s
c_1	the first Planck's constant, $c_1 = 3.74 \times 10^{-16}$ W m^2
c_2	the second Planck's constant, $c_2 = 1.4388 \times 10^{-2}$ m K
c_3	the third Planck's constant, $c_3 = 2.8976 \times 10^{-3}$ m K
c_4	the fourth Planck's constant, $c_4 = 1.2866 \times 10^{-5}$ W/(m^3 K^5)
E	emission of radiation, W
e	density of radiation emission, W/m^2
h	Planck's constant, $h = 6.625 \times 10^{-34}$ J s.
i	directional radiation intensity, W/(m^2 sr)
J	radiosity, representing a total radiation from a body, W
j	radiosity density, W/m^2
k	Boltzmann constant, $k = 1.3805 \times 10^{-23}$ J/K
n	refraction index
n, m	denotation of different surfaces
r	radius, distance, m
x	auxiliary value

Greek

α	absorptivity
β	flat angle (declination), deg
ε	emissivity of surface
ε_β	directional emissivity of surface in direction determined by angle β

φ	view factor
ρ	reflectivity
σ	Boltzmann constant for black radiation, $\sigma = 5.6693 \times 10^{-8}$ W/(m² K⁴)
τ	transmissivity
λ	wavelength, m
ν	oscillation frequency, Hz, (1/s)
ω	solid angle, sr

Subscripts

b	black surface
m	maximum
n, m	denotation of different surfaces
0	for $\beta = 0$
α	absorption
β	flat angle
λ	wavelength
ν	frequency
ρ	reflection
τ	transmission

CHAPTER 4

The Laws of Thermodynamic Analysis

4.1 Outline of Thermodynamic Analysis

4.1.1 Significance of Thermodynamic Analysis

The significance of thermodynamic analysis is that it can be applied to the investigations of all energy conversion phenomena. Such analysis provides different (energy, entropy, and exergy) views of the same phenomenon. Typical analysis is based on the material conservation equations that are used for developing energy balances, calculation of entropies, and, in recent decades, also for providing supplementary exergy balances.

Energy balance, based on the First Law of Thermodynamics, is developed to better understand any process, to facilitate design and control, to point at the needs for process improvement, and to enable eventual optimization. The degree of perfection in the energy utilization of the process, or its particular parts, allows comparison with the degree of perfection, and the related process parameters, to those in other similar processes. Comparison with the currently achievable values in the most efficient systems is especially important. Also, priorities for the required optimization attempts for the systems, or its components, can be established. Such priorities can be carried out either based on the excessive energy consumptions or on the particularly low degree of perfection.

However, the energy approach has some deficiencies. Generally, energy exchange is not sensitive to the assumed direction of the process, e.g., energy analysis does not oppose if heat is transferred spontaneously in the direction of the increasing temperature. Energy also does not distinguish its quality, e.g., 1 W of heat equals 1 W of work or electricity. Energy analyses can incorrectly interpret some processes; e.g., environmental air, when isothermally compressed, maintains its energy (e.g., enthalpy) equal to zero, whereas the exergy of the compressed air is larger than zero.

Entropy expresses the thermodynamic probability of a matter's state. According to the Second Law of Thermodynamics, the overall entropy growth in a process is required always to be positive. This requirement, applied even to an elemental step of any complex process, determines the only possible direction in which the step can occur. Entropy analysis allows for identification and location of the sources of irreversibility contributing to the overall unavoidable degradation of energy.

Entropy can be used for process optimization by minimization of entropy generation. The overall or local irreversible exergy loss can be calculated from the Guoy–Stodola law, equation (2.60), in which the respective entropy growth is applied. However, the Second Law has limited application for micro systems containing a countable number of independent particles. The smaller the number of particles, the less precisely the Second Law is fulfilled. For example, for any microbiological system containing only a few components, the Second Law may not be fulfilled.

The highest form of energy is mechanical energy, and work is the most valuable method of energy transfer. Therefore, exergy is defined as the maximum useful work obtainable from the considered matter (substance or field matter) in known environmental conditions. Exergy alone, not the energy, expresses the real ability to do work. The full classic definition of exergy, as a function of the states of matter and environment, is discussed in Section 2.6.

Exergy is a concept derived from simultaneous application of the First and Second Laws of Thermodynamics. Irreversibility destroys the exergy. There is no exergy conservation law, and exergy balance is completed with exergy loss; the greater the loss, the more irreversible is the process. Exergy balance allows for the development of exergy analysis according to a similar methodology for energy analysis. Exergy analysis applies to all applications mentioned for energy analysis. Whereas thermodynamic probability is expressed in units of entropy, exergy is expressed in units of energy. Consequently, exergy data are more practical and realistic in comparison to the respective energy values. Thus, exergy analysis provides a more realistic view of a process, which sometimes differs dramatically in comparison with the standard energy analyses. Exergy analysis can be compared to energy analysis, such as the second different projection in a technical drawing disclosing additional details of the subject seen from a different vantage.

Currently, there exist several different approaches to exergy analysis; see Moran and Shapiro (1992) or Bejan (1997). They all are largely mutually consistent, equally valid, and contribute to a better understanding of exergy analysis. Recently, the significance of exergy, used as a core thermodynamic variable for the investigation of biological systems, was presented by Jørgensen and Svirezhew (2004) in "mathematical biology."

However, the considerations in the present book are based on earlier pioneering approach to exergy analysis, which is the original monograph on exergy by Szargut and Petela (1965b, 1968), later developed also by Szargut et al. (1988). They apply exergy analysis to various processes, mostly to industrial processes. The analysis is based on classical thermodynamics and considerations are verified by numerous examples of applied engineering thermodynamics.

4.1.2 General Remarks and Definition of the Considered Systems

Knowledge about the environment in nature is continually being gained through many methods and observations. The scale of approach may be microscopic (e.g., a microscopic observation, or differential calculus) or macroscopic (phenomenological considerations, or integral calculus). Usually, studies are organized by focusing attention on a particular system that represents the targeted problem well.

Description and definition of the system is then a very important stage in any investigative approach. Consideration not based on a precisely defined system can lead to astonishing—but incorrect—results. The system has to be precisely determined by separating those elements included from those that are excluded. This is usually effectively rendered by applying the imaginary system boundary that tangibly separates the system from its surroundings. The best practical way is to draw a scheme of contents of the system indisputably separated from the surroundings by the drawn system boundary. Sometimes the investigated problem can be solved easily by introducing subsystems, also defined precisely.

The balance equations can be applied to each formulated system or subsystem. Below are discussed conservation equations for mass and energy. For the considered system one can also apply the equation of entropy growth and the balance equation of exergy. Each equation allows for determination of an unknown variable or for establishing a relation between variables.

The mass and energy conservation equations can be the basis for designing or exploiting the considered object. Complete data obtained from mass and energy considerations allows for development of entropy equations to verify the correctness of the mathematical model of mass and energy results from the viewpoint of the Second Law. The complete data can be also used to develop the exergy interpretation of the energy conversion process and mass transformation from the viewpoint of quality.

As discussed, the variables obtained from mass and energy analyses are very important; thus, they have to be prepared carefully. The variables can be measured, assumed, or calculated. If the system is *over-determined*, i.e., if the number of unknowns is smaller than the number of available independent equations, then all variables can be corrected based on the probability reconciliation calculus. For this reason the principles of exemplary reconciliation method are outlined in Section 4.7.

In practice, in the considered system the radiation processes are usually accompanied by processing on substances; therefore the laws of thermodynamics are considered below for systems in which radiation and substance play a role together.

4.2 Substance and Mass Conservation

The human brain understands matter either in the form of substance (material), i.e., as a collection of elementary chemical particles, or as field matter that appears as a field force of various kinds. Whereas substance is always connected with field matter (e.g., in the form of gravity), the field matter can exist independently of substance (e.g., matter of electromagnetic fields). Mass is the property of matter and is a measure of its inertia. The rest mass of field matter is zero, whereas the rest mass of substance is different from zero. For this reason the mass is commonly used as a measure of the amount of substance whose mass determines a weight in a gravitational field.

The change Δm of mass due to the increase Δe of energy is determined by Einstein's formula:

$$\Delta m = \frac{\Delta e}{c_0^2} \qquad (4.1)$$

where c_0 is the speed of light in vacuum.

For example, based on equation (4.1) the estimated mass m of the blackbody emission e_b calculated from formula (3.21) is shown in Figure 4.1 as a function of the emission temperature T. Even for the high temperatures T, the flux of the emission mass m of several $mg/(km^2\,s)$, as shown in Figure 4.1, is negligible in the common engineering mass flow rates.

It results from equation (4.1) that conservation laws for mass and energy are not independent and they both formulate differently the general law called the *law of matter conservation*. Thus, the mass conservation law corresponds to the energy conservation law.

The energy conservation law is commonly used in engineering considerations and calculations, whereas application of the mass conservation law is unnecessary and sometimes even not possible to apply. However, the substance conservation principle is commonly

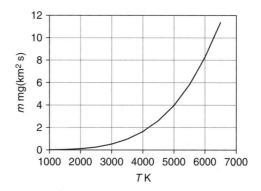

Figure 4.1 Mass m of the blackbody emission energy at temperature T.

applied, which is independent of the conservation laws of mass and energy.

The principle of conservation of substance claims that constant is always the number of molecules in physical processes, or the number of elements in chemical processes, or the number of nucleons in the processes of splitting and synthesis of nuclei.

The substance conservation equation does not need to account for radiation or any other form of matter except substance. Such an equation is developed for the system defined precisely by the system boundary. For an elementary process lasting very briefly:

$$dm_{in} = dm_S + dm_{out} \qquad (4.2)$$

where m_{in} and m_{out}, kg, is the elementary amount of substance delivered and extracted, respectively, from the system, and m_S is the elementary increase of the amount of substance within the considered system. Equation (4.2) can be appropriately modified for the steady state $(dm_S = 0)$, or for certain instants with use of mass flow rates, or for a certain period of time. The equation can be applied separately to particular compounds (if there is no chemical reaction) or elements. The amount unit can be kg, kmol, or the standard m^3 of the considered component.

The substance can be exchanged with the system by diffusion flux. For example, for gases the diffusion flux can be determined with use of Fick's law in a laminar situation or, with use of an equation with appropriately modified coefficients, in the case of turbulent diffusion. Due to the diffusion fluxes the enthalpy of the diffusing substance is carried out and the entropy effects occur, as discussed in Section 4.3.

A particular form of substance conservation equation can be the equation summarizing fractions of components in the considered composite material:

$$\sum_i f_i = 1 \qquad (4.3)$$

where f_i is the fraction of the ith component of the material.

4.3 Energy Conservation Law

4.3.1 Energy Balance Equations

The energy conservation equation is the result of observations and cannot be proved or derived. Throughout the long history of mankind there has not been recorded any phenomenon that disagrees with the First Law of Thermodynamics.

Energy balance based on the First Law of Thermodynamics is the basic method for solving problems of thermodynamics. If one wants to analyze any problem and helplessly does not know how, the general advice is to try to make an energy balance of the system that would represent the targeted problem. The First Law can be applied to a variety of problems, which, however, require a well-defined system for consideration. The system boundary should be the same for energy and matter balances because the matter balance is the basis for balance of energy. Sometimes only the specific definition of the system and particular tracing of the system boundary allows for the solution of the thermodynamic problem. In other cases the solution can be obtained by defining more of the different subsystems.

Generally, the energy E_{in} delivered to a system remains partly within the system as the increase ΔE_S of the system energy, and the rest is the energy E_{out} leaving the system. Thus, the general equation of energy balance is:

$$E_{in} = \Delta E_S + E_{out} \tag{4.4}$$

Usually, for better illustration of the balance equation, the particular terms of the equation are shown in the bands diagram. The principle of such a diagram is shown by a simple example (Figure 4.2) illustrating equation (4.4).

In principle, for energy considerations, the reference state for calculation of the energy of the matter included in the consideration can be defined arbitrarily; however, it is recommended to select this

Figure 4.2 Bands diagram of energy balance.

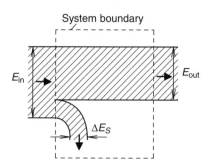

reference the same as for the exergy consideration in order to make a fair comparison of both the energy and exergy viewpoints.

Generally, application of the energy balance does not require analysis of processes occurring within the system boundary. It is sufficient only to know (e.g., from measurements) the parameters determining components of the energy delivered and leaving the system, as well as the parameters determining the initial and final states of the system. Obviously, if only the one unknown magnitude appears in the balance equation, then the equation can be used to calculate this magnitude.

Energy balance can be tailored differently depending on the considered viewpoint and actual conditions. For example, there are possibilities to categorize the case under consideration: (a) energy delivered is spent entirely for an increase of system energy with no energy leaving the system; (b) energy leaving the system comes entirely from the decrease of energy of the system with no energy delivered to the system; (c) there is neither delivered nor departing energy but only energy exchange within the system; (d) energy delivered is equal to energy leaving the system, with no change of the system energy. Other possibilities are that some components of energy can be neglected either due to relatively small changes or because they are not changed at all. The balance equation can be written for the steady or transient systems, for the system considered on a macro scale or a micro scale for which differential equations are applied, etc.

For example, for the elemental process lasting an infinitely short time, the balance equation (4.4) can take the form:

$$dE_{in} - dE_S + dE_{out} \tag{4.5}$$

In equation (4.5), only dE_S is the total differential and in order to demonstrate it clearly it is better to write equation (4.5) as follows:

$$\dot{E}_{in}(t)dt = dE_S + \dot{E}_{out}(t)dt \tag{4.6}$$

where $\dot{E}_{in}$ and $\dot{E}_{out}$ are the respective fluxes (e.g., in W) of energy delivered and extracted from the system, and t is the time.

Determination of dE_S requires not only accounting of the change of the intensive parameters of the system state but also of the eventual change of the substance amount in the system. If the considered system contains only the homogeneous substance, then

$$dE_S = d(m_S\, e_S) = m_S\, de_S + e_S\, dm_S \tag{4.7}$$

where m_S and e_S are, respectively, the amount of matter and its specific energy contained within the system.

Sometimes the subject of consideration can be recognized as moving in space (e.g., solar vehicle, radiometer vane, etc.). The simplest energy balance equation is then obtained by assuming that the coordinates system determining velocity and location is moving together with the system boundary. However, there are some consequences of such an assumption. Kinetic energy should be determined for the velocity relative to the moving system. The useful work done by the system does not appear in the energy balance because the forces acting on the system do not make replacements relative to the coordinates system. The useful work can be determined only for velocity and location relative to the earth.

Energy and mass calculations are the basis for engineering designs, whereas neither entropy nor exergy are. However, these latter have interpretative significance, as will be shown in the following chapters.

4.3.2   Components of the Energy Balance Equation

The *energy of a system* depends on its state. An increase ΔE_S in the energy system, changing from its initial to its final state, does not depend on the transition manner between these states, and is a difference between the final $E_{S,\text{fin}}$ and initial $E_{S,\text{inl}}$ energies of the system:

$$\Delta E_S = E_{S,\text{fin}} - E_{S,\text{inf}} \tag{4.8}$$

Generally, the system energy can consist of macroscopic components such as $E_{\text{macr},i}$ due to velocity (kinetic energy), surface tension (surface energy), gravity (potential energy), or any other energy of field nature (e.g., radiation). The remaining part of the system energy, containing microscopic components U_j, constitutes the internal energy (discussed in Section 2.3):

$$E_S = \sum_i E_{\text{macr}_i} + \sum_j U_j \tag{4.9}$$

where i and j are the successive numbers of the macro and micro components, respectively, of the system energy.

If the kind of substance before and after the process (e.g., physical process) is the same, then the reference state for calculation of the energy of the substance can be established with a certain degree of freedom. For example, the reference state can be assumed to be the state of the substance entering the system. Thus, the substance energy entering the system is zero, whereas the energy of the substance exiting the system is equal to the energy surplus relative to the reference state.

In addition, the components of negligible or constant value must not be taken into account. For example, the energy of the surface tension can be included only in consideration of the fluid mechanics process of liquid atomization or of the mechanical process of solid material comminution. Both processes have been analyzed, e.g., by Petela (1984a,b).

The energy exchange (E_{in} and E_{out}) with a system can occur in different ways.

Electrical energy can be delivered for heating the system, for driving an electric motor, or for generating an electromagnetic effect within the system (e.g., strong electric field effects combustion). In reverse processes the electric energy can be obtained; e.g., with the use of an electric generator, energy is obtained from the system. The energy flux of electric energy (power) is measured by a wattmeter.

Mechanical work can be exchanged with the system by means of a piston rod with reciprocal motion or with a rotating shaft.

The energy balance of a system should comprise mechanical work performed by all forces acting on a system boundary. Therefore, if a *substance flux* passes through the boundary, then the work performed by the force acting in the place of passing should be taken into account. Such transportation of a substance through the boundary is expressed by enthalpy (discussed in Section 2.3). For some kinds of substance, the enthalpy can be calculated with a specific formula; e.g., the formulae for plasma are discussed by Petela and Piotrowicz (1977).

If the considered system is moving relative to the coordinate system determining the location and velocity, then the work done by the forces causing the system displacement has to be considered. The energy balance should also include the work done by deformation of the system boundary if its shape changes during consideration.

Kinetic energy should be considered if the substance passes the system boundary with significant velocity relative to the boundary.

The *potential energy* of the substance exchanged with the system is included up to the energy balance if the substance has significant elevation above the reference level. This energy component results from the presence of the gravity field.

Energy transferred by *heat* occurs by direct contact of the system with the body at a temperature different from the system temperature, or can occur without contact, via radiation. The effect of contact during heat exchange appears in heat conduction as well as in heat convection. The model of *pure conduction* occurs when the particles of the contacted body do not change their location (solids). The energy is then transferred by free electrons and oscillations of atoms in the crystal lattice. Still, pseudo pure conduction can be recognized between fluids of very laminar flow; conduction occurs in the direction perpendicular to the ordered motion of particles at the component velocity only in the flow direction. In such a case, excluding the possibility of diffusion, there is no perpendicular substance flow and in spite of the medium flow this heat is transferred by conduction.

The essence of heat *convection* is the motion of substance (fluids), during which the mixing of hot and cold fluids occurs. However, the micromechanism of this mode of heat transfer also depends on the direct effective contacts (conduction) between the hot and cold fluids portions being replaced. If mixing is caused by the nonuniform distribution of density (temperature profile), then convection is called *natural convection.* In contrast, if the mixing is a result of the action of a pump or ventilator, etc., then a *forced convection* occurs.

Energy can be also exchanged with the system due to a *diffusive substance flux.* Then, the enthalpy of a diffusing substance has to be taken into account. For example, consider a system boundary demarcated over the laminar zone of a mixture of gases of the nonuniform temperature distribution. If it is assumed to be a laminar (no convection) mode of transparent gases (no radiation), then the energy E_L, W, exchanged through the boundary due to the heat conduction and enthalpy of the diffusing substance is composed of the two respective terms. The first term represents the heat conducted according to Fourier's law and the second term expresses the enthalpy of the diffusing gas according to Fick's law. Thus:

$$E_L = -A\left(k\frac{\partial T}{\partial y} + T\sum c_{p,i} D_{L,i}\frac{\partial c_i}{\partial y}\right) \qquad (4.10)$$

where A, m^2, is the surface area; k, $W/(m\,K)$, is the overall conductivity of the gas mixture; T, K, is the temperature of the gas at the boundary; y, m, is the space coordinate perpendicular to the system boundary surface and perpendicular to the gas flow direction, $c_{p,i}$, $J/(kg\,K)$; D_i, m^2/s, and c_i, kg/m^3, are, respectively, the specific heat at constant pressure, the laminar diffusion coefficient, and the concentration of the gas component, where i is the successive number of the gas mixture component.

Heat exchanged with the system (by convection, conduction, and radiation) is considered in related textbooks, e.g., by Holman (2009).

A friction work has to be spent in real processes which occur with friction. The friction work increases the energy of the system due to the absorption of heat in an amount equivalent to the friction work. Friction causes dissipation of energy, which can be only partly recovered. The *friction heat* does not appear as a member of the energy balance equation; however, it affects the final system energy and the components of the exiting energy.

The enthalpy and internal energy generally include physical and chemical components, both discussed in Chapter 2. *Chemical energy,* discussed in Section 2.7, is assumed to be the same for the substance considered as the component of the system and for the substance component separately exchanged with the system.

4.4 Entropy Growth

There are many articulations of the Second Law of Thermodynamics. They are based on various phenomena for which it has been noticed that they can occur only in one determined direction. For example, heat can flow only in the direction of the negative temperature gradient; real processes with friction are irreversible; every process occurring in nature is irreversible; a thermal engine cannot work without having available at least two heat sources at different temperatures, etc. These qualitative observations became possible for quantitative formulation with the use of entropy introduced by Clausius. Using entropy, the Second Law of Thermodynamics states that in nature there are possible only such phenomena for which the overall entropy growth, i.e., the sum $d\Pi$ of the elemental entropy increments dS_i of each ith matter participating in the considered phenomenon is larger than zero:

$$d\Pi = \sum_i dS_i > 0 \qquad (4.11)$$

For any theoretical model of reversible phenomenon $d\Pi = 0$; if, however, $d\Pi < 0$ then the phenomenon is impossible. Equation (4.11) expresses the overall entropy growth Π in integral form as follows:

$$\Pi = \sum_i \Delta S_i \qquad (4.12)$$

where $\Delta S_i = (S_{\text{fin}} - S_{\text{inl}})_i$ is the entropy increase of ith matter of the considered system or surrounding participating in the considered phenomenon.

Equation (4.11) expresses explicitly that the overall entropy growth has to be positive even in the smallest step $(d\Pi > 0)$ in the course of the process. For example, during the design of a heat exchanger, equation (4.12) for entering and exiting media can be fulfilled; however, in some particular cases within the exchanger can occur an unnoticed region, a so-called pinch point, for which locally $d\Pi < 0$, i.e., the whole process of heat exchange is impossible. Thus, entropy is very useful in verifying the design of new processes. The larger is the overall entropy growth, the more irreversible is the considered process.

There are some special cases for calculation of the overall entropy growth. If a certain substance remains unchanged during the process (e.g., a physical process), then only the respective increases of the substance entropy exiting and entering the system are taken into calculation of Π. In some cases one must take into account that the system can exchange substance on either a macro or micro scale (diffusion) as illustrated by equation (4.10). If a certain substance disappears (e.g., in a chemical reaction) then its absolute entropy has to be used with a negative algebraic sign. If a certain substance appears, then the positive sign of entropy should be used.

A heat source is defined as the body at given temperature T that can absorb or release infinitely large amounts of heat without a change in the heat source temperature T. Therefore, regarding the heat sources, the appropriate term based on equation (2.36) for the increase or decrease of entropy of the heat source should appear in equation (4.11) and (4.12), respectively.

The radiation entering or absorbed into the considered system has a negative sign of radiation entropy, whereas the radiation leaving the system, or being emitted, has positive entropy. The radiation entropy is recognized as absolute.

The components on the right-hand side of equation (4.12) can be the entropy of the substance, heat source, or radiation. The overall entropy growth Π does not include entropies of work (mechanical or electrical) nor the effect of fields such as the gravitational field or the surface tension of a substance. These magnitudes, although they contribute to the disorder, have no thermodynamic parameters and act only indirectly by changing parameters of involved matters.

4.5 Exergy Balance Equation

4.5.1 Traditional Exergy Balance

The exergy balance equation is the basis of the exergetic part of thermodynamic analysis. Exergy analysis can be applied to a range of problems that, like energy analysis, require an appropriately well-defined system for consideration. The system boundary should be the same as for the matter balance.

The exergy conservation equation can be applied only for reversibly occurring processes. For real processes the exergy conservation equation is fulfilled only when the unavoidable exergy loss, due to irreversibility of the process, is taken into account. Thus, correspondingly to energy equation (4.4), the following exergy balance equation is applied:

$$B_{\text{in}} = \Delta B_S + B_{\text{out}} + \delta B \tag{4.13}$$

where B_{in} and B_{out} are the respective sum of exergy delivered and released from the system, ΔB_S is the change in the exergy of the system, and δB is the exergy loss due to the process irreversibility, calculated from the Guoy–Stodola law, equation (2.60).

The bands diagram for exergy balance is shown in Figure 4.3. In comparison with the respective diagram for energy balance (Figure 4.2), the exergy diagram shows the exergy δB that disappears within the system.

Like the energy balance, the exergy balance can be tailored differently depending on the considered problem and the actual conditions. For example, some components of exergy can be neglected either due

Figure 4.3 Bands diagram of the exergy balance.

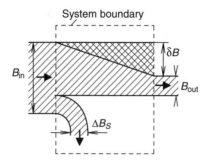

to relatively small changes, or because they are unchanged. The balance equation can be written for steady or transient systems, for a system considered on either a macro or micro scale using differential equations, etc. Obviously, for the calculation of exergies, there is no freedom in defining the reference state, which is only the environment, as determined by the definition of exergy.

In order to formulate the exergy balance it is required to know the parameters determining components of exergy delivered and leaving the system, the parameters determining the initial and final state of the system, and the parameters determining the environment. As in the energy balance equation, if only the one unknown magnitude appears in the exergy balance equation, then the exergy equation can be used to calculate this magnitude.

As discussed for the energy balance in Section 4.3, for the elemental process lasting an infinitely short time, the exergy balance equation can take the form:

$$\dot{B}_{in}\, dt = dB_S + \dot{B}_{out}\, dt + \delta B \qquad (4.14)$$

where $\dot{B}_{in}$ and $\dot{B}_{out}$ are the respective fluxes of exergy delivered and extracted from the system, and dB_S is the total differential exergy growth of the system.

The differential dB_S should be determined analogously to equation (4.7):

$$dB_S = d(m_S\, b_S) = m_S\, db_S + b_S\, dm_S \qquad (4.15)$$

where m_S and b_S are, respectively, the amount of matter and its specific exergy contained within the system.

If the subject of consideration is moving in space, then the simplest exergy balance equation is obtained by assuming, as for the energy balance equation, that the coordinates system determining velocity and location is moving together with the system boundary. The assumption of the moving system boundary requires some specific exergy interpretation as shown, e.g., by the exergy balance for a jet engine discussed by Szargut and Petela (1965, 1968).

4.5.2   Components of the Traditional Exergy Balance Equation

An increase ΔB_S of the exergy system, changing from its initial to its final state, does not depend on the path of change between these states and is equal to the difference of the final and initial components:

$$\Delta B_S = \left[\sum_i (B)_{\text{fin}, i} - \sum_j (B)_{\text{inl}, j} \right]_S \tag{4.16}$$

where the sum of the initial or the sum of the final components is:

$$(B)_S = B_k + B_p + B_S + B_b + \cdots \tag{4.17}$$

and where i and j are the successive numbers of the final and initial exergy components, respectively, B_k is the kinetic exergy, B_p is potential exergy, B_S is the thermal exergy of the system calculated with the use of formula (2.46), and B_b is the exergy of the photon gas (black radiation) calculated, e.g., based on equation (5.29). Also, the other eventual components in equation (4.17), as shown in Figure 2.1, can be added if necessary, e.g., the exergy of the surface tension which is equal to the energy of surface tension, etc.

The reference state for calculation of the exergy cannot be established arbitrarily as it can for the energy. The components of negligible or constant value may be not taken in the calculations.

Exergy exchanged with the considered system can occur in different ways described for the energy balance.

Electrical exergy is equal to electrical energy. *Exergy of mechanical work* is equal to work. Exergy of substance flux is calculated with the use of formula (2.45); however, *kinetic exergy* (calculated as the kinetic energy for absolute velocity) and *potential exergy* (equal to potential energy relative to the earth's surface level) should be calculated separately. The exergy of heat exchanged with the system is determined by formula (2.61).

Exergy can also be exchanged with the system by way of a *diffusive substance flux*. The exergies of diffusing substances are then taken into account as the exergy determined by formula (2.45) interpreted for the partial pressure of the substances.

The exergy loss δB_F caused by friction is determined by assumption that the friction heat Q_F, equal to the friction work, is entirely absorbed by the substance at temperature T. For the heat absorption process, assuming the entropy growth $\Pi_F = Q_F / T$, the exergy loss can be calculated from formula (2.60) as follows:

$$\delta B_F = Q_F \frac{T_0}{T} \tag{4.18}$$

The smaller is the exergy loss, the higher is the temperature of the absorbing substance. The exergy loss δB_F can be smaller or larger than the friction heat Q_F depending on the temperature ratio T_0/T. This observation is particularly important for refrigerating processes where often $T < T_0$.

The *chemical* exergy, discussed in section 2.7, is assumed to be the same for the substance considered as for the component of the system and for the substance component separately exchanged with the system.

4.5.3   Exergy Balance at Varying Environment Parameters

Exergy balance is usually carried out with an assumption of constant parameters of the environment during the time of consideration. The effects of varying environmental parameters are usually small, and the assumption of the mean environmental parameters is sufficient for the exergy analysis. The inclusion of the variations of the environmental parameters would make analysis more difficult because the considered exergy values should be taken for instantaneous environment parameters; i.e., the values used in the balance equation would need calculations by integration over the assumed time period.

Moreover, the balance equation would usually need the introduction of an additional member without which the equation could not be fulfilled. For example, such a need is shown by consideration of a perfectly insulated container that has been closed while being filled with a substance in equilibrium with the environment. Thus, the substance in the system (container) has zero initial exergy ($B_{S,inl} = 0$). If, meanwhile, the environment parameters are changed, then the enclosed substance gains the positive exergy ($B_{S,fin} > 0$) and the exergy of the system $\Delta B_S = B_{S,fin} - B_{S,inl} > 0$.

However, during the change in the environment there were no processes occurring, thus the overall growth of entropy $\Pi = 0$, which means $\delta B = 0$. No substance was exchanged with the system; $B_{in} = B_{out} = 0$. The above statements show that in the considered example the exergy balance equation (4.13) is not fulfilled; $\Delta B_S \neq 0$, and the environmental variation has generated a certain exergy ($B_{S,fin}$). However in another example, it can happen that also due to variations in the environment, there is no change in exergy or some exergy disappears. For example, ice stored during the summer has significant exergy, whereas the exergy of such ice in winter would be close to zero.

Therefore, generally, the exergy balance equation for the process occurring at the varying environment should contain the compensation term ΔB_e, which modifies equation (4.13) as follows:

$$\int_{inl}^{fin} \dot{B}_{in} \, dt = \Delta B_S + \int_{inl}^{fin} \dot{B}_{out} \, dt + \int_{inl}^{fin} \dot{\Pi} \, T_0 \, dt \; + \Delta B_e \qquad (4.19)$$

Figure **4.4** Bands diagram of the exergy balance with positive compensation term ($\Delta B_e > 0$).

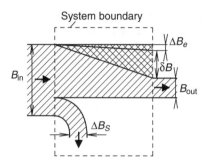

Figure **4.4** Bands diagram of the exergy balance with positive compensation term ($\Delta B_e > 0$).

where ΔB_e is the exergy gain due to variation of environment from initial state (inl) to the final (fin) state, $\dot{B}$ is the respective rates of exergy, and $\dot{\Pi}$ is the overall entropy growth. As mentioned, the ΔB_e can be positive or negative or zero. The bands diagram for the exergy balance at varying environment parameters is shown in Figure 4.4. Direct calculation of ΔB_e is not easy; thus, the best method is to calculate this value as the completion of the balance equation.

The Gouy–Stodola law, represented by equation (2.60), was derived for constant environment temperature T_0. If T_0 is varying, then the law can be applied only for the infinitely short process expressed by the presence of the appropriate integral term in equation (4.19). The Gouy–Stodola law cannot be applied to the processes that occur at the varying environment temperature if such variation is caused by the considered process.

The variation of environmental parameters can instantaneously generate or destroy exergy with no role in the internal processes of the examined system. The processes, wherever possible, should be organized to utilize the instantaneous positive value of ΔB_e. The effective prediction of change in the environment parameters would be helpful.

The fluctuation of parameters in the environment is one of the natural low-value resources, such as, e.g., waste heat at low temperature, etc. The fluctuation of these parameters has relatively insignificant influence on the high-value natural resources, e.g., natural fuels.

It is also possible to consider the variation of environmental parameters with altitude. Such consideration leads to the concept of mechanical exergy and to the determination of gravity influence on the exergy balance equation discussed in the following section 4.5.

A particular problem related to the temperature of the environment—with specific significance for radiation processes, especially occurring on the earth during night time—is the effective *sky temperature*, which can be used, e.g., for determination of the radiant heat lost from the earth's surface. Related problems are discussed, e.g., by Duffie and Beckman (1974).

Note that the energy balance is traditionally not considered at the varying reference states even if they are equal to the state determined by the environment parameters. As a result, consideration of the processes from an energy viewpoint does not show the interpretative features of the exergy approach with regard to the varying environment.

A calculation example illustrating the effect of the varying environment temperature is given in Section 6.9.

4.5.4 Exergy Balance with Gravity Input

Exergy analysis is an interpretative method for the study of energy conversion processes. The interpretive feature of exergy tolerates certain freedom in exergy application for disclosing as many as possible new viewpoints. Therefore, in some situations, the exergy balance equation requires the introduction of a special term to fulfill the traditional equation. For example, in Section 4.5.3, we discussed how to modify the traditional exergy balance equation for the situation where varying environment parameters are used as the reference states for the determination of exergy. The proposed solution in such a situation introduces a new compensation term to the exergy balance equation.

In another situation, when mechanical exergy (eZergy) is applied, the effect of gravity appears, which requires also an additional term in the exergy balance equation. Petela (2009a) proposed to insert an appropriate term, called *gravity input G*, as an additional exergy input in the left-hand side of the exergy balance equation. Thus, equation (4.13) for the constant environment parameters becomes:

$$B_{in} + G = \Delta B_S + B_{out} + \delta B \qquad (4.20)$$

The gravity input G can be positive, zero, or negative. The bands diagram for the exergy balance with included gravity input is shown in Figure 4.5. Usually the value of G is calculated from the exergy balance equation.

Prediction of the algebraic sign of gravity input is not discussed; however, interpretation of the sign from an exergy viewpoint can be proposed as follows. The gravity input can appear only when a substance is considered in the exergy balance and if eZergy is applied to the substance.

FIGURE **4.5** Bands diagram of exergy balance interpretation, including gravity input, in the case $G > 0$.

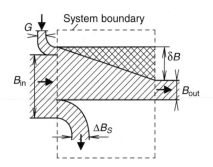

In the case $G < 0$, as a result of the effect of the gravity field on the considered process, the process product expressed by the total exergy value of the right-hand side of the exergy balance equation diminishes and has to be balanced by the negative gravity input G added to the left-hand side of the equation. The considered process can be recognized as opposing the effect of the gravity field.

In the case of $G > 0$, the presence of the gravity field during the considered process generates a certain "surplus" of exergy disclosed by the right-hand side of the exergy balance equation. This surplus has to be balanced by a positive gravity input G added to the left-hand side of the equation. The gravity field favors the process by contributing some exergy input.

In the case of $G = 0$, there is no change in the traditional exergy and this means that the work of the substance during theoretical expansion at altitude H (to obtain the equilibrium of densities), considered in Section 2.6.2, has no accountable importance.

Example 4.1 The gravity input significance is considered, e.g., by Petela (2009b). A chimney removes the hot waste gas (assume dry air) from a certain installation. The fresh air for installation is taken from the atmosphere at parameters T_0, p_0, ρ_0, and $x = 0$, where x is the altitude measured from the earth's surface. The parameters of air leaving the installation (the magnitudes at this point have subscript 1) are the parameters at the chimney inlet (bottom): T_1, p_1, ρ_1, and $x_2 = 0$. The air parameters at the chimney exit (the magnitudes at this point have subscript 2) are T_2, p_2, ρ_2, and $x_2 = H$, where H is the chimney height. The chimney is a cylindrical tube of constant inner diameter D; thus, the cross-sectional area of the chimney is also constant.

For a given H, the chimney diameter D is determined from the assumed ratio D/H. The hot air leaves the installation with velocity $w_1 = 10$ m/s and leaves the chimney at velocity $w_2 \approx w_1 \times \rho_1/\rho_2$. The pressure p_1 at the chimney bottom $p_1 = p_2 + g_x \times H \times (\rho_1 + \rho_2)/2$, where gravitational acceleration g_x is determined to be the arithmetic average of the values for $x = 0$ and for $x = H$. For the latitude assumed to be zero the approximation for gravitational acceleration g_x, m/s^2, is:

$$g_x = 9.780327 - 3.086 \times 10^{-6} x' \tag{a}$$

where x' is the altitude above sea level.

Air is assumed as the ideal gas with the individual gas constant $R = 287.04$ J/(kg K) and with the specific heat at constant pressure, $c_p = 1000$ J/(kg K). For the considered air the state equation $p = \rho \times R \times T$ can be applied. The density of the atmosphere is $\rho_0 = 1.225$ kg/m^3. The same reference state is assumed for calculations in analyses of energy and exergy; $p_0 = 101.325$ kPa and $T_0 = 288.16$ K.

Energy: Interpretation of the chimney process can be based on the following energy conservation equation:

$$E_1 + E_{w1} + E_{b1} = E_2 + E_{w2} + E_{b2} + E_Q \tag{b}$$

where E_1 and E_2 are the enthalpies at the chimney bottom and top, respectively, calculated as $E = m \times c_p \times (T - T_0)$, and where m is the mass flow rate of air.

Magnitudes E_{w1} and E_{w2} are the kinetic energies at points 1 and 2, respectively, calculated as $E_w = m \times w^2/2$.

Heat transferred from the chimney wall to the environment is:

$$E_Q = h\left(T' - T_0\right)\pi DH \tag{c}$$

where $T' = (T_1 + T_2)/2$ is the average temperature and h is the coefficient for the convection heat transferred from hot air to the environment.

The potential energy E_b of air is calculated as the possible work performed during the buoyant vertical replacement of the considered air from the actual locality to a certain equilibrium height H_b. The replacement occurs until the difference between the constant density, ρ (e.g., ρ_1 or ρ_2), of the actually considered air and the density of the atmospheric air, ρ_b, achieves zero ($\rho - \rho_b$). Such potential energy, which is equal to the respective potential exergy, can be expressed from equation (2.48) as:

$$E_b = m \int_{x=0}^{x=H_b} g_x \left(\frac{\rho_b}{\rho} - 1\right) dx \tag{d}$$

The reference altitude $x = 0$ is at the earth's surface. Using the solution of equation (d) in form of density ρ_b as function of altitude H, according to equation (f) (Example 2.1) is:

$$E_b = -\frac{m}{a_4\,\rho}\left[\frac{a_2}{6\,a_4}(\rho - a_3)^3 + \frac{a_1}{2}(\rho - a_3)^2\right] \tag{e}$$

where $a_1 = 9.7807\ \text{m/s}^2$, $a_2 = -3.086 \times 10^{-6}\ 1/\text{s}^2$, $a_3 = 1.217\ \text{kg/m}^3$, and $a_4 = -9.973 \times 10^{-5}\ \text{kg/m}^4$ are the constant values.

If all values are expressed as fractions of E_1 the normalized form of equation (b) can be written as:

$$100 + e_{w1} + e_{b1} = e_2 + e_{w2} + e_{b2} + e_Q \tag{f}$$

Exergy: balance equation for the considered chimney is:

$$B_1 + B_{w1} + B_{b1} = B_2 + B_{w2} + B_{b2} + B_Q + \Delta B \tag{g}$$

The subscripts of exergy streams B in equation (g) are, respectively, the same as in equation (b) and the additional term ΔB is the exergy loss due to irreversibility of the chimney process. According to the Gouy–Stodola law, equation (2.60), the exergy loss is calculated as the product of temperature T_0 and the overall entropy growth (entropies of heat and air):

$$\Delta B = T_0\left[\frac{E_Q}{T'} + m\left(c_p \ln \frac{T_2}{T_1} - R \ln \frac{p_2}{p_1}\right)\right] \tag{h}$$

The exergy of heat E_Q is calculated for the average temperature T' of air in the chimney:

$$B_Q = E_Q\left(1 - \frac{T_0}{T'}\right) \tag{i}$$

The exergy of the air (B_1 or B_2) is derived from the definition of the physical exergy of a gas:

$$B = m\left[c_p\,(T - T_0) - T_0\left(c_p \ln \frac{T}{T_0} - R \ln \frac{p}{p_0}\right)\right] \tag{j}$$

The potential exergy is equal to the potential energy, $B_b = E_b$, so that equation (e) may be used.

The exergy balance can be normalized as well:

$$100 + b_{w1} + b_{b1} = b_2 + b_{w2} + b_{b2} + b_Q + \Delta b \tag{k}$$

Notice that, although $B_{w1} = E_{w1}$, $B_{w2} = E_{w2}$, $B_{b1} = E_{b1}$ and $B_{b2} = E_{b2}$, the corresponding percentiles are not equal ($b_{w1} \neq e_{w1}$, $b_{w2} \neq e_{w2}$, $b_{b1} \neq e_{b1}$ and $b_{b2} \neq e_{b2}$) because the respective dimensional reference values are different: $E_1 \neq B_1$.

The eZergy balance equation for the considered chimney is:

$$Z_1 + Z_{w1} + G = Z_2 + Z_{w2} + Z_Q + \Delta Z \tag{l}$$

Equation (l) is used for calculation of G which supposedly is a measure of the effect of terrestrial gravity field on the considered process.

In dimensionless form, the eZergy balance becomes:

$$100 + z_{w1} + z_G = z_2 + z_{w2} + z_Q + \Delta z \tag{m}$$

Note again, that although $Z_{w1} = B_{w1} = E_{w1}$, $Z_{w2} = B_{w2} = E_{w2}$, $Z_Q = B_Q$ and $\Delta Z = \Delta B$, the correspondent percentiles are not equal ($z_{w1} \neq b_{w1}$, $z_{w2} \neq b_{w2}$, $z_Q \neq b_Q$, and $\Delta z \neq \Delta b$) because the reference Z_1 for percentage values is generally different than B_1 or E_1, ($Z_1 \neq B_1$). The terms corresponding to potential exergy do not appear in equations (l) and (m) because the potential exergy is already interpreted by eZergy.

Computation Results: While preparing results for Table 4.1 it was observed that moderate changes in T_0, p_0 or a similar kind of gas (varying R and c_p) have a negligible effect on the output data.

		Reference	Mono-variant changes of input parameters and resulting outputs			
Quantity	Units	value				
1	2	3	4	5	6	7
Input						
T_1	K	430	520	—	—	—
H	m	300	—	400	—	—
D/H	—	0.07	—	—	0.08	—
h	W/m^2 K	0.005	—	—	—	0.2
Output						
D	m	21	21	28	24	21
p_2	Pa	97385	97385	96106	97385	97385
T_2	K	428.85	518.58	428.46	428.86	428.72
ρ_2	kg/m^3	0.791117	0.654238	0.781451	0.791107	0.79136

Table 4.1 Output Trends Responsive to Change of Some Input Parameters; from Petela (2009b)

Quantity	Units	Reference value	Mono-variant changes of input parameters and resulting outputs			
1	2	3	4	5	6	7
p_1	Pa	99738	99326	99216	99738	99738
ρ_1	kg/m^3	0.808072	0.665458	0.803843	0.808072	0.80808
m	kg/s	2798.94	2304.89	4949.68	3655.63	2798.85
Energy						
e_1	%	100	100	100	100	100
e_2	%	99.192	99.387	98.912	99.196	99.098
e_{w1}	%	0.035	0.022	0.035	0.035	0.035
e_{w2}	%	0.037	0.022	0.037	0.0368	0.037
e_{b1}	%	7.151	9.663	7.338	7.151	7.151
e_{b2}	%	7.922	10.233	8.389	7.923	7.911
e_Q	%	0.035	0.043	0.035	0.031	0.141
Exergy						
b_1	%	100	100	100	100	100
b_2	%	90.66	96.229	87.312	90.672	90.490
b_Q	%	0.0652	0.074	0.067	0.057	0.261
b_{w1}	%	0.198	0.083	0.202	0.198	0.198
b_{w2}	%	0.207	0.086	0.214	0.207	0.207
b_{b1}	%	40.262	37.283	42.039	40.262	40.261
b_{b2}	%	44.605	39.481	48.057	44.608	44.539
Δb	%	4.918	1.497	6.591	4.917	4.963
eZergy						
z_1	%	100	100	100	100	100
z_2	%	101.161	100.573	101.566	101.166	101.036
z_{w1}	%	0.063	0.033	0.063	0.063	0.063
z_{w2}	%	0.066	0.034	0.067	0.066	0.066
z_Q	%	0.0208	0.0291	0.021	0.018	0.083
Δz	%	1.567	0.593	2.054	1.567	1.581
G	%	2.751	1.196	3.644	2.753	2.703
N	MW	5.4965	6.9208	12.9854	7.1791	5.4951

TABLE 4.1 Output Trends Responsive to Change of Some Input Parameters; from Petela (2009b) (*Continued*)

Column 3 of Table 4.1 represents an example of results for input data: $T_1 = 300K$, $H = 300$ m, $D/H = 0.07$ and $h = 0.05$ W/(m^2 K). In the results, the temperature drop in the chimney is relatively small ($T_1 - T_2 = 1.15$ K). The pressure decreases from $p_1 = 99.738$ kPa to $p_2 = 97.385$ kPa. The densities at point 1 and 2 differ insignificantly from each other although they both are clearly smaller than the density of atmosphere at $x = 0$, (ρ_0). For the considered chimney dimensions ($H = 300$ m and $D = 21$ m) the mass flow rate of air is $m = 2.799 \times 10^3$ kg/s and the power required for drawing air through the installation is $N = 5.4965$ MW.

For calculation of the percentage values for the balances of energy, exergy and ezergy the 100% bases are $E_1 = 397$ MW, $B_1 = 70.5$ MW and $Z_1 = 221$ MW, respectively. These values also illustrate the estimation of waste loss, which is the largest in terms of energy (397 MW), smaller in terms of ezergy (221 MW), and only 70.5 MW as interpreted by exergy. Column 3 shows also the split of the 100% input between other terms in the energy, exergy, and ezergy balances.

The terms representing the exit air differ a little from the terms for the inlet air: $e_1 - e_2 = 0.81\%$ and $b_1 - b_2 = 9.34\%$ for energy and exergy, respectively, whereas in the ezergy balance the difference is negative: $z_1 - z_2 = -1.16\%$.

The respective values of the potential energy and exergy expressed in W are equal, but their percentage values are different. As mentioned, in the ezergy considerations the altitudinal potential of air is interpreted by the ezergy value.

Factors such as a large mass rate of air, relatively small surface of the chimney wall, and low coefficient of heat transfer, all contribute to relatively small heat loss. The value of this loss is below 0.1% for all three balances. Exergy of lost heat (equal ezergy of this heat) is significantly smaller than the respective energy of this heat, obviously, because of its relatively low temperature.

Irreversibility loss ($\Delta B = \Delta Z = 3.47$ MW) of chimney process is disclosed obviously only in exergy and ezergy considerations; however, their percentiles are different ($\Delta b = 4.918\%$ and $\Delta z = 1.567\%$).

Gravity input, revealed only in ezergy considerations, is positive (2.751% or 6.09 MW), which reveals the extent to which the gravity field favors the chimney process.

Columns 4–7 of Table 4.1 illustrate the trends of the output data in response to changes in input parameters. The values in column 3 are considered as the reference values for studying the influence of the varying input parameters on the output data. Therefore, each of the next columns (4–7) corresponds to the case in which the input is changed only by the value shown in a particular column, whereas the other input parameters remain at the reference level.

Column 4 corresponds to a change in the air temperature T_1, which increases from 430 K to 520 K. The 90-K T_1 increase causes, e.g., a gravity input decrease from 2.752% to 1.196%. The change in temperature T_1 causes also an increase of T_2 and decreases of $p_1, \rho_1, \rho_2,$ and m. The increase in T_1 also remarkably changes some terms of the exergy balance, e.g., b_2 growing from 90.66% to 96.2%, Δb drops from 4.918% to 1.497%, and potential exergy grows from -4.343% to -2.198%.

The results shown in other column (5–7) can be analyzed similarly.

Column 5 shows the effect of increasing the chimney height from 300 m to 400 m. That change causes, e.g., the increases of power (from 5.4965 to 12.9856 MW) and gravity input (from 2.751% to 3.644%).

Column 6 shows that the increase of the D/H ratio (from 0.07 to 0.08) causes, e.g., growth of power (from 5.4965 to 7.18 MW).

Column 7 shows the influence of heat lost from chimney to the environment. An increase of coefficient h from 0.05 to 0.2 W/(m^2 K) (e.g., due to worsened

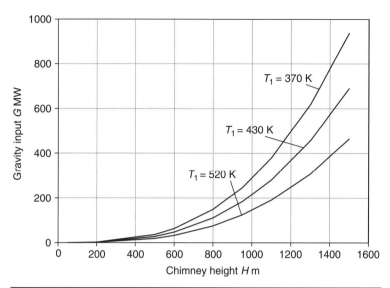

FIGURE 4.6 Gravity input as a function of chimney height H and air temperature T_1 at the chimney bottom ($D/H = 0.07$ and $k = 0.05$ W m^{-2} K^{-1}); from Petela (2009b).

insulation or strong wind), causes, e.g., a drop of gravity input (from 2.751 to 2.703) and the decrease of power.

The gravity input is shown (Figure 4.6) as function of chimney height H and air temperature T_1 at the chimney bottom. Gravity input increases significantly with increasing height H and with decreasing temperature T_1. Figure 4.6 takes into account very high chimneys, up to 1500 m, keeping in mind the solar chimneys for which rather lower temperature T_1 (e.g., ~370 K) would be considered.

In this example the three different thermodynamic interpretations (energy, exergy, and eZergy) were applied to the chimney phenomenon. The positive gravity input that represents the effect of the terrestrial gravity field was determined. The traditional exergy, contrary to eZergy, does not reveal the gravity input (G).

Other examples of the calculation of gravity input—adiabatic expansion of air in a turbine and drawing air through a throttling valve followed by a fan—are discussed by Petela (2009a). Further application of gravity input interpretation is also discussed in Chapter 11.

4.6 Process Efficiency

4.6.1 Carnot Efficiency

Work, or the efficiency of its generation, is one of the principal problems of technological progress being continually investigated by researchers. Some findings come from observations of nature. The continuous generation of a useful effect (e.g., work or heat) or conversion of energy is possible only in a situation when at least two heat sources with different temperatures are available. (The heat source has the feature that it can release or absorb an infinitely large amount of heat without changing temperature.)

The main mechanism for utilizing heat sources is by way of a working fluid, the parameters of which vary because the cyclical absorption of heat from the hotter source is followed by the release of heat to the colder source. One of the required heat sources can obviously be the freely available environment. Thus, practically, only one valuable heat source, different from the environment, is required for arranging the cyclic process. The parameters of the working fluid vary in successive subprocesses in such a way that the final state of the fluid cycle is identical to the initial state of the cycle. Illustration of the parameters varying in the cycle is a closed curve in the coordinate system of any two fluid parameters.

By searching for the most effective cycle process, which would occur reversibly without any losses, the ideal model was established by Carnot (1824). Real cycles can be designed close to this ideal model by applying different "carnotization" efforts. The model cycle of releasing and absorbing heat (at no entropy change) should consist of only ideal (reversible) processes. Thus, the cycle processes should occur with an infinitely small temperature difference between the heat source and the working fluid, and the flow of fluid should be frictionless. The other cycle processes, during which work is generated or consumed, should occur also reversibly (at constant entropy), which is possible if the fluid does not exchange heat (i.e., it is adiabatic) with its surroundings and, additionally, it expands or is compressed with no friction (i.e., it is isentropic).

For example, the parameters changing in the Carnot cycle with a photon gas as the working fluid is shown in the temperature–entropy (T, S) coordinates system in Figure 4.7. It is worth noting that the considerations of any cylinder–piston model system allows application of the obtained conclusions generally; not only to the cylinder–piston cases but also to the many other situations of the considered fluid and in different geometrical configurations. The piston bottom and the wall are generally mirrorlike except for the cycle phases during which heat is transferred from the heat sources of temperatures T_I (hot) or T_{II} (cold) to the photon gas within the cylinder. There is no substance in the cylinder. The considerations below use relations between the parameters of black radiation introduced later in Chapter 5. The four component processes of the cycle occur successively:

Process 1–2

Figure 4.7 presents the situation at the beginning of the first process 1–2; the piston is in the extremely left-hand position and the heat

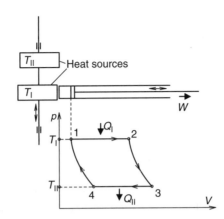

FIGURE **4.7** Carnot cycle.

source (T_I) is in contact with the cylinder bottom, so the cylinder is being gradually filled up with black emission of the cylinder bottom radiating into the cylinder space. The filling process occurs at temperature T_I equal to the photon gas temperature T_1 ($T_1 \approx T_I$) and, according to equation (5.21), at constant pressure. During this process, the piston moves to the right and performs work received through the piston rod.

Process 2–3

In the second process the piston moves continuously up to the extreme right-hand position performing work during expansion of the photon gas according to equation (5.26). In this process no heat source is in contact with the cylinder bottom, and the bottom is assumed to be mirrorlike inside.

Process 3–4

The heat source of T_{II} is in contact with the cylinder bottom through which heat transfer occurs at the infinitely small temperature drop $T_{II} \approx T_3$. Heat is released from the photon gas at a constant temperature and constant pressure; however, the volume occupied by the gas is decreasing until the gas state 4 is achieved.

Process 4–1

The cylinder bottom has no contact with any heat source and the compression of photon gas occurs up to the state of point 1.

The net work W performed in the cycle results from the energy conservation law:

$$W = Q_I - Q_{II} \tag{4.21}$$

where Q_I and Q_{II} are the amounts of heat exchanged, respectively, between the heat sources and the photon gas during processes 1–2 and 3–4. The efficiency η_C of the considered Carnot cycle is the ratio of work W to the cycle input Q_I; $\eta_C = W/Q_I$, where work W is expressed by formula (4.21). Additionally, as the exchanged heat is changing the energy of photon gas according to formula (5.13) and the processes at varying volume occur according to equation (5.25), the following relation can be derived: $Q_{II}/Q_I = T_{II}/T_I$ and thus the Carnot efficiency is:

$$\eta_C = 1 - \frac{Q_{II}}{Q_I} = 1 - \frac{T_{II}}{T_I} \tag{4.22}$$

The Carnot efficiency expressed by temperatures was already mentioned as formula (2.35).

The commonly called Carnot efficiency is in fact the efficiency of the Carnot cycle and is the most important efficiency in thermodynamics. All other defined efficiencies are less general, mostly arbitrary or specifically adjusted to the objects or situations.

One of the most significant properties of the Carnot efficiency is that it is valid independently of the nature of the working fluid and can be applied to any material or field matter used as the working fluid. For example, consider the two machines cooperating in Carnot cycles (Figure 4.8). In machine I the working fluid is photon gas and in machine II the working fluid is the ideal material gas. Both machines operate between the two heat sources at the constant temperatures T_I (hot) and T_{II} (cold), respectively. The machines are linked together

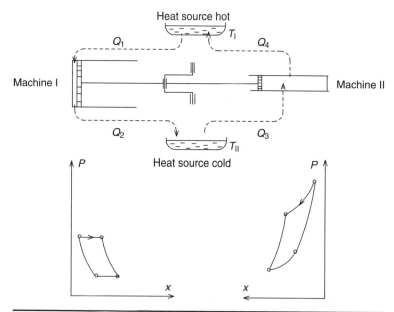

Figure 4.8 The two machines cooperating in two respective Carnot cycles.

and the unit does not exchange work with the surrounding. The two possibilities can be analyzed:

(i) Machine I is an engine whereas machine II plays the role of a heating pump. The directions of the heat fluxes for this possibility are shown in Figure 4.8. In the lower part of the figure the change of parameters in the cycle process is illustrated by diagrams in the p,x system of coordinates, where x is the distance of the piston motion proportional to the volume of working fluid in a respective cylinder. According to the Second Law of Thermodynamics, in the global effect of operating both machines, the cold heat source cannot lose heat and the hot heat source cannot gain. Thus the following inequalities result:

$$Q_1 \geq Q_4 \tag{a}$$

$$Q_2 \geq Q_3 \tag{b}$$

(ii) Machine I is a heating pump and machine II acts as an engine. The directions of the heat fluxes are opposite to those shown in Figure 4.8. Due to the assumed reversibility (Carnot cycle) of both machines the absolute amounts of heat remain unchanged. However, according to the Second Law of Thermodynamics the following inequalities result:

$$Q_1 \leq Q_4 \tag{c}$$

$$Q_2 \leq Q_3 \tag{d}$$

Relations (a) and (c) can be satisfied at the same time only when

$$Q_1 = Q_4 \tag{e}$$

and from relations (b) and (d):

$$Q_2 = Q_3 \tag{f}$$

Interpreting equation (4.21) for the Carnot efficiencies $\eta_{C,I}$ and $\eta_{C,II}$ of the considered machines and taking into account equations (e) and (f), one obtains:

$$\eta_{CI} = 1 - \frac{Q_2}{Q_1} = 1 - \frac{Q_3}{Q_4} = \eta_{CII} \tag{g}$$

Equation (g) shows that the Carnot efficiency does not depend on the nature of the working fluid and can be also applied for radiation.

The Carnot efficiency can be used as a reference value for calculation of exergy efficiency of a thermal engine. Consider the energetic efficiency of an engine:

$$\eta_{E,\,eng} = \frac{W}{Q_1} \tag{h}$$

and exergetic efficiency of the engine:

$$\eta_{B,\,eng} = \frac{W}{B_{Q1}} \qquad\qquad (i)$$

Based on formulae (2.35) and (2.61) the ratio of energetic and Carnot efficiencies is:

$$\frac{\eta_{E,\,eng}}{\eta_C} = \frac{W}{Q_1}\frac{T_I}{T_I - T_{II}} = \frac{W}{B_{Q1}} = \eta_{B,\,eng} \qquad\qquad (j)$$

The exergy efficiency of the engine demonstrates how much the real energy efficiency departs from the ideal efficiency represented by the Carnot efficiency. In the ideal case ($\eta_{E,\,eng} = \eta_C$) the exergy efficiency approaches 100%.

4.6.2   Perfection Degree of Process

Practically, process efficiency can be defined in different ways. For example, energy or exergy can be used for expressing the numerator and denominator of the efficiency. However, the best method for reviewing the process seems to be the application of the degree of perfection recommended by Szargut et al. (1988) for measuring the thermodynamic perfection of a process.

The energy and exergy degrees of perfection are defined analogously for convenient comparison. To determine the degree of perfection, all terms of the energy (or exergy) balance equation are categorized either as useful product, or process feeding, or loss. The perfection degree is then defined as the ratio of useful product to the process feeding. The loss is not disclosed in the perfection degree formula because it is a compensation of the perfection degree to 100%.

The losses can be of two kinds. The first loss appears in most processes during the unavoidable release of the waste heat or matter. The thermodynamic parameters of the waste usually differ from the respective parameters of the environment. Thus, the waste still has certain energetic or exergetic values that are dissipated in the environment unless utilized somehow beyond the considered system in an additional process of "waste recovery." The loss due to waste is called the *external loss* and such loss can be partially recovered. The second loss, noticed only in exergy analysis, appears within the system due to thermodynamic irreversibilities of component processes and such *internal loss* cannot be recovered even partially.

Energy balance can disclose only external loss, whereas the exergy balance can contain the terms of the external and internal losses. Internal loss is calculated from the Guoy–Stodola law. External loss is equal to the energetic or exergetic value of the waste. Internal losses in multiprocess systems can be summed in contrast to external losses, which theoretically can still be utilized in one of the other subsystems.

The concept of perfection degree can include exergy change due to the varying of environment parameters and the specific terms (e.g., gravity input). Thus, in the modified version it can be proposed that the denominator of the degree of perfection represents the feeding terms, gravity input, and exergy change due to the environment variation, whereas the numerator expresses the useful products. For example, for the steady process in which numerous fluxes of energy are exchanged, the exergy degree η_B of perfection can be proposed as follows:

$$\eta_B = \frac{\sum\limits_i B_{\text{use},i} + \sum\limits_k B_{Q,\text{use},k} + W_{\text{use}}}{\sum\limits_j B_{\text{feed},j} + \sum\limits_m B_{Q,\text{feed},m} + W_{\text{feed}} + G - \Delta B_e} \qquad (4.23)$$

where

i is the number of useful exergy fluxes B_{use} obtained from the process, including substance and radiation,
k is the number of useful exergy fluxes $B_{Q,\text{use}}$ of heat,
j is the number of entering exergy fluxes B_{feed}, including substance and radiation,
m is the number of entering exergy fluxes $B_{Q,\text{feed}}$ of heat,
W_{use} is the total work produced,
W_{feed} is the total work consumed,
G is the gravity input, considered if eZergy is applied and
ΔB_e is the exergy gain in case of variation of environment parameters.

Formula (4.23) can be applied also for combined processes in which more than one intended product is obtained (e.g., the combined generation of heat and power-cogeneration). A particular example of application of the energy and exergy perfection degrees, with no work, G and ΔB_e, is discussed, e.g., in Chapter 12, for photosynthesis.

Contrary to the not discoverable internal energy loss, the internal exergy losses have particularly practical significance. The exergy balance should be developed with possibly the most detailed distribution of the internal losses in order to obtain the most exact information about the possibility of perfection improvement of the considered system. For example, the internal exergy loss can be divided into the components corresponding to friction, heat transfer at a finite temperature difference, radiation emission and absorptions, etc.

If, in any part of the considered system, several irreversible phenomena occur, then, in principle, it is possible to calculate only the overall internal exergy loss caused by the phenomena. The splitting of the effects of these irreversible phenomena, occurring simultaneously at the same place and time, is impossible because these phenomena interact mutually (as is mentioned in the discussion of the Fourth Law of Thermodynamics). The splitting of the exergy loss in such a case can be based only on the assumed agreement. For example, for the combustion process the radiative heat exchange occurs between the flame and the surrounding wall. In order to split the effects of irreversible chemical reactions of combustion from the irreversible radiation exchange, it can be assumed that first combustion occurs and then heat exchange takes place. However, with such an assumption the temperature differences in the heat exchange are larger than they really are.

Therefore, it is better to split exergy losses according to the instant and site of occurrence, instead of according to the causes, unless the examined causes occur in different spots and different instants.

Theoretically, distribution of the exergy losses according to location can be also carried out even in a more detailed way than the common method. Application of the thermodynamic equation of the irreversible processes allows for calculation of the rate of entropy generation at a given location of the system, i.e., for calculation of the so-called entropy source. The entropy source can be used for calculation of the local exergy loss due to irreversible phenomena occurring at the given point of the system. However, the calculation of the local exergy losses based on the entropy source method is difficult because in practical cases the calculation of the entropy source of complex irreversible phenomena is difficult.

4.6.3 Specific Efficiencies

Generally, the efficiency of a process can be arbitrarily defined to expose the most important aspect.

For example, the exergy of the hot water generated from solar radiation can be related either to the exergy of heat Q at the sun's surface T_{sun}, $Q \times (1 - T_0/T_{sun})$, to the exergy b_{sun} of the sun's radiation, or to the exergy of heat Q absorbed at the water pipe temperature T_W, $Q \times (1 - T_0/T_W)$. The exergy efficiency increases successively through the above three possibilities due to the decreasing values of the denominators in the efficiency formulas: $Q \times (1 - T_0/T_{sun}) > b_{sun} > Q \times (1 - T_0/T_W)$. An exergy efficiency that relates the process effect to the decrease of the sun's exergy, $Q \times (1 - T_0/T_{sun})$, is unfair because the exposed surface of the water pipe obtains only the solar radiation exergy and the water pipe is independent of irreversible emissions at the sun's surface. Relating the process effect to the exergy of heat absorbed, $Q \times (1 - T_0/T_W)$, favors the exposed surface by neglecting its imperfectness during the absorption of heat Q. Thus, from these three possibilities, comparing the heated water effect to the exergy b_{sun} of the sun's radiation is the best estimation in this analysis.

Other examples of variously defined efficiencies are applied and discussed in the following chapters. However, from the comparative viewpoint of different processes, the best justified definition of the efficiency seems to be equation (4.23).

4.6.4 Remarks on the Efficiency of Radiation Conversion

The following discussion focuses on thermal radiation; however, among available sources of thermal radiation of significant value (high temperature), first of all is solar radiation. Until today there have been observed four possibilities of conversion of radiation (photon gas) into other forms of energy.

The conversion of radiation to *work*, so far not well developed, is one possibility. The theoretical efficiency of such a conversion is discussed in Section 6.4.1. An example of such conversion is the idea of sailing in space due to a photon wind. Another example is the concept of the light-mill, which is also used for measurement of the radiation pressure according to experiments by Lebedev (1901) as well as by Nichols and Hull (1901). The light-mill with its spinning action of a mirror placed on an arm and using the effect of radiation pressure is described by Halliday and Resnick (1967). There are also likely other photon devices or processes using the effect of radiation pressure that have not been invented yet.

Work performed directly by the photon gas can be obtained also within an enclosed space, which, e.g., explains the model considered in Section 5.7. Besides these already mentioned direct applications of the photon stream, some indirect utilizations of solar energy to perform work can also be achieved. An example of indirect utilization of radiation would be the effect combined with gravity and buoyancy observed in the solar chimney power plant. In Chapter 11, such a problem will be analyzed in more detail, using the concept of gravity input.

The conversion of radiation into *heat*, which, e.g., can increase the enthalpy of any working fluid, is based on the absorption of radiation on a surface exposed to solar radiation. The harvesting of heat from solar radiation is discussed in Chapter 10, which also discusses the parabolic solar cooker as a typical example of a device that absorbs solar radiation.

The conversion of radiation into *chemical energy* of substance occurs during the process of photosynthesis, the simplified model of which is discussed in Chapter 12.

The direct conversion of radiation into *electrical energy* occurs in photovoltaic devices; the simplified analysis of such conversion is discussed in Chapter 13.

4.6.5 Consumption Indices

Sometimes instead of efficiency, specially defined indices are used for the estimation of processes. For example, there are some processes that occur spontaneously due to interaction with the environment. Drying, cooling, vaporization, and sublimation are examples of such processes in which the self-annihilation of exergy takes place. Often these processes, especially in industrial practice, are accelerated with the use of the appropriate input. Exergy application for estimation of perfections of these processes reveals some problems.

For example, applying the common exergy efficiency definition— effect and input ratio—leads to the negative or infinite value of the efficiency. Therefore, instead of efficiency some specially defined criteria have to be used for the evaluation and comparison of processes perfection. For example, for drying processes the unit exergy consumption index is defined as the ratio of the exergy of the drying medium used to the mass of the liquid extracted in the form of vapor. In the case of the application of solar energy for drying, the index would express the exergy of absorbed radiation per mass of the vaporized moisture.

Another index can be used for the process occurring in a water cooling tower. Szargut and Petela (1968) propose the evaluation of the process with the index defined as the ratio of the sum of exergy lost in the tower and the heat extracted from the water. The typical value of the index for the cooling tower of a steam power station is about 0.088 kJ of exergy per kJ of heat.

Yet another example of processes which can occur spontaneously in the natural environment is desalination of sea water. In result of such desalination the separated salt and water vapor are obtained. However, desalination can be artificially accelerated, e.g. in a proper installation utilizing solar radiation, and the water vapor can be acquired in form of a condensate. Exergetic evaluation of such combined process can be based on a certain performance index taking into account the exergy input of utilized solar radiation. The exergetic effect of the process is the difference in chemical exergy of the sea water and condensate.

Petela (1990) proposed a specific approach to the exergy annihilation due to spontaneous processes. He considered the natural exergy annihilation rate that expresses the ability of the environment to spontaneously reduce the exergy of the substance or radiation. The natural wind velocity, the environment air temperature and composition, particularly humidity, as well as the solar radiation, the local surrounding surfaces' configuration, and its emissivities, all taken together into account can determine the available exergy effect for annihilation of exergy in the spontaneous processes of drying, cooling, etc. The so-called "wind chill factor" is an example of the concept expressing a certain ability of the environment air.

Therefore, the exergy B of any considered matter not being in equilibrium with the environment, exposed to interaction with the environment, experiences a reduction in its exergy at the natural exergy annihilation rate:

$$r_0(t) = -\left(\frac{\partial B}{\partial t}\right)_{\text{natural}} \tag{4.24}$$

where t is time. The rate r_0, always nonnegative ($r_0 \geq 0$), can be even recognized in some specific problems as the additional property of the environment together with environment temperature, pressure, etc.

Usually, for economic reasons, the natural approach to equilibrium with the environment is enforced by applying the rate:

$$r(t) = -\left(\frac{\partial B}{\partial t}\right)_{\text{forced}} \tag{4.25}$$

which also is always nonnegative ($r \geq 0$). Both rates can be used in the definition of the instant value e_r of the exergetic index (dimensionless) of process annihilation effectiveness:

$$e_r(t) = \frac{r(t) - r_0(t)}{B_{\text{in}}(t)} \tag{4.26}$$

where B_{in} is the driving exergy input flux of the considered process.

The cumulative exergetic index $\bar{e}_r$ of the annihilation process effectiveness within the time period from t_1 to t_2 can be determined as follows:

$$\bar{e}_r \big|_{t_1}^{t_2} = \frac{\int\limits_{t_1}^{t_2} [r(t) - r_0(t)] \, dt}{\int\limits_{t_1}^{t_2} B_{\text{in}}(t) dt} \tag{4.27}$$

An example of the application of formula (4.27) is discussed by Petela (1990) for the forced cooling of small balls that fill up the space with air flow. For the spontaneous processes without any technical input for acceleration of the process, the exergetic effectiveness e_r is the maximum and is equal to infinity. The idea of the index e_r can be developed further in some specific exergy problems related to the spontaneous annihilation of exergy.

4.7 Method of Reconciliation of the Measurement Data

From the balance equations of mass, energy, and exergy one can calculate some unknowns that, for different reasons, were not measured. Number u of such unknowns cannot be larger than the number r of available equations. If $u = r$, the solution obtained is unique. However, the best situation is when $u < r$, i.e., the problem is overdetermined, because there is the possibility of the introduction of new unknowns as the corrections to the measured values. Unavoidable errors of the measurements may cause the equations without introduced corrections to not be fulfilled accurately. Without using the corrections, called the *reconciliation*, it appears that the values of calculated unknowns depend on the calculation variant; i.e., they depend on the selected equations for the calculation procedure, whereas the not used equations are not fulfilled.

The purpose of the reconciliation of balance equations is also to calculate the unique and most probable values of the unknowns, verification of the assumed precision of measurements, and the estimation of real errors as well as their decrease. After the reconciliation the measurement errors decrease the most for the larger errors.

The convenient reconciliation method is proposed by Szargut and Kolenda (1968). The method theory is vast and only its outline is described and illustrated here with a simple calculation example.

The first step of the method is proper preparation of the equations set, called the *conditions set*, which is under consideration. The condition equations have to be independent and the unknowns have to be calculable. Then the following denotation is assumed:

$k = 1, 2, 3, \ldots, r$, is the successive number of total number r of condition equations,

α_i where $i = 1, 2, 3, \ldots, n$, is the successive magnitude being measured,

β_j where $j = 1, 2, 3, \ldots, u$, is the successive unknown.

Thus the condition equations are the set of functions:

$$F_k = F_k(\alpha_1, \ldots, \alpha_n, \beta_1, \ldots, \beta_u) \qquad (4.28)$$

where $u < r < n + u$. The next step is calculation of the *approximate values*, x_j, of unknowns from some condition equations or their combination. Then the following substitutions to the condition equations (4.28) are made:

- measured values (observations) z_i in place of α_i,

- approximate values x_j of unknowns in place of β_j.

Obviously, the obtained equations are not fulfilled:

$$F_k = (z_1, \ldots, z_n, x_1, \ldots, x_u) = -w_k \qquad (4.29)$$

where w_k is the discrepancy of the kth condition equation. To obtain the agreement of all the condition equations, i.e., to obtain $w_k = 0$ for every k, the following are introduced:

- corrections v_i for the observations,

- corrections y_j for the unknowns.

Thus from equations (4.29) is:

$$F_k = [(z_1 + v_1), \ldots, (z_n + v_n), \quad (x_1 + y_1), \ldots, (x_u + y_u)] = 0 \qquad (4.30)$$

For linearization of equations (4.30) the Taylor's series expansion is applied in the neighborhood of experimentally measured values z_i and calculated values x_j:

$$F_k = [(z_1 + v_1), \ldots, (z_n + v_n), (x_1 + y_1), \ldots, (x_u + y_u)]$$

$$\approx F_k(z_i, \ldots, z_n, x_i, \ldots, x_u) + \left(\frac{\partial F_k}{\partial \alpha_1}\right)_0 v_1 + \cdots, \left(\frac{\partial F_k}{\partial \alpha_n}\right)_0 v_n$$

$$+ \left(\frac{\partial F_k}{\partial \beta_1}\right)_0 y_1 + \cdots, \left(\frac{\partial F_k}{\partial \beta_u}\right)_0 y_u \tag{4.31}$$

where

$$\left(\frac{\partial F_k}{\partial \alpha_i}\right)_0 \equiv a_{k,i} \tag{4.32}$$

$$\left(\frac{\partial F_k}{\partial \beta_j}\right)_0 \equiv b_{k,j} \tag{4.33}$$

are the partial derivatives of functions F_k for the measured magnitudes α_i and unknowns β_j calculated at point $(z_1, \ldots, z_n, x_1, \ldots, x_u)$.

Relation (4.29) and the abbreviations (4.32) and (4.33) are introduced into (4.31) and the linearized equations are as follows:

$$\sum_{i=1}^{i=n} a_{k,i} v_i + \sum_{j=1}^{j=n} b_{k,j} y_j = w_k \tag{4.34}$$

Relation (4.34) represents the set of r equations with number n of unknown v_i and number u of unknown y_j, thus $n + u > r$. Therefore, the number $(n + u - r)$ of additional conditions can be introduced for determining the way to choose the corrections. The most logical policy is the method of least squares applied to corrections v_i with a normalization factor of m_i (the standard deviation of the experimental data) as the experimental uncertainty in the measurement of the independent magnitudes y_i. Thus:

$$\sum_{i=1}^{i=n} \left(\frac{v_i}{m_i}\right)^2 = \min \tag{4.35}$$

The minimum expressed by equation (4.35) is conditioned by equations (4.34). The conditional minimization can be solved by the method of Lagrange's multipliers λ_k with the Hamiltonian H. Thus, the simultaneous fulfilling of conditions (4.34) and (4.35) occurs when the following relations are fulfilled:

$$H = \sum_{i=1}^{i=n} \left(\frac{v_i}{m_i}\right)^2 - 2\sum_{k=1}^{k=r} \lambda_k \left(\sum_{i=1}^{i=n} a_{k,i} v_i + \sum_{j=1}^{j=u} b_{k,j} y_j - w_k\right) = \min$$

(4.36)

Applying the necessary conditions for the extreme:

$$\frac{\partial H}{\partial v_i} = 0 \quad \text{and} \quad \frac{\partial H}{\partial y_j} = 0$$

one obtains:

$$\frac{v_i}{m^2} = \sum_{k=1}^{k=r} a_{k,i} \lambda_k$$

(4.37)

$$\sum_{k=1}^{k=r} b_{k,j} \lambda_k = 0$$

(4.38)

The sets of linear equations (4.34), (4.37), and (4.38), whose number is $n + u + r$, allow for calculations of the same amount of unknowns which are v_i, y_j, and λ_k. Accuracy of measurements and the effect of reconciliation of the balance equations can be estimated based on the values of w_k, v_i, and y_j. For example, the correction v_i should not be larger than the respective standard deviation m_i; $(v_i \leq m_i)$.

Example 4.2 Heat is exchanged by radiation between two parallel, infinitely large black surfaces with a vacuum between the surfaces. As shown in Table 4.2 measured were temperature T_1 and T_2 of the surfaces, heat flux q and temperature T_0 of the environment. The exergy loss δb due to irreversible heat transfer has to be calculated.

The considerations are based on the two equations, one determining the heat flux q per $1m^2$, as:

$$q = \sigma \left(T_1^4 - T_2^4\right)$$

(a)

Symbol	Units	Measured or calculated value	Calculated correction	Corrected value
T_1	K	$\alpha_1 = 720$	$v_1 = 0.128958548$	720.128959
T_2	K	$\alpha_2 = 320$	$v_2 = -0.01132146$	319.988679
T_0	K	$\alpha_3 = 291$	$v_3 = 0$	291
q	kW/m²	$\alpha_4 = 13.1$	$v_4 = -1.5241912$	11.5758088
δb	kW/m²	$\beta_1 = 6.853$	$y_1 = -0.76866431$	6.08433569

TABLE 4.2 Data for the Considered Heat Exchange

where $\sigma = 5.6693 \times 10^{-11}$ kW/(m^2 K^4) is the Boltzmann constant for black radiation, and another equation for exergy loss δb as:

$$\delta b = \left(\frac{q}{T_2} - \frac{q}{T_1} \right) T_0 \qquad (b)$$

The considered problem is overdetermined because two independent equations are available and only one unknown has to be calculated; there are two condition equations ($r = 2$), one unknown ($u = 1$) and four measured values ($n = 4$). Equation (a) and (b) are rewritten as function F_1 with the use of measured data:

$$F_1 = C_b \alpha_1^4 - C_b \alpha_2^4 - \alpha_4 = w_1 = 1.53518618 \qquad (c)$$

and as function F_2, from which the preliminary value of unknown β_1 is calculated, thus the discrepancy $w_2 = 0$:

$$F_2 = \frac{\alpha_4 \alpha_3}{\alpha_2} - \frac{\alpha_4 \alpha_3}{\alpha_1} - \beta_1 = w_2 = 0 \qquad (d)$$

For reconciliation of the measurement data the following relations have to be formulated. From equation (4.34), ($k = 1$):

$$C_b 4\, \alpha_1^3 v_1 - C_b 4\, \alpha_2^3 v_2 - v_4 = w_1 \qquad (e)$$

and for ($k = 2$):

$$\frac{\alpha_4 \alpha_3}{\alpha_1^2} v_1 - \frac{\alpha_4 \alpha_3}{\alpha_2^2} v_2 + \left(\frac{\alpha_4}{\alpha_2} - \frac{\alpha_4}{\alpha_1} \right) v_3 + \left(\frac{\alpha_3}{\alpha_2} - \frac{\alpha_3}{\alpha_1} \right) v_4 - y_1 = w_2 \qquad (f)$$

From equation (4.37) we have:
for $i = 1$, (α_1):

$$\frac{v_1}{m_1^2} = C_b 4\, \alpha_1^3 \lambda_1 + \frac{\alpha_4 \alpha_3}{\alpha_1^2} \lambda_2 \qquad (g)$$

$i = 2$, (α_2):

$$\frac{v_2}{m_2^2} = -C_b 4\, \alpha_2^3 \lambda_1 - \frac{\alpha_4 \alpha_3}{\alpha_2^2} \lambda_2 \qquad (h)$$

$i = 3$, (α_3):

$$\frac{v_3}{m_3^2} = 0 + \frac{\alpha_4}{\alpha_2} \lambda_2 \qquad (i)$$

$i = 4$, (α_4):

$$\frac{v_4}{m_4^2} = -\lambda_1 + \left(\frac{\alpha_3}{\alpha_2} - \frac{\alpha_3}{\alpha_1} \right) \lambda_2 \qquad (j)$$

From equation (4.38):

$$-\lambda_2 = 0 \qquad (k)$$

Seven equations (e)–(k) contain seven unknowns: $v_1, v_2, v_3, v_4, y_1, \lambda_1,$ and λ_2. For simplification the standard deviations are assumed to be equal: $m_1 = m_2 = m_3 = m_4 = m = 2$. On the other hand, if standard deviations are equal, then they eliminate themselves from calculations; thus their values are not important. The corrected values of measured magnitudes are shown in the last column of Table 4.2. The Lagrange's multipliers are $\lambda_1 = 0.381047794$ and $\lambda_2 = 0$.

Nomenclature for Chapter 4

A	surface area, m^2
a	abbreviation, formula (4.32), in a reconciliation procedure
a_1	$= 9.7807\ m/s^2$, constant
a_2	$= -3.086 \times 10^{-6}\ 1/s^2$, constant
a_3	$= 1.217\ kg/m^3$, constant
a_4	$= -9.973 \times 10^{-5}\ kg/m^4$, constant
B	exergy, J
$\dot{B}$	exergy rate, W
b	specific exergy, J/kg
b	exergy percentile, %
b	abbreviation, formula (4.33), in a reconciliation procedure
c	concentration of component, kg/m^3
c_p	specific heat at constant pressure, J/(kg K)
D	diffusion coefficient, m^2/s
D	diameter, m
E	energy, J
$\dot{E}$	energy rate, W
e	emission density, W/m^2
e	specific energy of substance, J/kg
e	energy percentile, %
e_r	exergetic index of process effectiveness
$\bar{e}_r$	cumulative exergetic index of the annihilation process effectiveness
F	function in reconciliation procedure
f	composition fraction,
G	gravity input, J
g	gravitational acceleration, m/s^2
H	height of chimney, m
H	Hamiltonian
h	convection heat transferred coefficient, $W/(m^2\ K)$
k	thermal conductivity, W/(m K)
m	mass, kg, or mass flow rate, kg/s
m	standard deviation of the experimental data in reconciliation procedure
N	power, W
n	number of measured unknowns in reconciliation procedure
p	absolute static pressure, Pa

Q	heat, J
q	heat flux per 1 m², W/m²
R	individual gas constant for air, $R = 287.04$ J/(kg K)
r	number of available equations in a reconciliation procedure
r	exergy annihilation rate, W
S	entropy, J/K
T	absolute temperature, K
t	time, s
U	internal energy, J
u	number of unknowns in reconciliation procedure
V	volume, m³
v	correction of observation in reconciliation procedure
W	work, J
w	flow velocity, m/s
w	equation discrepancy in a reconciliation procedure
x	altitude measured from the earth's surface, m
x	distance, m
x	approximate value in place of β in a reconciliation procedure
$x\prime$	altitude above sea level, m
y	coordinate, m
y	correction of unknown in a reconciliation procedure
Z	eZergy, J
z	eZergy percentile, %
z	measured observation on α in a reconciliation procedure

Greek

α	measure variable in reconciliation procedure
β	unknown variable in reconciliation procedure
Δ	increment
δ	loss
η	efficiency
λ	Lagrange's multiplier
Π	overall entropy growth, J/K
ρ	density, kg/m³
σ	Boltzmann constant for black radiation $\sigma = 5.6693 \times 10^{-11}$ kW/(m² K⁴)

Subscripts

B	exergetic
b	black
b	buoyant replacement
C	Carnot
E	energetic
e	environment compensation
eng	engine

F	friction
fin	final
i	successive number
j	successive number
in	inlet
inl	initial
k	successive number
L	laminar
macr	macroscopic
out	outlet
p	pressure
Q	heat
S	system
sun	sun
w	velocity
W	water
x	altitude
0	environment
1, 2, 3, 4	denotations
I, II	denotations

CHAPTER **5**

Thermodynamic Properties of Photon Gas

5.1 Nature of Photon Gas

Modern physics is founded on two theories: general relativity and quantum mechanics. Both theories are defined by Einstein's postulates and are supported experimentally. Although these theories do not directly contradict each other theoretically, they are resistant to being incorporated within one cohesive model. General relativity is the most successful gravitational theory, whereas quantum mechanics has had enormous success in explaining many of the features of our world. The individual behavior of subatomic particles (electrons, protons, neutrons, photons, etc.) appearing in all forms of matter can often be described satisfactorily only by using quantum mechanics.

Radiation is one of the main phenomena appearing in surrounding nature and is described by quantum theory. During radiation of substantial bodies (solids, liquids, and some gases) a part of their energy (e.g., internal energy or enthalpy) trasforms into the energy of electromagnetic waves at a length theoretically from 0 to ∞. Radiation does not require a medium for its propagation. The radiation energy is noncontinuously emitted in the form of the smallest indivisible energy portions, called *photons*. If the energy of a body is not simultaneously supplemented from an external source, then the temperature of the body decreases. The phenomenon of such radiation is called *emission*.

Electrons orbit the nucleus of an atom at fixed orbital distances (called *orbital shells*) which for each atom are different and discrete. In a certain atom the electrons can orbit only at particular distances which are different from those for atoms of other chemical elements. In a stable state the electrons remain at a so-called *ground state*, which is the lowest energy level of an orbital distance.

An orbital shell is associated with a certain energy level. The greater the distance is from the nucleus, the greater is the energy level. Electrons, when excited by the absorption of energy, jump to a higher shell. Photons must have the exact amount of energy to replace electrons in the next shell and, e.g., a nonexact amount of energy in the photons cannot move electrons part of the way between shells. The excited atom stays in such an unstable state until the excessive energy is taken out and then the electron returns to a ground state. Because the amount of energy carried by a photon depends on the wavelength, the atoms of gas can absorb, or emit, energy only at a particular wavelength.

A theoretical model of a perfectly reflecting surface, discussed in Section 2.2.2, reflects 100% of the incident radiation. This means that the surface emits one photon of the same frequency and energy per each such photon absorbed. If a system is surrounded by a boundary with such a perfectly reflecting surface, then the system is radiatively adiabatic. However, the perfectly reflecting surface can participate in heat exchange by conduction and convection if the surface is in contact with any substance.

According to quantum mechanics an arbitrary potential may be approximated with the analogue of the classical harmonic oscillator in which the potential oscillates at the vicinity of a stable equilibrium point. In a one-dimensional harmonic oscillator the particle momentum is specified by a single position coordinate. In an N-dimensional analogue of an oscillator the momentum is considered for N position coordinates.

From a theoretical consideration of oscillators it may be concluded that the energies are quantized and may take only the discrete value multiplied by 1/2, 3/2, 5/2, etc. The lowest achievable energy is not zero but half of such a discrete value. This lowest energy is called the *zero point energy* (or the ground-state energy). In the ground state an oscillator performs null oscilations and its average kinetic energy is positive, although this zero energy is not perceptible as a meaningful quantity. Oscillators make no noise, probably because they have no rest mass. The ground-state energy has many implications, particularly in the quantum gravity problem.

Another conclusion is that the energy levels are equaly spaced, unlike in the *Bohr model* of an atom (which is a positively charged nucleus surrounded by electrons traveling in circular orbits around the nucleus) or in the *"particle in a box"* problem of a particle moving in a straight line, always at the same speed, until it reflects from a wall. For a randomly behaving oscillator the ground-state probability is concentrated at the origin. This means the particle spends most of its time at the bottom of the potential well at the state of little energy. When any energy is supplied, the probability density becomes concentrated at the *classical turning point*, at which the energy coincides with the potential energy.

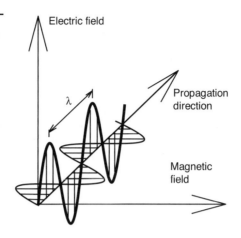

FIGURE 5.1
Simplified scheme of a polarized electromagnetic wave.

Thus, as noted, electromagnetic radiation consists of discrete packets of energy (photons). Each photon consists of both an oscillating electric field component and an oscillating magnetic field component. The electric and magnetic fields are perpendicular to each other; they are also orthogonal to the direction of propagation of the photon.

Thermal radiation is a case of mathematical description by the famous Maxwell equations for the general relation between electrical and magnetic fields in nature. The main Maxwell's equation, with three equation terms, expresses a magnetic field, respectively, to set up: (a) a changing electric field, (b) a current, and (c) the use of magnetized bodies.

The photon electric and magnetic fields flip direction as the photon travels. Figure 5.1 shows the recorded hypothetical history of a photon traveling over some distance and leaving a trail of electric and magnetic fields. The number of flips, or oscillations, that occur in one second is called the *frequency* (v) and is measured in hertz (1/s). The distance in the direction of wave propagation over which the electric and magnetic fields of a photon make one complete oscillation is called the *wavelength* λ, m, of the electromagnetic radiation. As mentioned in Section 3.1, the radiation propagation velocity c is equal $c = \lambda \times v$. The energy E_{ph} of a photon depends on its frequency:

$$E_{ph} = hv \tag{5.1}$$

where h is Planck's constant. The rest mass of a photon is zero. However, considering a photon as a relativistic particle, its energy can be equalized as $h \times v = m \times c_0^2$ and the mass m multiplied by speed c_0 determines the particle momentum P as follows:

$$P = \frac{hv}{c} \tag{5.2}$$

The electromagnetic nature of all photons is the same; however, photons can have different frequencies. The record of energy corresponding to a particular frequency or wavelength is the electromagnetic *spectrum* of radiation. Radiation can be examined by a spectrometer in which the radiation is dispersed by a spectrometer prism into its intensities, which are then measured. The spectrum represents the intensity components of radiation arranged in order of wavelengths or frequencies.

Because of a dual nature assigned to the radiation, the product of radiation can be considered either as the energy of photons or the energy of electromagnetic waves. The concept of the photon, introduced in quantum theory, leads to a certain interpretation of the space between surfaces exchanging heat by radiation. A radiation process can be understood either as a macroscopic effect of heat transfer considered in engineering thermodynamics or as a process of energy exchange by the energy carrier which are the photons. One can perceive not only a radiative heat transfer between surfaces of different temperatures but also the radiation product (photon gas) existing between the surfaces.

Therefore, the radiant heat transfer interpreted as the phenomenon of the electromagnetic wave propagation, can be also described as the behavior effect of a collection of particles (photons) within the space between the surfaces. Such photonics perception allows consideration of the thermodynamic properties of a radiation product, recognizing this product as a collection of energy quanta similar to a substance, which is recognized as a collection of molecules. The quanta generated in the oscillating way can be described by the theory of oscillation. The space with "photon gas" can be studied analogously to a substantial gas and the properties of such a nonsusbtantial working fluid (photon gas) and its behavior in space is one of the subjects of this book on the engineering thermodynamics of thermal radiation.

During emission of radiation by a body the absorption of radiation can occur simultaneously. During absorption of electromagnetic radiation the energy of photons is taken up by the electrons of a body atom. The photon is destroyed when absorbed and its electromagnetic energy is then transformed to other forms of energy, e.g., either electric potential energy, or emission of radiation, or to heat the body and raise its internal energy or enthalpy. It is also possible that due to absorption electrons can be freed from the atom as in the photoelectric effect (photo electrochemical or photovoltaic, discussed in Chapter 12) or in the Compton scattering of energy. The body absorptivity, discussed in Section 3.2, quantifies how much of the incident radiation is absorbed; the remaining amount of incident radiation can be reflected or transmitted. The absorption of radiation during its propagation through a medium is often called *attenuation*.

Usually the absorptivity of substances varies with the wavelength of the radiation because the energy of the incident photon must be similar to an allowed electron transition. As a consequence, a substance can absorb radiation in a range of selected wavelengths. For example, if a substance absorbs radiation in the wavelengths corresponding to the colors blue, green, and yellow, then the substance appears red (i.e., in the unabsorbed radiation wavelength range) when viewed under white light.

Absorption spectroscopy permits identification of a substance by precise measurements of absorptivity at different wavelengths if a substance is illuminated from one side and the intensity of the exiting radiation from the substance is measured in every direction.

From the thermodynamic viewpoint the photon gas can be considered as black radiation. There is no such concept as the emissivity of radiation because radiation is always black and emissivity is related only to the radiating surface, which emits black radiation at a rate determined by the surface property such as the surface emissivity. Therefore, the photon gas cannot have assigned properties such as the emissivity, which should be applied only for surfaces.

This observation can be supported by the following consideration. In the vacuum enclosed within white walls exists an elemental mass dm of black substance. In an equilibrium state the space is filled up with photon gas of black radiation. There is no possibility to exchange energy between the space and outer environment. Therefore, based on the energy conservation law, the energy of the gas will not change if, theoretically, the emissivity of the elemental mass decreases. As a result, the same photon gas (black radiation) will exist in the space in the presence of an elemental mass even at a very low value of emissivity. The element dm of emissivity smaller than one will absorb and emit radiation, but in the space the same black photon gas will always exist.

5.2 Temperature of Photon Gas

The key thermodynamic parameter is temperature, which has been used already in previous considerations (Chapter 2), although in terms of radiation it requires a more detailed interpretative discussion.

A photon is a modern physics model of a single energy quantum—an electromagnetic wave—which appears as a disturbance in the geometric properties of space. The concept of temperature does not apply to a single energy quantum, because temperature is a macro property of matter. The Boltzmann constant, which couples the kinetic energy and temperature of a substance molecule ($\sim kT$) in the suggestive relation $T = \sim h\nu/k$, should not be applied for assigning a temperature to a photon. The concept of temperature in radiation problems can be applied only to a batch of photons.

Geometrically, the simplest model space for consideration of an enclosed photon batch is the space between two flat, parallel, and infinitely large surfaces facing each other, separated by a distance that is large enough to accommodate the radiation with the longest meaningful wavelengths. In such a model space the history of reflections is simple: the whole radiosity of one surface arrives at the other surface, and vice versa. There is no need to involve the complex geometry of surfaces because this does not affect the final results in consideration of the radiation mechanism.

Thus, in such a simple model space the method of the test body and the Zeroth Law of Thermodynamic, discussed in Section 2.1.2, can be used for determination of the temperature of a certain photon population. The temperature of thermal radiation can only be determined indirectly, i.e., by measuring the temperature of the substance with which the radiation is in equilibrium.

There are three illustrative and instructive examples of equilibirum radiation, as suggested by Bejan (1997). The surfaces of the model space are perfectly white and initially the space does not contain any substance. Thus, the system of any considered photon batch within the space would be adiabatic.

First, assume that into the model space is inserted a substance I with the property that it can emit and absorb only a certain single frequency ν_I (to be exact, in a narrow frequency band from ν_I to $\nu_I + d\nu$). This means that if photons of many other frequencies are present in the considered space, then substance I will behave as completely transparent to those photons. After a sufficiently long time, the space will fill with the monochromatic radiation of frequency ν_I and the system (substance and photon gas) will achieve equilibrium at the initial substance temperature T_I. Substance I does not lose its energy; thus, the substance maintains a constant temperature, because any substance emission is reflected back and no energy is lost to the outside of the system. The initial emperature T_I was measured and the radiation can be determined as having temperature T_I.

Second, substance I is replaced by substance II, which is perfectly black. In time, the space fills with photons of all frequencies and the state of equilibrium is achieved. Similar to case I, the considered system with substance II also does not lose energy. The initially measured temperature T_{II} of substance II remains unchanged, and the radiation, which is an instantaneous collection of photons, can be determined as having tempetaturue T_{II}.

Third, both substances I and II, with initially measured temperatures, respectively, T_I and T_{II}, are inserted into the space. At total equilibrium, both substances achieve the same temperature, T_{III}. The equilibrium at temperature T_{III} prevails also between substance I and the monochromatic radiation in the space. Moreover, the equilibrium between substance II and black radiation prevails also at temperature T_{III}. Therefore, for temperature T_{III}, the monochromatic radiation (of substance I) and the sample of photons in the frequency band (from ν_I

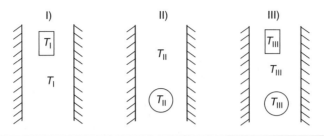

Figure 5.2 Three considered cases (I, II, and III) of equilibrium.

to $\nu_1 + d\nu$) of the black radiation spectrum (of substance II) became the same. Since the frequency ν_1 was taken to be arbitrary, it is possible, with the same result, to vary and analyze in a similar way, band-by-band, the frequency dependence of blackbody radiation. The result is that the batches of radiation, which contain photon collections of different numbers and frequencies, achieve the same temperature at the equilibrium state.

Summing up, the three considered cases are schematically shown in Figure 5.2. In case I, the substance I and monochromatic photon gas at the equilibrium state have the same temperature T_I. In case II, the substance II and the black photon gas at the equilibrium have the same temperature T_{II}. However, in case III, substance I, substance II, and the black photon gas, all at the joint equilibrium state, have the same equilibrium temperature T_{III}.

In practice, radiation temperature can be determined based on the measured radiosity density. It is assumed that based on equations (3.17), (3.20), (3.28), and (3.35) the radiosity density j can be interpreted as the emission density of the gray surface, $j = e = \varepsilon \times \sigma \times T^4$. Therefore, based on measured value of j and assumed value of emissivity ε of the examined surface, the temperature T can be determined as follows:

$$T = \left(\frac{j}{\varepsilon\sigma}\right)^{\frac{1}{4}} \qquad (5.3)$$

The idea of measured temperature based on equation (5.3) is applied in a pyrometer. The radiosity j of the target object is used to deduce the object temperature T as the output signal. The pyrometer, in very simple terms, consists of an optical system and detector. The optical system focuses the energy radiated by an object onto the radiation detector. The output of the detector is proportional to the amount of energy radiated by the target object (decreased by the amount absorbed by the optical system), and to the response of the detector to the specific radiation wavelengths.

For example, the infrared pyrometer measures the energy being radiated from the target only in the 0.7–20 μm wavelength range. To increase the output and make it more readable, the detector can be a thermopile consisting of a number of thermocouples connected in series or parallel. The emissivity of the examined surface has to be known because it is an important variable in converting the detector output into an accurate temperature signal. There is no need for direct contact between the pyrometer and the object. A pyrometer is also suitable for measuring the temperature of moving objects. A chamber such as the cavity radiator (Figure 3.4), submerged in a bath of the known constant temperature, can be used for the calibration of pyrometers.

The radiation temperature allows determination of the emitted energy. For example, frequently considered is the radiation energy only within a specified wavelength range. Such energy can be calculated in a fashion similar to the calculation of the mean specific heat, equation (2.23). In the case of radiation, both sides of equation (3.29) can be divided by T^5:

$$\frac{e_{b\lambda}}{T^5} = \frac{c_1 (\lambda T)^{-5}}{\exp\left[\frac{c_2}{(\lambda T)}\right] - 1} \tag{5.4}$$

Function (3.29) with two variables, λ and T, is now changed into function (5.4) of only one variable, ($\lambda \times T$). For the given temperature T, the fraction of the total energy radiated between 0 and λ can be introduced:

$$\left(\frac{e_{b,\,0-\lambda}}{e_{b,\,0-\infty}}\right)_T = \left(\frac{\int\limits_0^\lambda e_{b\lambda}\, d\lambda}{\int\limits_0^\infty e_{b\lambda}\, d\lambda}\right)_T = \frac{e_{b,\,0-\lambda,T}}{\sigma T^4} \equiv e_{\lambda T} \tag{5.5}$$

The dimensionless values $e_{\lambda T}$ are tabulated, e.g., by Holman (2009). Table 5.1 shows some exemplary values of $e_{\lambda T}$ for the ($\lambda \times T$) ranging from 1000 to 50,000 μm K. For any wavelength range (λ_1, λ_2)

$\lambda \times T$ μm K	$e_{\lambda T}$	$\lambda \times T$ μm K	$e_{\lambda T}$
1000	0.321×10^{-3}	6000	0.73777
2000	0.06672	7000	0.80806
3000	0.27322	8000	0.85624
4000	0.48085	10,000	0.91414
5000	0.63371	50,000	0.99889

TABLE 5.1 Radiation Function

the energy radiated from the black surface is:

$$e_{b,\,(\lambda_1-\lambda_2),\,T} \equiv e_{bT}\Big|_{\lambda_1}^{\lambda_2} = \sigma\, T^4\left(e_{(\lambda T)_2} - e_{(\lambda T)_1}\right) \tag{5.6}$$

Example 5.1   A furnace interior, assumed to be a black surface at temperature of 2000 K, is viewed through a small window in the furnace wall. The window glass plate has the surface area $A = 0.006$ m². The emissivity ε and transmissivity τ of the glass are considered for different two wavelength ranges. For $\lambda < 3.5$ μm: $\varepsilon = 0.3$, $\tau = 0.5$; and for $\lambda > 3.5$ μm: $\varepsilon = 0.9$, $\tau = 0$. Energy absorbed in the glass and energy transmitted can be calculated as follows.

The values $(\lambda \times T)_1 = 0.2 \times 2000 = 400$ μm K and $(\lambda \times T)_2 = 3.5 \times 2000 = 7000$ μm K. From Table 5.1 $(e_{\lambda T})_1 \approx 0$ and $(e_{\lambda T})_2 = 0.8081$. Using formula (5.6) the energy of incident radiation within wavelengths from 0.2 to 3.5 μm is:

$$A\sigma\, T^4\left(e_{(\lambda T)_2} - e_{(\lambda T)_1}\right) = 0.006 \times 5.6693 \times 10^{-8} \times 2000^4 \times (0.8081 - 0)$$
$$= 4.398 \text{ kW}$$

The energy radiation transmitted is: $0.5 \times 4.398 = 2.199$ kW. Radiation energy absorbed within the wavelength from 0.2 to 3.5 μm is $0.3 \times 4.398 = 1.319$ kW and within wavelength from 3.5 μm to ∞, is:

$$A\sigma\, T^4\left(1 - e_{(\lambda T)_2}\right) = 0.006 \times 5.6693 \times 10^{-8} \times 2000^4 (1 - 0.8081) = 2.199 \text{ kW}$$

The total absorbed radiation energy is $1.319 + 0.94 = 2.26$ kW.

5.3 Energy of Photon Gas

Now, with knowledge of the radiation temperature, the energy of the photon gas can be considered. As mentioned in Section 3.4, the cavity radiator model shown in Figure 3.4 is usually the basis for consideration of radiation. Now, it can be assumed that (a) the cavity is filled only with radiation in the form of photon gas, i.e., there is no substance within the cavity and the refractive index is $n = 1$; (b) under conditions prevailing in the cavity the gas remains in thermodynamic equilibrium; (c) the gas has the temperature of a blackbody surface; and (d) the principle of classic phenomenological thermodynamics can be applied to the photon gas. If the dimensions of the cavity are significantly larger than the meaningful radiation wavelengths, the radiation can be recognized as isotropic, i.e., the state of the gas is the same at each cavity point.

A photon gas has rest mass equal to zero. Therefore, energy U, in J, of the gas cannot be related to its mass but rather to its volume V. Thus, the photon gas energy density u, J/m³, i.e., density of radiation energy, is:

$$u = \frac{U}{V} \tag{5.7}$$

The normal radiation intensity i_0 is discussed in Section 3.7. Now, let us introduce into consideration the black normal radiation intensity $i_{b,0}$, W/(m² sr), which is the product of u and the light velocity c_0:

$$i_{b,0} = uc_0 \tag{5.8}$$

When considering radiation in a volume (not radiation flux emitted from a surface), the intensity $i_{b,0}$ can be interpreted as being free to vary within the spherical solid angle ω equal 4π sr; thus:

$$u = \frac{1}{c_0} \int_0^{4\pi} i_{b,0} d\omega \tag{5.9}$$

Using equation (3.28) with interpretation of radiosity j as a blackbody radiation (emission) present in the cavity ($j = e_b$), one obtains:

$$u = \frac{1}{c_0} \int_0^{4\pi} \frac{e_b}{\pi} d\omega \tag{5.10}$$

Integrating (5.10):

$$u = \frac{\sigma T^4}{\pi c_0} \int_0^{4\pi} d\omega = \frac{4\sigma}{c_0} T^4 \tag{5.11}$$

and using (3.21), the internal energy of photon gas is:

$$u = a T^4 \tag{5.12}$$

Equation (5.12) leads us to the conclusion that if the temperature T of the photon gas is held constant, then the radiation energy density u also remains constant. The total energy U of the photon gas within volume V can be calculated by using relations (5.12) and (5.7):

$$U = aVT^4 \tag{5.13}$$

Formula (5.12) can be derived also in a different fashion as shown in Section 5.4. Radiation energy u is a similar quantity to the specific internal energy of a substance.

5.4 Pressure of Photon Gas

Electromagnetic waves may transport linear momentum, i.e., it is possible to exert radiation pressure on an object by shining a light on it. Obviously this pressure is relatively very small. The existence of radiation was predicted theoretically by Maxwell, and was confirmed experimentally by several researchers many years later. For example, solar radiation is so feebly weak that it can be detected only by allowing the radiation to fall upon a delicately poised vane of metal of high reflectivity (a Nichols radiometer). Thus, radiation pressure is a real effect of exerting a positive force due to momentum given up during the interaction of electromagnetic waves with substance matter. Using quantum terminology, radiation pressure is an effect of the photons hitting a target.

Electromagnetic radiation pressure is proportional to the energy intensity of the electromagnetic field and inversely proportional to the speed of light. The pressure acts in the same direction as the wave propagation represented by the Poynting vector (which is the instantaneous vector cross-product of the electric and magnetic fields). While the electric and magnetic fields oscillate in transverse mode, the Poynting vector oscillates in longitudinal mode. The vector can travel through a vacuum, and the vector's magnitude is always positive.

However, saying that electromagnetic waves are transverse waves is not exactly true, because the electric and magnetic fields may randomly alternate their polarity, although then the Poynting vector varies its amplitude in at the unchanged vector direction. Thus, the momentum transfer between a wave and a material target is only due to the existence of this vector.

Radiation pressure in pascals (N/m^2) is equal to the time-averaged Poynting vector magnitude divided by the speed of light. The Poynting vector describes the rate of energy flowing through a surface and has the dimensions of power per unit area. The pressure within the electromagnetic field is considered here as being independent of the properties of the target eventually hit. The nature of the Poynting vector can be considered as the radiosity density j. The radiation pressure p of electromagnetic radiation can be defined either as the force F per unit area A, or using momentum P, or by the radiosity density j, as is shown by the following multi-equation:

$$p = \frac{F}{A} = \frac{\frac{dP}{dt}}{A} = \frac{j}{c} \qquad (5.14)$$

where t is time and c is the speed of light.

The above explanations suggest that the radiation mechanism responsible for the pressure effects can be analyzed from two viewpoints. First, the pressure interaction between radiation and substance can be considered, and second, the pressure effect only within the internal structure of the radiation field can be examined. Both viewpoints are outlined as follows.

The pressure exerted on an object by a given amount of radiation depends on whether the radiation is absorbed, reflected, or transmitted. If radiation is reflected, then the object that reflects must recoil with enough momentum to stop the incoming wave and then send it

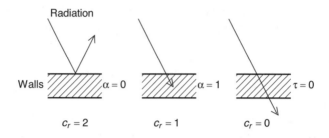

FIGURE 5.3 Examples of walls of typical values c_r of the radiation pressure coefficient.

back out again. When radiation is absorbed, then the object must only stop it. If radiation is transmitted through the object, then there is no pressure effect on the object.

Consider a parallel beam within radiation that neighbors a wall for a time t and that the incident radiation is totally absorbed by the wall (i.e., the wall is perfectly black, $\alpha = 1$). This means the momentum P delivered to the wall is equal to energy divided by radiation speed, $P = U/c$, where U is the radiation energy in J. However, if the radiation energy is totally reflected (i.e., the wall is perfectly white, $\alpha = 0$), the momentum delivered to the wall is twice that given during absorption, $P = 2 \times U/c$. If the incident radiation is partly absorbed and partly reflected ($0 < \alpha < 1$), the delivered momentum lies between U/c and $2U/c$. The transmission of radiation through the wall occurs without momentum effect. Thus, generally, the momentum can be expressed with use of the radiation pressure coefficient c_r as follows:

$$P = c_r \frac{U}{c} \qquad (5.15)$$

The value of c_r can vary theoretically from 0 to 2. Typical examples of momentum are shown schematically in Figure 5.3.

Example 5.2 A parallel beam of radiation with energy flux of 4 W/m^2 falls for 10 s on a white surface ($c_r = 2$) of 3 m^2 area. The energy arriving at the surface is:

$$U = 4 \times 10 \times 3 = 120 \, \text{J}$$

From equation (5.15) the momentum delivered after 10 s of illumination (at $c \approx 3 \times 10^8$ m/s) is:

$$P = c_r \frac{U}{c} = 2 \frac{120}{3 \times 10^8} = 8 \times 10^{-7} \, \text{N s}$$

Based on Newton's law, the average force on the surface is equal to the average rate at which momentum is delivered to the surface:

$$F = \frac{P}{t} = \frac{8 \times 10^{-7}}{10} = 8 \times 10^{-8} \, \text{N}$$

whereas the pressure exerted by radiation on the surface is:

$$p = \frac{F}{A} = \frac{8 \times 10^{-8}}{3} = 2.7 \times 10^{-8}\,\text{Pa}$$

The radiation pressure within the electromagnetic radiation field, with no material target, can be now considered for comparison.

In the quantum theory viewpoint, although photons have no mass, they do have momentum and can transfer that momentum to other particles upon impact. The pressure observed within the electromagnetic field, regardless of any hit target—thus regardless of the properties of the target (α or τ)—can be considered, however, based on the interaction of photons only with an exceptional wall, which is a perfectly white wall.

Then, similar to substance, the equipartition theorem is also applied to the number N of the photons. Analogously to equation (2.6) for a substance gas, the gas energy expression $\rho \times w^2$ in this equation can be substituted by the radiation energy density u, (J/m^3), determined with use of equation (5.1):

$$u = NE_{ph} = Nh\nu \tag{5.16}$$

and according to the equipartition theorem, equation (2.6) changes the interpretation respectively for black radiation as follows:

$$p = \frac{u}{3} \tag{5.17}$$

Using formula (5.12) in (5.17) to determine the internal energy u, the pressure can be expressed as a function of temperature T.

However, the same result can be obtained if the internal energy u is derived from one of the general mathematic relations applied for thermodynamic functions, which are discussed in Section A.3. Taking into account that u is independent of volume:

$$\left(\frac{\partial u}{\partial v}\right)_T = u$$

equation (A.16) is:

$$u = T\left(\frac{\partial p}{\partial T}\right)_v - p \tag{5.18}$$

From equation (5.17) the derivative in (5.18) can be determined as:

$$\left(\frac{\partial p}{\partial T}\right)_v = \frac{1}{3}\left(\frac{\partial u}{\partial T}\right)_v$$

and, substituted together with (5.17), to (5.18):

$$u = \frac{T}{3}\frac{du}{dT} - \frac{u}{3} \tag{5.19}$$

After separating variables in (5.19) and rearranging to:

$$\frac{du}{u} = 4\frac{dT}{T}$$

one can integrate:

$$\ln u = 4\ln T + \ln C \tag{5.20}$$

The integration constant in equation (5.20) can be interpreted as the constant $C = a$, and thus, from (5.20), equation (5.12) can be obtained.

Substituting (5.12) into (5.17) the pressure of the photon gas can be expressed by temperature:

$$p = \frac{a}{3}T^4 \tag{5.21}$$

The considerations of pressure from both viewpoints allow for some additional comments and conclusions. For example, from equation (5.21) we obtain the result that if the temperature T of photon gas is held constant, then the constant remains not only the radiation energy density u, according to equation (5.12), but also the pressure p. Thus, the thermodynamic state of the photon gas is completely determined by arbitrarily choosing one of the possible parameters: u, T, and p. For comparison, determination of the thermodynamic state of substance according to equation (2.1a) requires two parameters.

Radiation pressure can be extremely different in special situations. For example, the solar radiation power incident on the earth's surface is about 1370 W/m². Thus, from equation (5.15), at $c_r = 1$, the radiation pressure at the earth's surface is only $1370/c_0 = 4.57$ μPa. However, the flux density from the NOVA experiment laser beam is about 10^8 W/m², which corresponds to the radiation pressure of the target, 3.3 GPa.

Interaction between solar radiation pressure and the earth's gravity can be observed for various dust particle sizes suspended in the air at a small particle density—about 2 g/cm³. From the whole particle size range only particles of diameter about 0.4–μ.8 μm remain stably suspended, whereas smaller and larger particles drop. For this particular size range of particles the forces caused by gravity and solar radiation pressure are equal. Explanation of this lies in the relation between particle diameter and wavelength. The 0.5-μm-diameter particles result in the best momentum transfer from the 0.5-μm wavelength of the solar radiation spectrum (discussed in Chapter 7) for which the peak of radiation intensity occurs. For this wavelength there appears the maximum radiation pressure, lowest dissipation, and maximum of radiation pressure coefficient ($c_r \approx 2$).

The radiation pressure for the condition of a particle diameter equal to the wavelength is usually referred to as the Mie-scattering regime of a maximum momentum transfer. The particle of a diameter close to the wavelength of a single laser beam can levitate. The dimensions of the actual body or its constituents are very important when considering the radiation pressure coefficient that relates to properties such as absorptivity and transmissivity. A large body can have a large surface area; however, if its diameter, or atomic diameter, is not close to the incident radiation wavelength, then the body has a lower radiation pressure coefficient, and thus a lower momentum is imparted to the body. The Mie's scattering mechanism describes the problem of heating up or increase in mass, and, based on equation (4.1), it explains the source of internal energy of matter that comes from the relatively small difference in energy between the incoming radiation wave and the outgoing reflected wave. The theory for determination of the conditions for radiation pressure coefficient to equal the gravity is beyond the scope of the present book.

Two other examples of radiation pressure aspects are both the radiation pressure used in the design of solar sails and that used in analysis of the so-called *Casimir effect*.

Solar sails, or any light sails using sources other than the sun, are considered for spacecraft propulsion with the use of mirrors with a large surface area. Relatively small solar thrust can be powered by a laser from the earth.

The Casimir effect appears as a small attractive force between the two conducting uncharged plates, which are close to and parallel to each other. According to modern physics, a vacuum is full of oscillating electromagnetic waves of all possible wavelengths that fill the vacuum with a vast amount of invisible energy. With a gradually narrowing gap between the plates, the larger wavelengths are eliminated and fewer waves can contribute to the vacuum energy, which falls below the energy density of surroundings. As a result of the different pressure, a tiny force pulls the plates together. The idea can be developed for different applications.

The nature of the radiation pressure of a photon gas is that radiation with an extremely high frequency can be confronted with the gravity effect especially so that such radiation penetrates any material and acts all over its constituent particles, not just over the material surface. The shadow of a shower of such radiation, imparting impulses of momentum to all bodies in space, can be imagined as the carriers of the gravitational force assigned to the new theoretical concept of *gravitons*.

Whereas photons represent the luminance of electromagnetic radiation, the gravitons represent shadowing and are considered to be negative energy waves, i.e., having no photons or photon-holes. Such considerations, not related to solar radiation, which does not affect gravity, can be carried out for the source of rays in the upper gamma region called *cosmic radiation*. Such considerations represent the sample of a new scientific avenue in which the concept of radiation exergy would be enigmatic. Many problems of radiation pressure can be novel subjects for exergy analyses. It is also noteworthy that if radiation propagates in space filled with substance, then the total pressure in the space is the sum of radiation and substance pressures.

5.5 Entropy of Photon Gas

The concept of entropy, discussed in Secrion 2.5 for a substance and heat, applies also to radiation. The entropy of heat transferred by conduction or convection, in the case of fluids, is calculated as the exchanged heat divided by the appropriate temperature; thus many thermodynamic texts imply erroneously that the entropy of radiation can be determined similarly, i.e., as the transferred radiant heat divided by a surface temperature. However, such an interpretation neglects the effects of the growth of entropy during the irreversible processes of emission and absorption, which are unavoidable in the mechanism of radiant heat transfer. Entropy of the photon gas is different from the entropy of exchanged heat.

Entropy can be derived in different ways. The simplest is the derivation of the entropy density s_S, J/(K m^3), of a photon gas in an equilibrium state residing in a system. Based on Maxwell's relation (A.11c):

$$s_S = \frac{1}{v} \int \left(\frac{\partial p}{\partial T} \right)_v dv \qquad (5.22)$$

The derivative in equation (5.22) can be determined from equation (5.21) and then, after calculation of the obtained integral, one obtains from (5.22):

$$s_S = \frac{4}{3} a T^3 \qquad (5.23)$$

Analogously to the emission density e determined by formula (3.22) for the gray surface of emissivity ε, the entropy of the emission density s, W/(m^2 K), can be introduced. The formula on s is obtained by multiplying the entropy s_S by the factor $c_0/4$. Using relation (3.21): $a \times c_0/4 = \sigma$, as well introducing the surface emissivity ε, the entropy s of the surface emission density can be determined as:

$$s = \varepsilon \frac{4}{3} \sigma T^3 \qquad (5.24)$$

It is noteworthy that the emissivity ε, defined for the surface emission of energy is applied in formula (5.24) for determination of the emission of entropy. Motivation of such application is discussed in Section 8.1. The entropy of radiation can be used in analyses of the irreversibility of different radiation processes.

5.6 Isentropic Process of Photon Gas

One of the possible processes of photon gas is the isentropic process during which the photon gas does not exchange heat with its surroundings. The isentropic process occurs reversibly and the entropy in each elemental process stage remains constant. The entropy in J/K of the gas occupying volume V is determined based on formula (5.23), and the condition of constant entropy in the process is:

$$V \frac{4}{3} a T^3 = \text{const.}$$

or also

$$VT^3 = \text{const.} \tag{5.25}$$

Eliminating temperature T by pressure p with use of formula (5.21), the following condition for other pairs of parameters (V and p) results:

$$pV^{\frac{4}{3}} = \text{const.} \tag{5.26}$$

Dividing side-by-side equations (5.25) and (5.26), the conditions for the third possible pair of parameters, p and T, can be obtained:

$$pT^4 = \text{const.} \tag{5.27}$$

Relations (5.25)–(5.27) can be used for determining one of the unknown parameters. For example, if the initial pressures p_1 and T_1 are known, then at the end of the considered isentropic process of the photon gas at the known temperature T_2, the unknown temperature p_2 can be determined based on relation (5.27) as follows: $p_2 = p_1 \times (T_1/T_2)^4$.

5.7 Exergy of Photon Gas

In engineering thermodynamics, the exergy of radiation, like e.g., energy or entropy, is considered for macro objects that consist of multielement populations of photon gas. Traditionally, to make the consideration easier, the model of the cylinder–piston system with the considered medium is usually applied. The conclusions established

FIGURE **5.4**
Adiabatic process
of a photon gas
within the cylinder
with a piston.

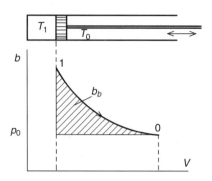

with the use of such a model can have a general meaning and can be applied to the medium in other appropriate situations.

Based on the cylinder–piston system (Figure 5.4), the formula for the exergy of photon gas was first derived by Petela (1964). The cylinder–piston system, from which work can be received through the piston rod, is situated in a vacuum and contains only the trapped black radiation at an equilibrium state at the initial absolute temperature T_1. Heat exchange by conduction or convection does not occur because there is no substance. The cylinder walls and piston base are white; thus there is no heat exchanged by radiation and the process occurring in the cylinder is perfectly adiabatic. On the outer side of the piston there is only black radiation in equilibrium at the constant environmental absolute temperature T_0. The initial pressure p_1 of the photon gas in the cylinder and the pressure p_0 of the environment radiation are determined by formula (5.21). The piston moves frictionlessly to the right if $T_1 > T_0$ or to the left if $T_1 < T_0$, due to different pressures p_1 and p_0.

Hence it is evident that the system performs work regardless of whether the considered photon gas has a higher or a lower temperature than the environmental temperature. In other words, for all temperatures T_1 different from the environmental temperature, the work performed by the system is positive. The objective is to determine the exergy $b_{b,s}$ of radiation enclosed within the system.

Consider the adiabatic (isentropic–frictionless) process (1-0) in which the final pressures on both sides of the piston become equal and the photon gas approaches the final state at pressure p_0. The process occurs according to equation (5.26), as shown in the p, V diagram (Figure 5.4).

According to the definition of exergy the useful work w, J/m^3, performed in the considered process is equal to the exergy $b_{b,s}$ of the photon gas (black radiation) at the initial state 1. The useful work $w \equiv b_{b,s}$, shown in Figure 5.4 by the shadowed area, is determined as the absolute work with subtracted work $p_0 \times (V_0 - V_1)$ spent on the

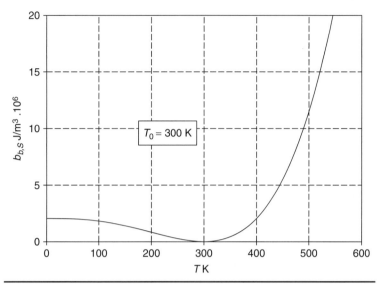

FIGURE 5.5 Exergy of a photon gas as a function of temperature.

compression of the environment:

$$b_{b,S} = \int\limits_{V=V_1}^{V=V_0} p\,dV - p_0(V_0 - V_1) \tag{5.28}$$

Assuming $V_1 = 1\ \text{m}^3$ and using (5.21) and (5.26) in (5.28) the following formula for exergy $b_{b,S}$ of the photon gas within the system is obtained:

$$b_{b,S} = \frac{a}{3}\left(3\,T_1^4 + T_0^4 - 4\,T_0 T_1^3\right) \tag{5.29}$$

The exergy of the photon gas, calculated from formula (5.29) for $T_0 = 300\ \text{K}$ is shown in Figure 5.5. With growing temperature T_1, equal to any arbitrary temperature T, the exergy $b_{b,S}$ diminishes to the 0 value for environment temperature T_0 and then continuously grows.

The correctness of the discussed model is confirmed by the fact that the derived formula is exactly the same as the formula derived by the method of exergy balance of the radiating surface. Such a method is discussed in Chapter 6. Unfortunately, the authors ignore the method of surface exergy balance for verification of their consideration.

Certain modification of the discussed cylinder–piston model is possible in order to make it more realistic and convincing. For example, the enclosed photon gas can be imagined as existing at the presence of the gaseous substance, which is nonradiative (i.e., not emitting and not absorbing). The amount of this substance can be continuously adjusted appropriately to maintain the substance pressure constant and equal to the constant environment pressure. As the substance pressures inside and outside the cylinder always counterbalance, so work is performed only due to the isentropic expansion of the photon gas.

5.8 Mixing Photon Gases

One of the possible processes to be considered is mixing n portions of photon gas of different temperatures T_i and occupying respective volumes V_i. Initially, the portions are separated and enclosed within white walls. Then, the separating walls are removed and all the portions are brought together within a white-wall enclosure of volume V that is the sum of the volumes of the separated portions:

$$V = \sum_{i=1}^{i=n} V_i \tag{5.30}$$

The energy conservation equation for the process of bringing the portion together is:

$$Vu = \sum_{i=1}^{i=n} V_i u_i \tag{5.31}$$

where u is the energy density of the photon gas (black radiation) at a resultant temperature T, determined by using equation (5.12) in equation (5.31):

$$T = \sqrt[4]{r_i T_i^4} \tag{5.32}$$

where r_i is the volume fraction $r_i = V_i/V$. Using (5.21) in (5.31) the formula for resultant pressure p can be also obtained as follows:

$$p = \sum_{i=1}^{i=n} r_i p_i \tag{5.33}$$

where p_i is the pressure of the ith separated portion.

The irreversibility of the mixing can be measured by the overall entropy growth Π, J/(m^3 K), using formula (5.23) as follows:

$$\Pi = \frac{4}{3}a \left(T^3 - \sum_{i=1}^{i=n} r_i T_i^3 \right) \tag{5.34}$$

Formulation of the exergy balance equation is discussed in detail in Chapter 4. In the considered case the exergy balance applied for

Component i	Volume, m^3	Temperature, K	Energy, J	Pressure, μPa	Entropy, mJ/K	Exergy, J
1	30,000	400	0.5809	6.455	1.937	0.0613
2	40,000	600	3.9212	32.677	8.714	1.3888
3	55,000	350	0.6243	3.784	2.378	0.0231
Total	125,000	—	5.1264	—	—	—
Resultant	—	482.5	—	13.670	14.165	1.1322

TABLE 5.2 Input and Output Data for Example 5.2

the considered mixing process of the photon gases consists only of decreases of the exergy of photon gases, before ($b_{b,S,i}$) and after ($b_{b,S}$) mixing, and of the exergy loss δb due to process irreversibility:

$$b_{b,S} = \sum_{i=1}^{i=n} r_i b_{b,S,i} - \delta b \qquad (5.35)$$

The exergy loss δb can be determined by formula (2.60) with applied formula (5.34). The exergy of radiation portions $b_{b,S,i}$ can be determined from formula (5.29) used for respective temperature T_i. The exergy $b_{b,S}$ of the radiation mixture results from formula (5.35) but can be also determined from (5.29) in which temperature T is determined by formula (5.32).

Example 5.3   Within the white walls the three portions of photon gas are mixed together. The initial data were used in calculations according to formulae (5.29)–(5.34), and the output data are shown in Table 5.2.
As shown in Table 5.2 the overall entropy growth is 14.165 – 1.937 – 8.714 – 2.378 = 1.136 > 0; thus the process is irreversible.

5.9 Analogies Between Substance and Photon Gases

Some analogies noticed in thermodynamics problems can be helpful in better understanding and interpreting problems. One example is the analogy amongst mass, heat, and momentum transfer. Obviously, acknowledgment of the analogy is not necessary for effective consideration. The manifestation of analogies originates from a certain formalism, although consideration of the analogies provides a basis for mutual verification of analogous processes and satisfaction. Thermodynamics of the gaseous phase of a substance and of the photon gas also indicates some analogies. Some such analogies for the ideal substance gas and the black radiation can be discussed as follows.

One feature suggesting the analogy is that the substance consists of small elements (molecules), whereas the photon gas is a population of small indivisible portions of energy (photons).

Five *typical reversible processes* are considered in the thermodynamics of substance. Each such process has a constant value of a characteristic parameter: a temperature T for isothermal, a pressure p for isobaric, a volume V for isochoric, entropy S for isentropic, and the fifth process, polytropic, being a kind of process generalization, occurs at the constant exponent n in the constant value expression $p \times V^n$. For a properly taken value of n the polytropic process becomes appropriately one of the four other processes.

The corresponding processes for the photon gas are trivial for isotherm, isobar, and isochoric because only one parameter of the possible three, T, p, and V, is required to be known in order to determine the other two parameters. However, the isentropic process, which for a substance is determined by equation (2.33), has its analogical counterpart in the form of equation (5.26) for radiation.

The kind of process affects the *specific heat* of the processed matter. The relations for radiation can be derived from the general thermodynamic relations discussed in Section A.3. As for the unit of the amount of matter, instead of kg for substance, the unit of m^3 is applied for radiation. The specific heat c_v, J/(m^3 K), of a photon gas at constant volume is derived from relation (A.12) in which the derivative is calculated from equation (5.23):

$$c_v = T\left(\frac{\partial s}{\partial T}\right)_v = 4aT^4 \tag{5.36}$$

The radiation process at constant pressure occurs at constant temperature; thus the isobar simultaneously is the isotherm and $dp = dT = 0$. The specific heat c_p of the photon gas at constant pressure is determined from relation (A.13) as follows:

$$c_p = T\left(\frac{\partial s}{\partial T}\right)_T = \infty \tag{5.37}$$

The specific heat c_T of the photon gas, at constant temperature, can be determined from relation (2.21) as follows:

$$c_T = \left(\frac{dq}{dT}\right)_T = \infty \tag{5.38}$$

As mentioned in Section 2.3, in the thermodynamics of substance the internal energy represents the ability to do work by the substance, which remains during the consideration within the considered system, even if the substance would be in a local motion. Engineering thermodynamics of a substance introduces the concept of enthalpy, which expresses the energy exchanged with the system due to the exchange of the substance. The energetic effect of transportation of the substance through the system boundary is then included in the enthalpy value. Thus, the enthalpy H is defined, according to equation (2.13), as the internal energy U with the added work term $p \times V$ representing the *transportation* through the boundary system, $H = U + p \times V$.

Equation (2.13) interpreted for $V = 1 \ \mathrm{m}^3$ of radiation at temperature T, and after using equation (5.17), is

$$h_b = u + p = \frac{4}{3}u \qquad (5.39)$$

where h_b and u are the enthalpy and energy of the black radiation, respectively.

However, in practice, the energy brought into a system by black emission would be e_b expressed by formula (3.21), which, when using (5.12), becomes:

$$e_b = \frac{c_0}{4}u \qquad (5.40)$$

It results that for determination of energy exchanged with the system, the *enthalpy of radiation* h_b cannot be used as used is the enthalpy h of substance ($h_b \neq e_b$), not even mentioning that the dimensions of e_b and h_b are different. Eventually, other possibilities can be examined. The exchanged radiation energy, e.g., expressed as a certain radiation flux f_r, can be defined by the product of $u/3$ (division by 3 is according to the equipartition theorem) and radiation speed c_0:

$$f_r = \frac{c_0}{3}u \qquad (5.41)$$

The obtained value f_r is also different from e_b, ($f_b \neq e_b$), as well as is different from h_b; $f_b \neq h_b$. In conclusion; only the energy u should be used for expressing radiation within the system. and only emission e_b can be used to express energy of radiation exiting or entering the system. There is no direct analogy between the enthalpy of substance and any magnitude for radiation.

Searching for a radiation analogy to the state equation for a gas (2.1), Bosnjakovic (1965) formulated the following state equation for radiation:

$$pv_B = 0.9 \ RT \qquad (5.42)$$

where p is the pressure, v_B is Bosnjakovic's concept of the specific volume, $\mathrm{m}^3/\mathrm{kmol}$, R is the universal gas constant, $R = 8316 \ \mathrm{J}/(\mathrm{kmol\,K})$, and T is the absolute temperature. Some numerical data are used in Table 5.3 to illustrate the calculation with formula (5.42). For a given temperature, the pressure is determined by formula (5.21) and then v_B is calculated from (5.42). With growing temperature the radiation

T, K	p, kPa	1/v_B, kmol/m³	Comments
1000	2.52×10^{-9}	3.37×10^{-11}	
5700	2.66×10^{-5}	6.24×10^{-9}	Approximate temperature of sun's surface
10^4	2.52×10^{-4}	3.37×10^{-8}	
10^5	2.52×10^{-1}	3.37×10^{-5}	
10^6	2.52×10^3	3.37×10^{-2}	
2×10^7	4.03×10^8	2.7×10^2	Approximate temperature at the sun's center

TABLE 5.3 Some Values of Temperature T, Pressure p, and the Bosnjakovic's Radiation Density ($1/v_B$)

density represented by the value $1/v_B$ also grows. In practice, equation (5.42) can be considered as a simple function of a single variable, e.g., $v_B(T)$, determined as $v_B = 2.97 \times 10^{19}/T^3$.

A peculiar property of photon gas is the Gibbs free energy g, J/m³, which in contrast to substance is equal to zero. This value can be determined by expressing all the members of the formula on g:

$$g = u + p - Ts$$

by the temperature as follows:

$$g = aT^4 + \frac{a}{3}T^4 - \frac{4}{3}aT^4 = 0 \tag{5.43}$$

Equation (5.43) is valid for either monochromatic or black radiation. From equation (5.43) the *chemical potential* μ_r of the photon gas is also zero:

$$\mu_r = \left(\frac{\partial g}{\partial N}\right)_T = 0 \tag{5.44}$$

where N is interpreted as the number of photons in the considered volume of 1 m³. The expression dN is mathematically informal because N is an integer. Another comment is that even if the photon gas is perfectly isolated the number of photons is not conserved. Properties such as the Gibbs free energy, energy, or entropy are introduced for statistically predicted amount of photons in a considered volume, and these properties do not depend on the number of photons that instantaneously exist in the volume.

The influence of temperature T on exergy of photon gas, shown in Figure 5.5, can be compared to the physical exergy of substance varying with temperature. The specific physical exergy b, J/kg, based on formula (2.45) is:

$$b = h - h_0 - T_0 (s - s_0) \tag{5.45}$$

where T_0 is the environment temperature; h and s are the specific enthalpy and entropy, respectively; and h_0 and s_0 are the specific enthalpy and entropy, respectively, of gas in equilibrium with environment. In order to examine the value b near absolute zero temperature one can use the Third Law of Thermodynamics, which, according to Planck, states that the specific heat and entropy of all substances approaches zero for the temperature diminishing to absolute zero. The quantum considerations have led Debye to the conclusion that the specific heat c of crystals near absolute zero varies according to the cubic parabola:

$$c = C_D T^3 \tag{5.46}$$

where C_D is constant. Thus, near to the zero temperature the specific enthalpy is:

$$h = c T^4 \tag{5.47}$$

whereas the specific entropy is:

$$s = \int_0^T \frac{c}{T} dT = \frac{C_D}{3} T^3 \tag{5.48}$$

After inserting (5.46) and (5.47) into (5.45) the exergy of substance near absolute zero becomes:

$$b = C_D T^4 - h_0 - T_0 \left(\frac{C_D T^3}{3} - s_0 \right) \tag{5.49}$$

Thus, for the temperature T diminishing to absolute zero the physical exergy of substance approaches the finite positive value

$$\lim (b)_{T \to 0} = T_0 s_0 - h_0 > 0 \tag{5.50}$$

Figures 5.6 (left) shows how the substance temperature varies with entropy s at constant pressure p. Exergy of such a substance, as a function of temperature T, is shown in Figure 5.6 (right). The straight parts of the plots in Figure 5.6 correspond to the phase changes.

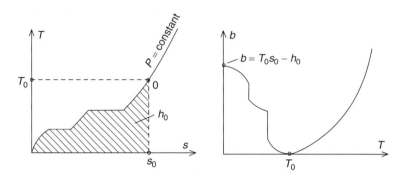

FIGURE 5.6 The isobar (left) and exergy (right) of a substance (after Petela 1964).

Generally, the variation of the exergy of substance and radiation with varying temperature is similar. Neglecting linear sections of the plot in Figure 5.6 (right) the analogy in variation of exergy with temperature can be noticed by comparison to Figure 5.5. In both cases (substance and radiation) the exergy is positive for any temperature different from T_0. The exergies assume zero value for $T = T_0$ and have finite values for temperature approaching absolute zero. For $T > T_0$ the exergies continuously grow with growing temperature. There are characteristic temperatures for substance and for radiation, respectively, for which the energy is larger than the exergy (for large temperature T) and smaller (for small T). This consistent analogy can be also recognized as a certain confirmation of the correctness of the derived formula for radiation exergy.

Nomenclature for Chapter 5

A	surface area, m^2
b	exergy of radiation, J/m^3
b	specific physical exergy of substance, J/kg
a	$= 7.564 \times 10^{-16}$ $J/(m^3\ K^4)$, universal constant
C_D	constant, $J/(kg\ K^4)$
c	speed of light, m/s
c	specific heat of solid, $J/(kg\ K)$
c_0	speed of light in a vacuum, m/s
c_1	$= 3.743 \times 10^{-16}$ $W\ m^2$, the first Planck's constant
c_2	$= 1.4388 \times 10^{-2}$ $m\ K$, the second Planck's constant
c_r	radiation pressure coefficient
c_p	specific heat of photon gas at constant pressure, $J/(m^3\ K)$
c_T	specific heat of photon gas at constant temperature, $J/(m^3\ K)$
c_v	specific heat of photon gas at constant volume, $J/(m^3\ K)$
E	energy, J

e density of emission energy, W/m^2

F force, N

f_r auxiliary concept of radiation flux, W/m^2

g Gibbs free energy for radiation, J/m^3

H enthalpy of substance, J

h $= 6.62 \times 10^{-34}$ J s, Planck constant

h specific enthalpy of substance, J/kg

$i_{b,0}$ black normal radiation intensity, $W/(m^2 \text{ sr})$

J radiosity, J

j radiosity density, W/m^2

k $= 1.3805 \times 10^{-23}$ J/K, the Boltzmann constant

m mass, kg

N number of photons

N number of dimensions of the oscillator analogue

n refractive index

n integer number

n polytropic exponent

P momentum, kg m/s

p static absolute pressure, Pa

q heat, J/m^3

r volume fraction

s entropy of emission density, $W/(m^2 \text{ K})$

s specific entropy of substance, J/(kg K)

s_S entropy of photon gas, $J/(K \text{ m}^3)$

T absolute temperature, K

T_0 environment temperature, K

t time, s

U black radiation energy, J

U internal energy of substance, J

u density of radiation energy, J/m^3

u specific internal energy of substance, J/kg

V volume, m^3

v_B Bosnjakovic's concept of the specific volume of radiation, $m^3/kmol$

w work, J/m^3

Greek

α absorptivity of surface

ε emissivity of surface

λ wavelength, m

μ_r "chemical potential" of the photon gas, J/m^3

ν oscillation frequency, 1/s

ρ mass density, kg/m^3

σ $= 5.6693 \times 10^{-8}$ $W/(m^2 \text{ K}^4)$, Boltzmann constant for black radiation

τ transmissivity of surface

Subscripts

b	black
i	successive number
ph	photon
S	system
T	temperature
1,2	different cases
I, II, III	different cases
0	environment
0, ∞	zero or infinity
λ	wavelength

## CHAPTER 6

Exergy of Emission

6.1 Basic Explanations

A photon gas trapped in a space surrounded by mirrorlike walls was considered in Section 5.7. The product of the emission process is the photon gas, which is black radiation with a temperature equal to the temperature of the emitter. Emissivity of the emitter, e.g., the emissivity of a solid surface, determines the surface ability measured by the rate at which the black radiation is produced. Thus, e.g., **the perfect gray surface of the emissivity ε emits black radiation in an amount determined by the emissivity ε.** In other words, the density of emission e_b expresses the amount of emitted black radiation energy from 1 m² of black surface at ε = 1, whereas density $e(e = ε × e_b)$, determined by formula (3.22), expresses the amount of emitted black radiation energy from 1 m² of gray surface, at a rate reduced by ε ≤ 1.

For example, the measured emission of radiation from any body allows for directly determining the temperature of the body only if it is black, as discussed in Section 5.2. However, if the examined body is gray, its real temperature can be determined if the emissivity ε of the body is known or guessed. Since the emissivity is smaller than one, the examined real temperature of the body is appropriately higher than the temperature resulting from the measured emission.

The black emission e_b has exergy b_b; however, the rate of emission e of the gray surface is smaller ($e = ε × e_b$) and has exergy b of emission e, also reduced by ε:

$$b = εb_b \qquad (6.1)$$

If the exergy b_b of the black surface emission at temperature T is known, then the exergy b emitted from the gray surface at temperature T and emissivity ε can be determined from equation (6.1).

The definition of exergy, given by equation (2.45), can be interpreted for the photon gas. Exergy is a function of an instant state of a matter (e.g., a photon gas at the considered instant) and of the state of this matter in the instant of equilibrium with the environment. Such equilibrium is the basis for determination of the reference state for the exergy of the photon gas.

The environment consists of many bodies at different temperatures and with different radiative properties (e.g., emissivities or transmissivities). The dominant temperature of the environmental bodies can be assumed to be the standard (averaged) environmental temperature T_0. As discussed previously, the surface always emits black radiation; thus the environment surface at temperature T_0, regardless of the surface properties, emits black emission at temperature T_0. The properties of the surface determine only the rate at which the emission occurs. Thus, the environment space permanently contains the black radiation at temperature T_0 and this radiation is in equilibrium with the environmental surfaces at T_0.

Such reasoning leads to the conclusion that **the exergetic reference state for a photon gas (black radiation) is its state at temperature T_0; such a reference state depends only on temperature T_0 and does not depend on diversified values of emissivities of the environmental bodies.** All surfaces emit only black emission and the emissivities of the surfaces (e.g., of any two surfaces x and y) determine only the effect of exchange of emission energy (e_{x-y}) or exergy (b_{x-y}) between the surfaces.

The black emission exergy b_b, expressed by formula (6.1), is always a function only of temperature T of the considered surface and of the environmental temperature T_0, $b_b = f(T, T_0)$. No pressure has to be considered for establishing the exergy reference state for radiation. The pressure of the environmental substance does not affect the radiation, which is not a substance, whereas the pressure of radiation is determined only by the radiation temperature.

It can also be stated that the concept of a perfectly black surface plays a basic role in the exergetic considerations of radiation, and **the concept of emissivity is applied only for a surface but not for a photon gas.**

Consideration of the interaction between surfaces can be significantly simplified in the case when the considered surfaces are models of black surfaces. The obtained results of such a consideration, although valid exactly only for the model surfaces, often allows for obtaining practical qualitative information with acceptable accuracy for situations with nonblack surfaces.

6.2 Derivation of the Emission Exergy Formula

Determination of the radiation exergy of surfaces is very important in practice. Radiation exergy allows evaluation of energy resources represented by the hot radiation of the sun, or by any other hot radiation (i.e., a surface hotter than the environment), and also by cold radiation (i.e., a surface colder than the environment). Thus, the radiation exergy of a surface is the pivotal problem in the engineering thermodynamics

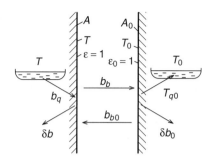

FIGURE **6.1** Radiating parallel surfaces.

of thermal radiation and deserves particular attention. The formula for exergy $b_{b,s}$ (J/m^3) of the photon gas enclosed in a system was derived in Section 5.7. Now, the exergy b_b (W/m^2) of the emission density of a black surface, or any flux of propagating radiation, will be determined.

Simple derivation of the emission exergy of a black surface, published for the first time by Petela (1961b) in Polish and then republished in English, Petela (1964), is based on the balance of the emitting surface according to the model shown in Figure 6.1. The two surfaces A and A_0 are black, flat, infinite, parallel, facing each other, and they enclose the space without substance (vacuum) and interchange heat by means of radiation. This model is often selected for consideration because the space is enclosed by the simplest possible geometry involving only two surfaces. Each surface is maintained at a constant temperature due to the exchange of the compensating heat with the respective external heat sources. Surface A_0 at temperature T_0 represents the environment, whereas surface A at arbitrary temperature T emits the considered radiation. The simplicity of the model of black surfaces is that there is no reflected radiation to be considered.

In order to derive the formula for the emission exergy density b_b of a black surface the following exergy balance for surface A, in the steady state, is considered:

$$b_0 + b_q = b_b + \delta b \tag{6.2}$$

where the terms in equation (6.2) or in Figure 6.1, all in W/m^2, are:

b_b, b_0 exergy of emission density of surfaces A and A_0, respectively;

b_q, b_{q0} change in exergy of respective heat source;

$\delta b, \delta b_0$ exergy loss due to irreversibility of simultaneous emission and absorption on the respective surface.

From the definition of exergy, the radiation of a surface at the environment temperature is:

$$b_0 = 0 \tag{6.3}$$

The change in exergy of the heat source, based on formula (2.61), is:

$$\delta b_q = q \frac{T - T_0}{T} \tag{6.4}$$

where q, W/m^2, is the heat delivered by the heat source of temperature T. This is the amount of heat that allows surface A to emit and maintain its constant temperature T. This is also the heat exchanged by radiation between surfaces A and A_0 and calculated from the energy balance of surface A based on formula (3.21):

$$q = \sigma \left(T^4 - T_0^4 \right) \tag{6.5}$$

The overall entropy growth Π due to simultaneous emission and absorption of heat taking place at surface A is:

$$\Pi = -\frac{q}{T} + s - s_0 \tag{6.6}$$

where $-q/T$ is the decrease in entropy of the heat source at temperature T, based on formula (2.39); and s and s_0 are the entropy of emission densities from surface A and A_0, respectively, determined by formula (5.24) (at $\varepsilon = 1$).

Based on the Gouya–Stodola law (2.60), the exergy loss δb is:

$$\delta b = T_0 \Pi \tag{6.7}$$

Making use of (5.24) and (6.3)–(6.7) in equation (6.2), and after some rearranging, the formula for the exergy of emission density b_b, W/m^2, of the black surface A is obtained:

$$b_b = \frac{\sigma}{3} \left(3T^4 + T_0^4 - 4T_0 T^3 \right) \tag{6.8}$$

The term in parentheses in equation (6.8), characteristic for radiation exergy, is discussed in Section 6.3.

The convenient form of equation (6.8) for practical calculation of the total exergy B_b (W) of emission for the whole black surface area A can be obtained by application of constant C_b, used already in formula (3.23), as follows:

$$B_b = AC_b \left[3 \left(\frac{T}{100} \right)^4 + \left(\frac{T_0}{100} \right)^4 - 4 \left(\frac{T_0}{100} \right) \left(\frac{T}{100} \right)^3 \right] \tag{6.9}$$

where $C_b = 5.6693$ W/(m^2 K^4).

Based on formula (6.1) as well as on formulae (6.8) and (6.9), the exergy of the gray surface with emissivity ε can be determined as follows:

$$b = \varepsilon \frac{\sigma}{3} \left(3T^4 + T_0^4 - 4T_0 T^3 \right) \tag{6.10}$$

$$B = A\varepsilon C_b \left[3 \left(\frac{T}{100} \right)^4 + \left(\frac{T_0}{100} \right)^4 - 4 \left(\frac{T_0}{100} \right) \left(\frac{T}{100} \right)^3 \right] \tag{6.11}$$

Formula (5.29), derived based on study of the expansion of a photon gas in the cylinder with a piston, also contains the characteristic expression in parentheses in formula (6.8). The derivation of formula (5.29) may give rise to some doubts because the work carried out by the considered system during the filling of the cylinder with photon gas was not taken into account. However, such filling is achieved by means of emitting the radiation of a certain body, at the cost of energy of that body, or of some heat that is conducted to that body. Therefore, the process of filling cannot be taken into account if the exergy of a photon gas, already existing within the cylinder, is considered. The derived formula (6.8) confirms the correctness of the above discussion in neglecting the filling cylinder in the derivation procedure of formula (5.29). The exergy values from formulae (5.29) and (6.8) differ only by a factor of $c_0/4$, which results from purely geometrical consideration.

The method of exergy balance of radiation surface allowed for undisputed derivation of formula (6.8) on the exergy of black emission. However, it is noteworthy that the possible application of such a method to any considered exergy radiation problem is usually missed by many authors. The exergy balance method may be used not only for the surface of known temperatures and properties but also for any arbitrary radiation reaching certain surfaces and coming from an unknown source. For example, in Chapter 7 such a case is analyzed to determine the exergy of arbitrary radiation of an irregular spectrum, e.g., determined by measurement.

Finally, it can be shown that the exergy of the density of black emission can be derived also based on the exergy definition equation (2.45) in which enthalpy has to be interpreted as the emission density, according to (3.21), and the respective entropy of the emission density, according to (5.24). Substituting appropriately to formula (2.45) the following equation is obtained:

$$b_b = \sigma T^4 - \sigma T_0^4 - T_0 \left(\frac{4}{3}\sigma T^3 - \frac{4}{3}\sigma T_0^3 \right) \tag{6.12}$$

which can be rearranged to the exact formula (6.8). This method of derivation of b_b, applied by Petela (1974), confirms again the correctness of formula (6.8), although the correctness of this method was previously uncertain until it was disclosed that the substance enthalpy and entropy in formula (5.24) can be replaced, respectively, by emission and its entropy of black radiation. The method is also applied for any arbitrary radiation, as shown in Section 8.4.

6.3 Analysis of the Formula of the Exergy of Emission

Analysis is carried out on the exergy of emission from the perfectly gray surface expressed by equation (6.10). As the subject of the following analysis, equation (6.10) is now rewritten:

$$b = \varepsilon \frac{\sigma}{3} \left(3T^4 + T_0^4 - 4T_0 T^3 \right) \tag{6.13}$$

where ε is the emissivity of the considered gray surface, T is the temperature of this surface, and T_0 is the environment temperature.

The mathematical analysis published for the first time by Petela (1964) reveals first of all that exergy b determined by equation (6.13) is always positive and has the lowest value zero when $T = T_0$:

$$(b)_{T=T_0} = 0 \qquad (6.14)$$

The above conclusions result even more explicitly after transformation applied by Planck for the entropy considerations. According to this transformation the expression in brackets of equation (6.13) can be presented in another form:

$$3T^4 + T_0^4 - 4T_0T^3 \equiv (T - T_0)^2 \left(3T^2 + T_0^2 + 2T_0T\right) \qquad (6.15)$$

The right-hand side of relation (6.15) is the product of two always-positive expressions. For any different temperatures $T \neq T_0$, the exergy of radiation is positive.

The exergy b also reaches zero if the considered surface is white (i.e., perfectly reflecting, $\varepsilon = 0$). From (6.13) it results:

$$(b)_{\varepsilon=0} = 0 \qquad (6.16)$$

It results from formulae (6.13) and (3.22) that the exergy of emission, for the environment temperature approaching absolute zero, is equal to the emission:

$$\lim_{T_0 \to 0} (b) = \varepsilon \sigma T^4 = e \qquad (6.17)$$

It is noticed that the characteristic term in brackets in formula (5.29), appearing also in formula (6.13), was derived by Petela (1964) from consideration of the work done by the cylinder–piston system and without using the Stefan–Boltzmann law (3.21). The obtained equation (6.13) can be recognized as being independent of equation (3.21). Therefore, the energy of emission e can be interpreted as the particular case of the exergy of this emission at the theoretical condition $T_0 = 0$, or that the Stefan–Boltzmann law expresses the exergy of emission when the environmental temperature equals absolute zero.

In the conditions of cosmic space, interpretation of the environment becomes specific and significantly different from the earth's environment. The environmental temperature in such conditions, considered within the large range from zero to infinity ($0 < T_0 < \infty$), is justified, whereas under earth's conditions the environmental temperature ranges only a little.

As the surface temperature T approaches absolute zero, the exergy of emission expressed by formula (6.13) approaches the finite value:

$$\lim_{T \to 0} (b) = \varepsilon \frac{\sigma}{3} T_0^4 \qquad (6.18)$$

This means that the so-called "cold" radiation (discussed also in Section 6.8) emitted by the surface at temperature smaller than T_0 represents a certain practical value.

Equation (6.13) changes to the form of equation (5.29) if the photon gas enclosed within a system is considered. Petela (1964) mentioned for the first time that there is a peculiar theoretical case of the lack of radiation in a so-called "empty tank." It appears that the exergy of radiation matter in the case when its amount is zero, which corresponds also to the theoretical case of the photon gas temperature $T = 0$, has an exergy value larger than zero, similarly to the exergy for the substance matter.

We obtain this result from the following reasoning. Neglecting all field matter (e.g., gravity) and neglecting interpretation of a ground-state energy, discussed in Section 5.1, according to which even in a vacuum there still exist the null oscillations ("idling" photon oscillations), one can imagine the empty tank, with all mirrorlike walls, as the case of no radiation, thus no radiation temperature, $T = 0$. For such interpretation of the empty tank situation the exergy b_{ET} of the radiation vacuum results from equation (5.29) for $T = 0$ as follows:

$$\lim_{T \to 0} (b_{b,S}) = \frac{a}{3} T_0^4 \equiv b_{ET} > 0 \qquad (6.19)$$

Equation (6.19) expresses a certain exergetic value for the theoretical situation in which the exergy of an "empty tank," from the radiation viewpoint, is proportional to the environment temperature in the fourth power. This peculiarity of radiation exergy was mentioned later also by Parrott (1979).

Based on equation (6.13), Figure 6.2 illustrates the exergy b_b (solid thick line) of emission density of a black surface ($\varepsilon = 1$) at the constant

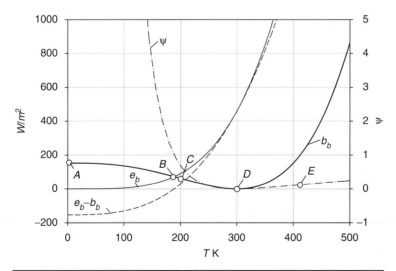

Figure 6.2 Emission e_b, exergy b_b, difference $(e_b - b_b)$, and the exergy/energy ratio ψ as a function of surface temperature T, at $T_0 = 300$ K.

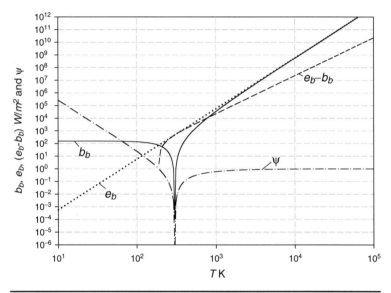

FIGURE 6.3 Emission e_b, exergy b_b, difference $(e_b - b_b)$, and the exergy/ energy ratio ψ as a function of surface temperature T in a very large range, at $T_0 = 300$ K.

value of the environment temperature $T_0 = 300$ K. As shown, the exergy b_b has its minimum at point D. There are the inflexions at points A and C. For comparison, Figure 6.2 presents also energy e_b (solid thin line) of emission density according to equation (3.21). At point B the energy emission equals the exergy of emission. Point B can be determined by comparison of equation (6.13) and (3.21) ($b_b = e_b$). The exergy of black emission is larger than the energy of such emission if the radiation temperature T is small enough ($T < T_B$), and the temperature at point B is $T_B = T_0/4^{1/3}$.

Figure 6.2 presents also the exergy/energy radiation ratio (dotted line) and the difference $e_b - b_b$ (dashed line) which, together with other peculiarities of radiation exergy, are discussed in the following. Figure 6.3, in comparison with Figure 6.2, presents the considered variables for the wide range of temperature T.

6.4 Efficiency of Radiation Processes

6.4.1 Radiation-to-Work Conversion

Thermal radiation can be converted by different processes. Work is a process of energy transfer during which the energy does not degrade. For this reason the work is used to define the exergy. A real *energy conversion efficiency*, η_E, of thermal radiation into work can be defined as the ratio of work W, performed due to utilization of the radiation, to the energy e of this radiation:

$$\eta_E \equiv \frac{W}{e} \qquad (6.20)$$

In an ideal (reversible) process the maximum work W_{max} can be obtained from radiation energy. Then, such work is the exergy of the radiation, $W_{max} \equiv b$, and efficiency η_E changes to the maximum conversion efficiency $\eta_{E,max}$ which is equal to the so-called *exergy/energy radiation ratio* ψ defined for the first time by Petela (1964):

$$\frac{b}{e} = \eta_{E,max} \equiv \psi \qquad (6.21)$$

If the emission density e from formula (3.22) and exergy b of emission density from formula (6.10) are used in (6.21), then:

$$\psi = 1 + \frac{1}{3}\left(\frac{T_0}{T}\right)^4 - \frac{4}{3}\frac{T_0}{T} \qquad (6.22)$$

where T is the temperature of the considered radiation.

Therefore, ψ represents the relative potential of maximum energy available from radiation. This characteristic ratio ψ has the significance similar to that of the Carnot efficiency for heat engines. The term ψ is reluctantly called the *efficiency* because, as also Parrott (1979) showed later, it can have values larger than unity.

The *exergy conversion efficiency* η_B of thermal radiation into work can be defined as the ratio of the work W, performed due to utilization of the radiation, to the exergy b of this radiation:

$$\eta_B = \frac{W}{b} \qquad (6.23)$$

Introducing (6.20) to (6.23) to eliminate the work W, and then using equation (6.21) to eliminate the exergy b, one obtains that the exergy efficiency of conversion of radiation exergy to work is equal to the ratio of the real and the maximum energy efficiencies:

$$\eta_B = \frac{\eta_E}{\eta_{E,max}} \leq 1 \qquad (6.24)$$

Using equation (6.21) in (6.24) to eliminate $\eta_{E,max}$, the ratio ψ is equal also to the ratio of energy-to-exergy efficiency of the radiation conversion to work:

$$\psi = \frac{\eta_E}{\eta_B} > 1, = 1, \text{ or } < 1 \qquad (6.25)$$

which can be larger than, equal to, or smaller than unity.

It is noteworthy that, in the consideration of the heat engine cycle with a working fluid, an interpretation the same as in (6.25) can be derived.

The ratio ψ is a function of the two temperatures

$$\psi\,(T, T_0) = \left(\frac{b}{e}\right)_{T, T_0} = 1 + \frac{1}{3}\left(\frac{T_0}{T}\right)^4 - \frac{4}{3}\frac{T_0}{T} \qquad (6.26)$$

Figure 6.2 presents the example of the ratio ψ (dotted line) for $T_0 = 300$ K. With the growing temperature T from zero to infinity the value ψ decreases from infinity to the minimum value zero for $T = T_0$ and then increases, with inflexion point E, to the unity:

$$\lim_{T \to \infty}(\psi) = \lim_{T \to \infty}\left[1 + \frac{1}{3}\left(\frac{T_0}{T}\right)^4 - \frac{4}{3}\frac{T_0}{T}\right] = 1 \qquad (6.27)$$

However, in spite of ψ approaching unity for infinite temperature T the difference $(e_b - b_b)$ between energy e and exergy b does not approach expected zero, but it does approach infinity:

$$\lim_{T \to \infty}(e - b) = \lim_{T \to \infty}\left\{\varepsilon\sigma\left[T^4 - \frac{1}{3}\left(3T^4 + T_0^4 - \frac{4}{3}T_0 T^3\right)\right]\right\} = \infty \qquad (6.28)$$

The difference $(e_b - b_b)$ from the growing temperature T of the negative values grows indefinitely as shown (dashed line) in Figure 6.2. For the large values of temperature T, the influence on the difference and ψ is shown also in Figure 6.3.

The ratio ψ is dimensionless because energy and exergy are expressed in the same units; however, for some interpretations ψ can be recognized as the amount of kJ of exergy per amount of kJ of energy for any radiation at given temperatures T and T_0. The ratio ψ has a certain practical significance: although the ψ was not defined as efficiency, it can be recognized in the same way as the efficiency of the maximum theoretical conversion of radiation energy to radiation exergy. For example, for any arbitrary radiation of the known energy and at certain presumable temperature T, the exergy of this radiation can be approximately determined as the product of the considered energy and the value ψ taken for this temperature T. Table 6.1 presents some data for the characteristic values of temperatures.

The ratio ψ can be expressed in a more general way as a function only of one variable, by using the temperature ratio $x \equiv T/T_0$ in equation (6.26) as follows:

$$\psi\,(x) = 1 + \frac{1}{3x^4} - \frac{4}{3x} \qquad (6.29)$$

T	e_b	b_b	$e_b - b_b$		Point in
K		W/m²		ψ	Fig. 6.2
0	0	153.1	–153.1	∞	A
100	5.67	136.1	–130.4	24.0	—
189	72.3	72.3	0	1	B
200	90.7	62.4	28.3	0.6875	C
300	459.2	0	459.2	0	D
407	1556	179.8	1376	0.1156	E
1000	56,693	34,169	22,524	0.6027	—
3000	4,592,000	3,980,000	612,000	0.8667	—
6000	73,474,000	68,570,000	4,898,000	0.9333	—
∞	∞	∞	∞	1	—

TABLE 6.1 Some Data on e_b, b_b, $(e_b - b_b)$, and ψ for Different T ($T_0 = 300$ K)

Figure 6.4 shows that with growing x from 0 to 1 the ratio ψ decreases from infinity to zero, and then with growing x from 1 to infinity the ratio ψ increases asymptotically to 1.

Example 6.1 The value $\psi = 0.2083$ for a black emission at temperature $T = 473$ K (200°C) can be calculated from formula (6.26) at $T_0 = 300$ K and $x = 473/300 = 1.577$. In Section 7.3.5, Example 7.1, the ψ_{wv} value for water vapor at 473 K and $T_0 = 300$ K is calculated as $\psi_{wv} = 0.185$. The smaller value of ratio ψ_{wv} for water vapor, in comparison with black surface radiation ($\psi_{wv} < \psi$) results from

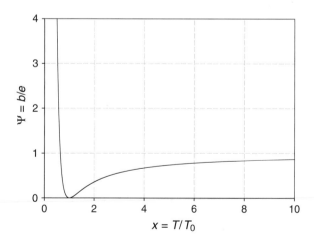

FIGURE 6.4 Ratio ψ as a function of ratio x.

significant difference in spectra of the water vapor and the black surface. In Example 7.5 (Section 7.6.1) for solar radiation, the difference between calculated $\psi_S = 0.9326$ for the considered solar spectrum and the value $\psi = 0.9333$ for a black surface at 6000 K is insignificantly smaller because the solar spectrum is not much different from the black surface spectrum.

6.4.2   Radiation-to-Heat Conversion

Besides work, heat is another process of radiation conversion. For example, such conversion can be considered based on a scheme depicting the absorption of incident radiation (Figure 6.5).

Figure 6.5 shows schematically the fluxes of energy, exergy, and entropy. To simplify considerations it is assumed that the surface of temperature T is black ($\varepsilon = 1$) and the emission e of this surface arrives at the absorbing gray surface of temperature T_a and emissivity ε_a. The heat receiver at temperature T_a absorbs the radiation heat q exchanged between the surfaces:

$$q = \varepsilon_a (e - e_a) \tag{6.30}$$

where e and e_a are the black emission densities calculated, respectively, for temperatures T and T_a. The energy conservation equation for the balanced system (i.e., absorbing surface) is:

$$e = (1 - \varepsilon_a)e + \varepsilon_a e_a + q \tag{6.31}$$

The *energy conversion efficiency* η_E can be interpreted as an output-to-input ratio q/e, and after using relations (6.30) and (3.21), is determined as:

$$\eta_E \equiv \frac{q}{e} = \varepsilon_a \left[1 - \left(\frac{T_a}{T} \right)^4 \right] \tag{6.32}$$

The energy efficiency does not depend on the environment temperature T_0. The higher is the emissivity ε_a of the absorbing surface,

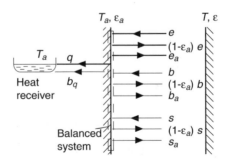

Figure 6.5 Scheme of emission and absorption by the surface at temperature T_a (from Petela, 2003).

the higher is the efficiency. The highest efficiency is for the black absorbing surface ($\varepsilon_a = 1$). The smaller is the surface temperature T_a, the higher is the efficiency. For example, for temperature $T_a = T_0$ the efficiency is high, whereas the practical value of heat absorbed at T_0 is zero.

In contrast to the work, heat is marked by temperature, which determines the quality of the heat. A conversion of radiation energy into heat occurs at the exergy loss during irreversible absorption accompanied by emission, and the value of the radiation matter is degraded to the level marked by the temperature of heat.

The effectiveness of conversion of the incident radiation into heat q can be evaluated by the *exergy conversion efficiency* η_B. Again interpreting appropriately the terms in the following exergy balance equation for a balanced system (Figure 6.5), completed by exergy loss δb due to irreversibility:

$$b = (1 - \varepsilon_a) b + \varepsilon_a b_a + b_q + \delta b \qquad (6.33)$$

the exergy efficiency η_B of conversion, as the ratio of the useful effect expressed by the exergy b_q of heat, and the exergy of incident radiation b, is determined as follows:

$$\eta_B \equiv \frac{b_q}{b} \qquad (6.34)$$

where exergy b_q of the heat receiver is:

$$b_q = q \frac{T_a - T_0}{T_a} \qquad (6.35)$$

Using equations (6.10), (6.30), and (6.35) in equation (6.34) and expressing the energy density emissions with formula (3.22), one obtains:

$$\eta_B = 3\varepsilon_a \left(1 - \frac{T_0}{T_a}\right) \frac{T^4 - T_a^4}{3T^4 + T_0^4 - 4T_0 T^3} \qquad (6.36)$$

In contrast to the energy efficiency, the exergy efficiency does depend on environment temperature T_0. The lower is the environment temperature, the higher is the efficiency. The higher is the emissivity ε_a of the absorbing surface, the higher is the efficiency, and the highest efficiency is achieved for the black absorbing surface ($\varepsilon_a = 1$).

For the given emissivity ε_a of the absorbing surface, environment temperature T_0, and the temperature T of the arriving black emission exergy b, the efficiency η_B depends on the temperature T_a of the absorbing surface. The temperature T_a can be controlled in the

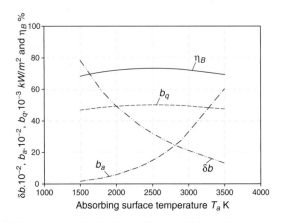

Figure 6.6   Effect of varying temperature T_a of the adsorbing surface at constant $T = 6000$ K, $T_0 = 300$ K, and $\varepsilon_a = 0.8$.

determined range by a proper arrangement of the withdrawn heat q, and it can be shown that the efficiency η_B has its maximum. The condition:

$$\frac{\partial \eta_B}{\partial T_a} = 0 \tag{6.37}$$

allows for derivation of the following equation from which the temperature $T_a = T_{a,\text{opt}}$ can be determined:

$$4T_{a,\text{opt}}^5 - 3T_0\, T_{a,\text{opt}}^4 - T^4 T_0 = 0 \tag{6.38}$$

For example, for $(T/1000) = 6$, and $(T_0/1000) = 0.3$, one obtains $(T_{a,\text{opt}}/1000) = 2.544$ which corresponds to $T_{a,\text{opt}} = 2544$ K.

The calculated results were used to illustrate (Figure 6.6) the values of some terms in equation (6.33). With the growing temperature T_a of the adsorbing surface, the exergy b_a of emission of this surface increases, whereas the exergy loss δb that occurs on this surface decreases. Both the efficiency η_B and the exergy b_q of heat delivered to the receiver have their maxima.

Therefore, the optimal utilization of any radiation arriving at the absorbing surface occurs at the determined temperature of this surface. This means that the heat extraction should be arranged in such a way that the temperature of this surface would be maintained at the level of the optimal (exergetic) temperature $T_{a,\text{opt}}$. This conclusion should be used for designing systems utilizing any hot radiation, e.g., solar radiation, by a device in which the exergy of solar radiation is harvested by an adsorbing surface.

The exemplary comparison of the considered formulae for the energetic and exergetic radiation conversion efficiencies is summarized

Efficiency	Radiation to work conversion	Radiation to heat conversion
Energetic η_E	$\eta_E = \dfrac{W}{e}$, $\eta_{E,\max} \equiv \psi = \dfrac{b}{e}$	$\eta_E = \dfrac{e - e_a}{e} = 1 - \left(\dfrac{T_a}{T}\right)^4$
Exergetic η_B	$\eta_B = \dfrac{W}{b}$	$\eta_B = \dfrac{b_q}{b}$

TABLE 6.2 Comparison of Some Exemplary Radiation Conversion Efficiencies

in Table 6.2. The exergetic efficiency problems for various different processes were discussed also in Section 4.6.

6.4.3 Other Processes Driven by Radiation

Besides processes of the conversion of radiation to work or heat, there are also other various processes in which thermal radiation (not necessarily solar radiation) is meaningful or plays the driving role. The thermodynamic analysis and exergy efficiency for such processes, mentioned already in Section 4.6.4, has to be considered individually according to the process specificity, although a certain general methodology exists and can be applied. Examples of such other processes can be the conversion of solar radiation energy to:

- heat, particularly from solar energy
- power in process combined with the buoyancy effect
- chemical energy of green plant substance
- electricity

The methodology for these processes was outlined respectively in Chapters 10–13.

Whereas Section 6.4.2 concerned mainly conversion of any thermal radiation to heat, Chapter 10 presents analysis particularly for solar radiation with its specific geometry and spectrum. and it also discusses the strategic viewpoint about the example of a cylindrical–parabolic cooker.

Chapter 11 develops exemplary thermodynamic analysis for the conversion of solar radiation into power, but in contrast to the general evaluation shown in Section 6.4.1, it presents more detailed aspects, in particular including the effect of the gravitational field of the earth.

A simplified approach to the energy and exergy analyses of photosynthesis is outlined in Chapter 12 by considering the conversion of solar energy into the substance of the green plant represented by the model of a leaf.

Chapter 13 briefly outlines the photovoltaic in which, specifically for the photovoltaic, a conversion of solar radiation energy into electricity has unavoidably to occur with simultaneous conversion of part of this energy into heat.

The energy and exergy analyses of these processes show the different values of energetic and exergetic magnitudes compared to each other based on the common background of the same reference states.

6.5 Irreversibility of Radiative Heat Transfer

Whereas the irreversibility of processes with a substance is caused by friction, diffusion, and heat exchange at a finite temperature, the irreversibility of radiation processes occurs due to basic phenomena such as emission and adsorption. The irreversibility mechanism of many radiation processes, e.g., the dilution or attenuation of propagating radiation, can be explained based on the irreversibility of emission and absorption. Obviously, in the combined processes in which substance and radiation take place, all the sources of irreversibility should be considered—those for substance as well as those for radiation.

The irreversibility of radiative heat transfer was considered by Petela (1961b). Consideration of a radiation system can be carried out conveniently based on a vacuum enclosed within solid surfaces. Again, the simplest configuration is the system with two parallel flat surfaces, x and y, infinitely large and facing each other as shown in Figure 6.7. The system is in the steady state and the heat exchanged by radiation between the surfaces occurs at constant rate q, W/m². The considered surfaces are black ($\varepsilon_x = \varepsilon_y = 1$) at the respective temperatures, T_x and T_y, uniformly distributed over these surfaces, and the heat sources, at the respective temperatures T_x and T_y are in direct contact with the respective surface x and y. The irreversibility will be examined based on the second law of thermodynamics by determination of the overall entropy growth Π for the examined object, which, e.g., is surface x.

The heat q exchanged by radiation between surfaces x and y is equal to the difference in the surfaces emission e_x and e_y, and with use of formula (3.21):

$$q = e_x - e_y = \sigma \left(T_x^4 - T_y^4 \right) \tag{6.39}$$

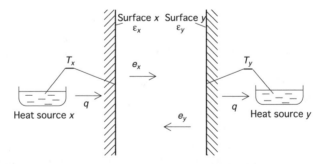

FIGURE **6.7**   Radiative heat transfer between two surfaces, x and y.

The respective entropies s_x and s_y of emissions are calculated from formula (5.24) as follows:

$$s_x = \frac{4}{3}\sigma T_x^3 \quad \text{and} \quad s_y = \frac{4}{3}\sigma T_y^3 \tag{6.40}$$

The overall entropy growth Π_x for surface x consists of the entropy decrease $(-q/T_x)$ of heat source x, entropy of the generated emission (s_x), and entropy of the disappearing emission $(-s_y)$; thus:

$$\Pi_x - \frac{q}{T_x} + s_x - s_y \tag{6.41}$$

Using equations (6.39) and (6.40) in (6.41):

$$\Pi_x = \frac{\sigma}{3}\left(T_x^3 + 3\frac{T_y^4}{T_x} - 4T_y^3 \right) \tag{6.42}$$

The entropy growth is a function of temperatures: $\Pi_x(T_x, T_y)$. To find the extreme of the function the first partial derivatives are assumed zero:

$$\frac{\partial \Pi_x}{\partial T_x} = \frac{\sigma}{3}\left(3T_x^2 - 3\frac{T_y^4}{T_x^2} \right) = 0 \quad \text{and} \quad \frac{\partial \Pi_x}{\partial T_y} = \frac{\sigma}{3}\left(12\frac{T_y^3}{T_x} - 12T_y^2 \right) = 0 \tag{6.43}$$

Both results of (6.43) indicate that the function reaches extreme $T_x = T_y$ at which $\Pi_x = 0$. The second derivative of function (6.42) at the extreme point $(T_x = T_y)$, e.g.:

$$\frac{\partial^2 \Pi_x}{\partial T_x^2} = \frac{\sigma}{3}\left(6T_x + 6\frac{T_y^4}{T_x^3} \right)_{T_x=T_y} = 4\sigma T_x \geq 0 \tag{6.44}$$

is always nonnegative which means that the extreme is a minimum (at $T_x = T_y$ and $\Pi_x = 0$). Thus, for temperatures $T_x \geq T_y$ the heat transfer is possible and irreversible, except the case of $T_x = T_y$ when $\Pi_x = 0$ and $q = 0$. Figure 6.8 illustrates function $\Pi_x(T_x, T_y)$ for exemplary ranges of T_x and T_y. The plane on the left-hand side from the line of $\Pi_x = 0$ represents the case when $T_x < T_y$. Such a case occurs also at $\Pi_x = 0$ (is possible although irreversible); however, the effective heat exchange is from surface y to surface x ($q < 0$).

Analogically, the considerations can be carried out also for surface y and the overall entropy growth Π_y can be calculated as:

$$\Pi_y = -\frac{q}{T_y} - s_x + s_y \tag{6.45}$$

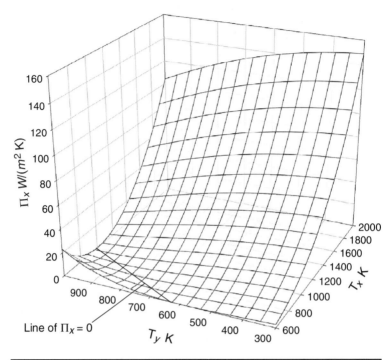

FIGURE 6.8 Overall entropy growth Π_x as function of temperatures T_x and T_y.

from which:

$$\Pi_y = \frac{\sigma}{3}\left(T_y^3 + 3\frac{T_x^4}{T_y} - 4T_x^3\right) \tag{6.46}$$

The sum of the overall entropy growths Π_x and Π_y for both surfaces is equal to the overall entropy growth Π for the whole process of heat transfer:

$$\Pi = \Pi_x + \Pi_y = q\left(\frac{1}{T_y} - \frac{1}{T_x}\right) \tag{6.47}$$

Equation (6.47) confirms correctness of the consideration of irreversibility.

The entropy growth does not depend on the environment temperature T_0. However, any exergy loss due to irreversibility does depend on and can be calculated according to formula (2.60) as the product of T_0 and the respective overall entropy growth. For example, the exergy loss δb_x at the considered surface x is:

$$\delta b_x = T_0 \Pi_x \tag{6.48}$$

In conclusion, the process of simultaneous emission and absorption occurring on any surface during radiative heat transfer is possible although irreversible. The irreversibility grows with the growing temperature difference during transfer of radiative energy. The larger is the irreversibility, the larger is the degradation of transferred energy. The reversible process of simultaneous emission and absorption occurs only in case of equal temperatures ($T_x = T_y$), but then the net heat transferred is zero.

6.6 Irreversibility of Emission and Absorption of Radiation

As previously mentioned, one of the possible methods of utilization of the radiation exergy is application of any absorbing surface exposed to this radiation. From the exergy viewpoint, the possible processes that then occur on such a surface were first considered by Petela (1961b). Due to the irreversibility of the processes of emission and absorption of radiation, the loss δb of exergy appears, which can be calculated according to the Gouy–Stodola law (2.60) as the product of the overall entropy growth Π and the environment temperature T_0:

$$\delta b = \Pi T_0 \qquad (6.49)$$

Consideration can be based on the scheme (Figure 6.5) of the emission and the absorption of the surface of emissivity ε_a. It can be assumed that the absolute temperature T_a of the considered surface is constant due to the appropriate amount q of heat exchanged between the surface and a heat sink or a heat source. (In the case of a source, heat q in Figure 6.5 has the opposite direction.)

During *alone emission* ($e = b = s = 0$) it is assumed that the emitted energy, due to heat q delivered from the heat source, according to the energy conservation law for the steady state, is equal to this heat q ($e_a = q$), and is calculated with the use of formula (3.22). The entropy s_a of this emission, based on formula (5.24), is:

$$s_a = \varepsilon_a \frac{4}{3} \sigma T_a^3 \qquad (6.50)$$

The overall entropy growth for the considered emission process consists of the entropy drop (−) of the heat source and of the entropy of the produced (+) emission:

$$\Pi = -\frac{q}{T_a} + s_a \qquad (6.51)$$

Using equations (3.22) and (6.50) in (6.51), one obtains the expression for calculation of Π which appears to be always larger than zero

$$\Pi = \varepsilon_a \frac{\sigma}{3} T_a^3 > 0 \qquad (6.52)$$

and this proves that the emission alone (not accompanied by any adsorption) is possible; however, it is irreversible. An approximate example of the emission without absorption is the radiation of the sun. If one assumes, that the sun is surrounded by a vacuum at temperature practically $T_0 \approx 0$ and emissivity $\varepsilon_0 \approx 1$, then no radiation comes to the sun, and its surface represents the case of alone emission from the surface at temperature T_a equal to the sun temperature T_S ($T_a = T_S$). At the sun's surface the conversion of heat transferred from the sun's interior to its radiative emission occurs at an exergy loss determined by formula (6.49). The percentage value of the exergy loss can be determined by the exergy loss divided by the sun radiation exergy. Using equations (6.10), (6.49), and (6.52), one obtains:

$$\left(\frac{\delta b}{b}\right)_{\text{sun}} = \frac{1}{3\dfrac{T_S}{T_0} + \left(\dfrac{T_0}{T_S}\right)^3 - 4} \qquad (6.53)$$

As a result of equation (6.53), the larger is the environmental temperature, the larger is the exergy loss. For $T_0 \to 0$, we obtain $\delta b \to 0$. For example, for the sun's surface temperature, $T_S = 6000$ K and from the viewpoint of the earth's conditions (assuming, e.g., $T_0 = 300$ K), equation (6.53) gives $\delta b/b = 1.786\%$. This conversion of heat from the sun to its radiative energy is relatively very effective because it occurs at a high temperature.

During *alone absorption* of any incident emission e by the surface with temperature T_a (Figure 6.5), it is assumed that $e_a = b_a = s_a = 0$ and the absorbed heat q ($q = \varepsilon_a e$) is transferred to the sink of temperature T_a. Again, to simplify considerations the incident emission is assumed to be black ($\varepsilon = 1$). The overall entropy growth for the considered absorption process consists of the entropy increase ($+$) of the heat sink and the entropy disappearing ($-$) during absorption:

$$\Pi = \frac{q}{T} - s \qquad (6.54)$$

Using equations (3.22) and (6.50) in (6.54), one obtains:

$$\Pi = \varepsilon_a 4\sigma T^3 \left(\frac{1}{4}\frac{T}{T_a} - \frac{1}{3}\right) \qquad (6.55)$$

Analysis of the alone absorption phenomena should be carried out at the same temperatures of the absorbing surface and of the incident emission ($T = T_a$). Otherwise, if $T \neq T_a$, analysis takes into account not only the pure absorption process but also the consequences of the degradation of energy. Thus, assuming the condition $T = T_a$ in (6.55), one obtains the expression on Π that happens to be always smaller than zero:

$$\Pi = -\varepsilon_a \frac{\sigma}{3} T_a^3 < 0 \tag{6.56}$$

and this proves that the absorption alone, without accompanying emission of the considered surface, is *impossible*. This conclusion is in agreement with the Kirchhoff's identity, stating that the emission ability of any surface is equal to its absorption ability at the same temperature. In contrast to this conclusion, De Vos and Pauwels (1986) argue that the absorption without emission is *irreversible*.

The *simultaneous emission and absorption* can be considered based also on the scheme depicted in Figure 6.5. An emissivity e of a black radiation ($\varepsilon = 1$) from a surface of temperature T arrives at the considered surface of emissivity ε_a and temperature T_a. Between the two surfaces the heat q is exchanged:

$$q = \varepsilon_a \sigma \left(T^4 - T_a^4 \right) \tag{6.57}$$

The overall entropy growth in such a case consists of the entropy increase (+) of the heat receiver, of disappearing (−) entropy of absorbed radiation, and of the entropy produced (+) due to emission of the considered surface:

$$\Pi = \frac{q}{T_a} - \varepsilon_a s + s_a \tag{6.58}$$

where s, based on formula (5.24), is calculated as follows:

$$s = \frac{4}{3} \sigma T^3 \tag{6.59}$$

Using equations (6.50) and (6.57)–(6.59) one obtains:

$$\Pi = \varepsilon_a \frac{\sigma}{3} T_a^3 \left[3 \left(\frac{T}{T_a} \right)^4 - 4 \left(\frac{T}{T_a} \right)^3 + 1 \right] \geq 0 \tag{6.60}$$

The expression in the quadratic brackets is always nonnegative, except for $T = T_a$ when there is a minimum that amounts to zero. This means that the simultaneous emission and absorption for $T \neq T_a$ is always possible, although irreversible. If $T = T_a$ the process is reversible, but there is no heat exchange.

In conclusion, the exergy of radiation reaching any surface can be reflected (i.e., re-radiated) and the reflected radiation has its exergy at the temperature of the original radiation, which is not utilized by the absorbing surface. If the reflection process does not change the radiation temperature, then this process is reversible and does not generate any exergy loss. However, the radiation emitted by the absorbing surface has its own exergy determined by the emissivity and temperature of the absorbing surface. This is the problem of the efficiency of the absorbing surface, or any other device utilizing the radiation somehow, in how much of the whole incident exergy b the surface, or the other device, can grasp and utilize. The efficiency of the absorbing device or surface is an entirely different thing and does not depend on the theoretical potential represented by b. Acceptance of such an interpretation is very important in correct reasoning about the theory of radiation exergy, because if not noticed by some researchers, this can mislead to strange conclusions.

6.7   Influence of Surroundings on the Radiation Exergy

As mentioned earlier, the exergy of radiation matter, which is the emitted photon gas, is a function of the instant state of the matter and of the state of such a photon gas in the case of radiative equilibrium with the environment. Such a function of radiation exergy does not depend on the history of the considered matter or on the way in which the matter was created. It can be shown that radiation exergy does not depend on any external contemporary factors such as the environmental emissivity, configuration of the emitting surfaces under consideration, or the presence of any other surfaces at a temperature different from the environment and radiating on the considered surface.

In other words, besides parameters of the considered radiation, the only factor that counts in determination of the radiation exergy is environment temperature T_0. Except T_0 there are no other environmental characteristics (geometrical configuration, properties, or any nonuniformity) that could influence the work used to measure the exergy. This rule can be commonly experienced in practice.

Additionally, as mentioned in Section 6.1, the emissivity of the considered surface does not affect the exergy of black emission radiated from the surface at its temperature.

The same formula for exergy emission can be derived based on the exergy balance of a surface of any temperature T or based on exergy balance of any environment surface (at T_0) at which arrives the exergy emission of the temperature T.

The independence of the radiation exergy on some environmental factors is discussed in the following section.

6.7.1   Emissivity of the Environment

As emphasized before, the practical observations reveal that any surfaces at different emissivities, facing each other, but at the same temperature T_0, remain in thermodynamic equilibrium, and the presence of such emissivities diversity does not demonstrate any practical (exergetic) significance, i.e., any possibilities to perform work. Even in the case of the close neighboring surfaces of extreme emissivities, such as white snow and black soot, the surfaces remain in thermal equilibrium if their temperatures are remained equal to T_0.

However, at temperature $T \neq T_0$, for any surface of arbitrary properties (gray surface) the properties of the considered surface and the environment (e.g., ε and ε_0) play a role in determining the exchanged exergy of emission at temperature T.

In other words, from a radiation viewpoint, the diversified emissivity values of the environment bodies have no exergetic significance. However, if any environment body would change its temperature, then the other bodies immediately disclose their diversified abilities to affect the rate of exchange of emission exergy between bodies of different temperatures.

It is noteworthy that the diversified emissivity of the bodies of the environment can be compared to the problem of the reference substances applied in calculation of the chemical exergy of the considered substance. At different locations in the environment the same reference substance can appear under a different concentration although at the same temperature T_0. There is the problem of which of the concentration values should be chosen for exergy calculations as the reference because in case of a substance, the difference of concentration can be theoretically utilized to do work in an isothermal theoretical process of equalization of the concentrations. However, as only one concentration value can be taken for exergy calculations, thus, to solve the problem for a substance, only one of the existing concentrations of the reference substance in the environment, is defined as the *standard concentration reference*. Such choice is based on the agreement or any specific reasoning or just selecting the one that dominates.

The problem of choosing the reference emissivities of environmental bodies does not exist in the consideration of radiation. If we recognize that the environment space contains the black radiation that is in equilibrium with the environment surfaces, then the only problem left is to establish the standard environmental temperature representing the only sufficient reference for calculation of radiation exergy.

6.7.2 Configuration of Surroundings

Regarding the effect of the configuration on the radiation exergy, the Prevost law (Section 3.1) can be quoted, according to which the surface radiates independently of the presence of other surfaces existing in the surroundings. Confirmation of this law for exergetic considerations can be illustrated with the following example.

Consider the model shown in Figure 6.9. The two spherical surfaces A_1 and A_0, at thermodynamic steady state, are exchanging heat Q by radiation through a vacuum. The constant temperatures T_1 and

FIGURE 6.9 Two
concentric surfaces.

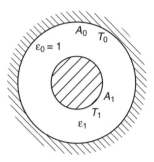

T_0 are maintained due to the connection of the surfaces with the respective heat sources. The outer surface A_0 simulates environment of the temperature T_0.

The concentric spherical shapes of the surfaces are assumed for convenience of consideration in which the local view factors are uniform over the whole respective surfaces. The view factor was introduced by formula (3.9) and is discussed with some more detail in Section 7.5.1. The configuration of the surface is described by the following values of the view factors: $\varphi_{1-0} = 1$ (the whole radiation of surface A_1 arrives in surface A_0), $\varphi_{1-1} = 0$ (surface A_1 is nonconcave), $\varphi_{0-1} = A_1/A_0$ (based on the reciprocity rule), and $\varphi_{0-0} = 1 - \varphi_{0-1}$ (based on the complacency rule).

The value of the environmental surface emissivity ε_0 has no influence on the exergy of the radiation of the considered surface A_1; thus, for convenience, it is assumed that $\varepsilon_0 = 1$ and the consideration of reflected radiation of surface A_1 is avoided. Emissivity ε_1 of the considered surface A_1 is arbitrary. Based on the exergy definition, the radiation exergy of surface A_0 is $b_0 = 0$.

First, the *exergy balance equation for surface A_1* is considered. The exergy balance input consists of the exergy of heat Q determined by formula (2.61). The exergy output consists of the emission exergy of surface A_1 and of the exergy loss of irreversibility at surface 1, determined by the Gouy–Stodola law (2.60). Thus:

$$Q\left(1 - \frac{T_0}{T_1}\right) = A_1 b_1 + T_0 \left(A_1 s_1 - A_0 \varphi_{0-1} s_0 \varepsilon_1 - \frac{Q}{T_1}\right) \qquad (6.61)$$

where b_1 is the unknown in the consideration and represents the emission exergy of surface per 1 m² of this surface. The entropies of the emission densities s_1 and s_0, respectively, for surfaces 1 and 0, are determined from formula (5.24).

Heat Q exchanged between the two surfaces can be determined from the energy balance equation, e.g., for surface 1. The energy input consists of heat Q and of the emission energy arriving from surface A_0 in surface A_1 and absorbed at absorptivity equal to emissivity ε_1.

The energy output is represented only by the emission of surface 1, entirely absorbed by surface A_0. Thus:

$$Q + A_0 \varphi_{0-1} e_0 \, \varepsilon_1 = A_1 e_1 \qquad (6.62)$$

where e_1 and e_0 are the emission density of surfaces A_1 and A_0, respectively, and they are determined by formula (3.22).

After using (6.62) in solving equation (6.61) on b_1, the exact formula (6.10) is obtained. The geometrical magnitudes such as A_1, A_0, or shape factor φ_{0-1} have been cancelled out and do not appear in the derived result, which means that these magnitudes of configuration have no effect on the universal formula (6.10).

Now, the *exergy balance of surface A_0* will also be applied to confirm correctness of the derived formula (6.10). The exergy balance equation is:

$$A_1 b_1 + A_1 \varphi_{1-0} \varepsilon_1 b_0 + A_0 \varphi_{0-0} \varepsilon_0 b_0 = A_0 \varepsilon_0 b_0 + Q \left(1 - \frac{T_0}{T_0} \right) + \delta B_0 \qquad (6.63)$$

If all terms containing emission exergy of surface A_0 ($b_0 = 0$) are neglected, as well as the exergy of heat Q at temperature T_0, then the exergy balance equation is reduced to $A_1 b_1 = \delta B_0$. Thus, expressing the exergy loss δB_0 at the surface A_0 by formula (2.60), one obtains:

$$A_1 b_1 = T_0 \left[\frac{Q}{T_0} - A_1 s_1 + A_0 s_0 - A_0 s_0 \varphi_{0-1} (1 - \varepsilon_1) - A_0 s_0 \varphi_{0-0} \right] \qquad (6.64)$$

After solving equation (6.64) on b_1 again the exact formula (6.10), containing no configuration parameters, can be obtained.

6.7.3 Presence of Other Surfaces

From interpretation of the Prevost law it results that the radiation exergy of the considered surface in the system of other surfaces of different temperatures is not affected by the other surfaces. The Prevost law for radiation exergy can be illustrated by the following example of the four different surfaces of the system shown in Figure 6.10.

A very large flat surface of area A, split into two parts (1 and 2), faces other flat parallel surfaces of area A that is also split into two parts (3 and 4). All the four surfaces are black and remain in thermal equilibrium at uniform and constant respective different temperatures (T_1, T_2, T_3, and T_4). The areas of the surfaces are expressed with the factors: $a_1 = A_1/A, a_2 = 1 - a_1, a_3 = A_3/A, a_4 = 1 - a_3$ where $A = A_1 + A_2 = A_3 + A_4$.

FIGURE **6.10** The parallel surfaces.

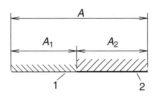

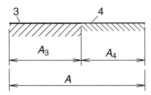

The exergy balances are considered for the four surfaces. Each surface receives the exergy of heat at the surface temperature and the radiation exergies from the two opposite surfaces. The output of the exergy balance equation consists of the radiation exergy from the considered surface and of the exergy loss due to irreversibility at the considered surface. Thus the four exergy balance equations are:

$$b_{q1} + a_1 a_3 b_3 + a_1 a_4 b_4 = a_1 b_1 + T_0 \Pi_1 \tag{i}$$

$$b_{q2} + a_2 a_3 b_3 + a_2 a_4 b_4 = a_2 b_2 + T_0 \Pi_2 \tag{ii}$$

$$b_{q3} + a_3 a_1 b_1 + a_3 a_2 b_2 = a_3 b_3 + T_0 \Pi_3 \tag{iii}$$

$$b_{q4} + a_4 a_1 b_1 + a_4 a_2 b_2 = a_4 b_4 + T_0 \Pi_4 \tag{iv}$$

where the values of heat delivered to surfaces:

$$q_1 = a_1 (e_1 - a_3 e_3 - a_4 e_4) \tag{v}$$

$$q_2 = a_2 (e_2 - a_3 e_3 - a_4 e_4) \tag{vi}$$

$$q_3 = a_3 (e_3 - a_1 e_1 - a_2 e_2) \tag{vii}$$

$$q_4 = a_4 (e_4 - a_1 e_1 - a_2 e_2) \tag{viii}$$

exergy of heat:

$$b_{q1} = q_1 \left(1 - \frac{T_0}{T_1}\right) \tag{ix}$$

$$b_{q2} = q_2 \left(1 - \frac{T_0}{T_2}\right) \tag{x}$$

$$b_{q3} = q_3 \left(1 - \frac{T_0}{T_3}\right) \tag{xi}$$

$$b_{q4} = q_4 \left(1 - \frac{T_0}{T_4}\right) \tag{xii}$$

and the total entropy growth for the surfaces:

$$\Pi_1 = a_1 (s_1 - a_3 s_3 - a_4 s_4) - \frac{q_1}{T_1} \qquad \text{(xiii)}$$

$$\Pi_2 = a_2 (s_2 - a_3 s_3 - a_4 s_4) - \frac{q_2}{T_2} \qquad \text{(xiv)}$$

$$\Pi_3 = a_3 (s_3 - a_1 s_1 - a_2 s_2) - \frac{q_3}{T_3} \qquad \text{(xv)}$$

$$\Pi_4 = a_4 (s4 - a_1 s_1 - a_2 s_2) - \frac{q_4}{T_4} \qquad \text{(xvi)}$$

Radiation energies e and entropies s are determined, respectively, from formula (3.22) and (5.24) (for $\varepsilon = 1$).

The numerical solution of the system of equations (i)–(xvi) gives the values of the emission exergies b_1, b_2, b_3, and b_4 identical with the respective values obtained on the other hand from equation (6.8) applied for temperatures T_1, T_2, T_3, and T_4 at given T_0. This sameness occurs independently on surfaces temperature values (T_1, T_2, T_3, or T_4), areas factors (a_1, a_2, a_3, or a_4) and environment temperature T_0. Thus, it results that radiation exergy determined by formula (6.8) for any considered black surface does not depend on the presence of other surfaces of arbitrary temperatures, configuration of the surfaces, and the level of environment temperature.

Example 6.2 The four very large black surfaces (Figure 6.10) have temperatures $T_1 = 6000$ K, $T_2 = 1200$ K, $T_3 = T_0 = 300$ K, and $T_4 = 200$ K. The surfaces areas are determined by factors $a_1 = 0.1$ and $a_3 = 0.7$. From the system of equations from (i)–(xvi) the following values of emission exergies can be obtained: $b_1 = 68.576$ MW/m^2, $b_2 = 78.525$ kW/m^2, $b_3 = 0$, and $b_4 = 62.36$ W/m^2. Identical values are obtained from formula (6.8) used respectively. The obtained values do not change with varying a_1 and a_2, but they do change at the changed surface temperatures.

6.8 "Cold" Radiation

"Cold radiation" can be distinguished from "cold light" emitted by luminescence, discussed in Section 1.1. By cold radiation one can understand radiation at temperature T_c smaller than the environment temperature, $T_c < T_0$. Exergy of cold radiation is positive and for sufficiently low temperature is even larger than energy of this radiation (e.g., Figure 6.2). However, the exergy of cold radiation is relatively small. For example, the significance of the cold emission, relative to the emission at a temperature larger than the environment, can be estimated based on formula (6.8), which is used to equate the exergy of black emission at $T > T_0$ to the exergy of black emission at $T_c < T_0$:

$$3T^4 + T_0^4 - 4T_0 T^3 = 3T_c^4 + T_0^4 - 4T_0 T_c^3 \qquad \text{(6.65)}$$

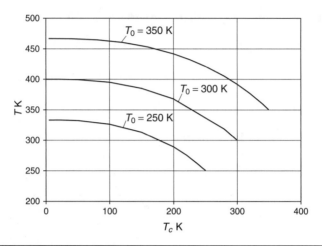

Figure 6.11   Equivalent radiation temperature T (at $T > T_0$) to the cold radiation temperature T_c.

Figure 6.11 illustrates equation (6.65) for the three different environment temperatures T_0 (250, 300, and 350 K). The higher is the environment temperature T_0, the higher is the equivalent temperature T.

Example 6.3   Based on Figure 6.11, it results that exergy of radiation at $T_c = 200$ K (at $T_0 = 300$ K) is equivalent to the exergy of radiation at $T = 367.9$ K. However, for the unchanged $T_c = 200$ K but at the increased environment temperature to $T_0 = 350$ K, the equivalent temperature $T = 442$ K.

Generally, the exergy of the radiation flux at temperature T is, e.g., decreasing as it travels through the environment at decreasing temperature T_0 within range $T > T_0$. There could be mentioned the case, nowadays recognized as very academic, in which, e.g., a human exists (obviously within a perfectly insulated capsule), on the surface of Venus, thus in an environment with a temperature of about 740 K. Such a human might recognize the exergy of emission of the Venusian surface as being equal to zero; however, the exergy of emission of the earth's surface at about 300 K would be interpreted on Venus according to formula (6.8) as: $b_{b,\text{Venus}} = (1/3) \times 5.6693 \times 10^{-8} \times (3 \times 300^4 + 740^4 - 4 \times 740 \times 300^3) = 4.615$ kW/m². For comparison, a human on the earth's surface might recognize the exergy of emission of the surface of Venus as $b_{b,\text{Venus}} = (1/3) \times 5.6693 \times 10^{-8} \times (3 \times 740^4 + 300^4 - 4 \times 300 \times 740^3) = 7.964$ kW/m².

A similar comparative analysis for the earth and sun would give the values: $b_{b,\text{earth}} = (1/3) \times 5.6693 \times 10^{-8} \times (3 \times 300^4 + 6000^4 - 4 \times 6000 \times 300^3) = 24.48$ GW/m² and, respectively, $b_{b,\text{sun}} = (1/3) \times 5.6693 \times 10^{-8} \times (3 \times 6000^4 + 300^4 - 4 \times 300 \times 6000^3) = 68.576$ GW/m².

In practice, the example of cold radiation can be radiation of the sky, which can be determined by the effective temperature of the sky. The concept of the sky temperature T_{sky} arises, e.g., when considered is a radiative heat transferred to space from objects (at temperature T_0) on earth. The characteristic difference $T_0^4 - T_{sky}^4$ is a measure of such heat, which can be sometimes negative if $T_0 < T_{sky}$.

As mentioned by Duffie and Beckman (1991), many models of the sky have been considered that take into account the beam and reflected radiation. The sky is usually considered as a blackbody at effective temperature T_{sky}. In fact, the atmosphere is not at a uniform temperature and it radiates only in some wavelength bands. For example, the atmosphere is practically transparent in the band from 8 to 14 μm, and beyond this range absorbs much of the infrared radiation. Determination of temperature T_{sky} is proposed to be based on the measured meteorological parameters such as water vapor content or dew point temperature. For example, for a clear sky the Swinbank (1963) formula, based on the environment temperature T_0, is applied:

$$T_{sky} = 0.0552\, T_0^{1.5} \qquad (6.66)$$

Obviously, the presence of clouds would increase the sky temperature in comparison to a clear sky. Experiments show that the sky temperature T_{sky} can be lower than the environmental temperature T_0 by about 5 K in hot and humid conditions and by about 30 K in cold and dry conditions.

Distinguishing between T_0 and T_{sky} in thermodynamic processes is illustrated, e.g., in Chapter 11 on the thermodynamic analysis of a solar chimney power plant.

6.9 Radiation Exergy at Varying Environmental Temperatures

Variation in the average environmental temperature T_0 can appear, e.g., for the local environment depending on time, or when different climate zones are considered, or eventually when the radiation flux travels through the remote environments of different temperatures.

The exergy of emission at constant temperature T varies with the varying environment temperature T_0 of which the effect of T_0 can be determined based on equation (6.10) by the partial derivative of the exergy b in regard to T_0:

$$\left(\frac{\partial b}{\partial T_0} \right)_{T=\text{const}} = \varepsilon \frac{4}{3} \sigma \left(T_0^3 - T^3 \right) \qquad (6.67)$$

The radiation exergy b, with the decreasing environment temperature T_0, proportionally to the third power of this temperature, grows for $T_0 < T$ and diminishes at $T_0 > T$.

As mentioned already in Section 4.5.3, the exergy balance equation for the process occurring at varying T_0 should contain the appropriate compensation term (ΔB_e) as is generally shown in equation (4.19) for processes in which radiation together with substance take place.

Sensitivity of radiation exergy to the varying T_0 can also be illustrated with the following example. An adiabatic tank of a volume 1 m^3 is fulfilled with black radiation ($\varepsilon = 1$) at temperature T_1 at the initial environment temperature $T_{0,1}$. After some time, the environment temperature changes to the value $T_{0,2}$. If nothing more happens (no work, no heat, no matter exchange so no exergy loss), then all the components of exergy balance equation (4.19) are equal to zero except the term Δb_S which expresses the system exergy increase due to the change of environment temperature. The increase Δb_S can be calculated by respective application of formula (5.29) as follows:

$$\Delta b_S = \frac{a}{3} \left[T_{0,2}^4 - T_{0,1}^4 - 4T_1^3 \left(T_{0,2} - T_{0,1} \right) \right] \tag{6.68}$$

For $T_{0,1} \neq T_{0,2}$ the term Δb_S is different from zero and thus, as shown by equation (5.19), the correction ΔB_e has to be included to complete the exergy balance equation for the radiation system considered at varying environment temperature.

In a particular case, when temperature T_1 would be equal to $T_{0,1}$, ($T_1 = T_{0,1}$), e.g., some radiation at $T_{0,1}$ would be trapped initially in the tank; then after changing the environment temperature to a value $T_{0,2}$, the initially worthless radiation would gain the value determined from (6.68) as follows:

$$\left(\Delta b_S \right)_{T_1 = T_{0,1}} = \frac{a}{3} \left(3T_{0,1}^4 + T_{0,2}^4 - 4T_{0,2}T_{0,1}^3 \right) \tag{6.69}$$

Referring to a radiation flux, the varying environment temperature can occur, e.g., when the flux travels through different spaces characterized by different local environment temperatures or when the local environment temperature varies due to atmospheric changes.

Determination of radiation exergy during traveling through different environments (e.g., on earth or in cosmic space) is obvious. One of the formulae discussed in Chapter 7 can be selected to fit the kind of considered flux and change in environment temperature taken into account. It is worth emphasizing that the energy of the considered radiation traveling through various environments remains constant and only exergy of this radiation expresses variation of the practical thermodynamic values of the traveling flux.

However, considering the variation of the environment during analysis of the determined system requires, as in the case of tank, the appropriate correction term in the exergy balance equation (4.19). The term can be different from zero and calculated as the completion of the balance equation.

When the environment temperature changes from $T_{0,1}$ to $T_{0,2}$, then the exergy radiation flux, considered as a black at temperature T_1, changes its exergy by

$$\Delta b_b = \frac{\sigma}{3} \left[T_{0,1}^4 - T_{0,2}^4 - 4T_1^3 \left(T_{0,1} - T_{0,2} \right) \right] \qquad (6.70)$$

The right-hand sides of formulae (6.70) and (6.68) differ only by the algebraic sign and the constant (a and σ, respectively).

It is worth noting that any variation in the environment temperature T_0 can be interpreted as a certain form of natural exergy resource. This variation can appear periodically, during a day, a month, or a year, and the amplitude of variation can be taken as a certain measure of this natural resource.

Example 6.4 The influence of varying environment temperatures can be illustrated by the following example. Consider a ball (Figure 6.12) with black surface at temperature T and at conductivity close to infinity, which motivates the consideration of the uniform distribution of temperature over the whole ball volume. The ball cools down in a vacuum (i.e., no convection and conduction beyond the ball) surrounded by a black wall of temperature T_0.

The surface area of the ball is 1 m²; thus the ball diameter is $D = (1/\pi)^{0.5} = 0.564$ m and the ball volume is $V = \pi \times D^3/6 = 0.094\,\mathrm{m}^3$. The density of ball material is $\rho = 7860\,\mathrm{kg/m^3}$ and the specific heat $c = 452\,\mathrm{J/(kg\,K)}$. The ball mass is $m = V \times \rho = 738.8$ kg. The ball absorbs emission $e_0 = \sigma \times T_0^4$ of environment and emits energy $e = \sigma \times T^4$, where $\sigma = 5.6693 \times 10^{-8}\,\mathrm{W/(m^2\,K^4)}$. Initial temperature of the ball is $T_{inl} = 400$ K.

(A) Environment temperature $T_0 = 280$ K is constant
For the system shown in Figure 6.12, the energy balance equation for an infinitely short time period dt is:

$$e_0\,dt = mc\,dT + e\,dt \qquad (a)$$

Assume the solution for T in the form:

$$T = T_{inl}\,e^{A_A t} \qquad (b)$$

where A_A is a constant for the considered case A. Dividing equation (a) by $m \times c \times dt$, expressing e_0 and e by respective temperatures T_0 and T, and

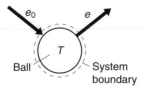

FIGURE **6.12** The cooling ball.

Ball — System boundary

T_{inl} K	A_A 1/s
350	-4.2973×10^{-6}
400	-8.2561×10^{-6}
500	-1.9133×10^{-5}
600	-3.493×10^{-5}
1000	-1.6872×10^{-4}

TABLE 6.3 Constant A_A for the Considered Example as a Function of Initial Temperature T_{inl} of the Ball

calculating the derivative $dT/d\tau$ based on equation (b), equation (a) becomes as follows:

$$T_{inl} A_A e^{A_A t} = \frac{\sigma}{mc} \left(T_0^4 - T^4 \right) \tag{c}$$

For the initial instant for which $t = 0$, the ball temperature is $T = T_{inl}$, thus from (c):

$$A_A = \frac{\sigma}{mc \, T_{inl}} \left(T_0^4 - T_{inl}^4 \right) \tag{d}$$

Substituting data for the considered example to equation (d) one obtains $A_A = -8.2561 \times 10^{-6}$ 1/s. However, as it comes from equation (d), the constant A_A depends on some parameters. For example, for m, c and T_0 assumed constant in the considered example the constant A_A depends on initial temperature T_{inl} of the ball as illustrated in Table 6.3.

Now, the emission terms e_0 and e in equation (a) can be expressed by the respective temperatures, whereas temperature T is given from (b). After integrating from 0 to t:

$$\sigma T_0^4 t = mc \left(T - T_{inl} \right) + \int_0^t \sigma \left(T_{inl} \, e^{A_A t} \right)^4 dt \tag{e}$$

After calculating integral and rearranging, equation (e) yields:

$$T = T_{inl} + \frac{\sigma}{mc} \left[T_0^4 \, t - \frac{T_{inl}^4}{4 A_A} (e^{4 A_A t} - 1) \right] \tag{f}$$

For example, from formula (f), temperature $T = 389$ K is obtained when the substitution is $t = 3600$ s.

Thus, consider other data for the cooling ball during the time period $t = 3600$ s from $T_{inl} = 400$ K to $T_{fin,A} = 389$ K. Based on calculated terms of equation (e) the drop of the internal energy of the ball (assumed as basic 100%) is equal to $E_{ball,A} = 3,671.5$ kJ and the absorbed environment emission (34.17%) are spent for the ball emission (−134.17%), as shown in column 2 of Table 6.4.

Item %	Energy balance $T_0 = 280$ K	Exergy balance $T_0 = 280$ K	Energy balance $T_0 \neq$ const	Exergy balance $T_0 \neq$ const
1	**2**	**3**	**4**	**5**
Environment emission	34.17	0	44.20	0
Internal energy of ball	100	100	100	100
Emission of ball	−134.17	−63.72	−144.20	−15.72
Irreversibility loss	—	−36.28	—	−75.25
Effect (Δb_e) of $T_0 \neq$ const	—	—	—	−9.03
Total	0	0	0	0

TABLE 6.4   Calculation Results from Example 6.4

For the considered system (Figure 6.12) the exergy balance equation for an infinitely short time period dt is:

$$\int_0^t b_0 \, dt = 0 = -b_{\text{ball}} + \int_0^t b \, dt + \delta b \tag{g}$$

where, based on the exergy definition: $b_0 = 0$ and δb is the exergy loss due to the irreversibility occurring during the considered time period. Interpreting formula (2.64) for the ball, the ball exergy b_{ball} drop is:

$$b_{\text{ball}} = mc \left(T_{\text{inl}} - 389 - T_0 \ln \frac{T_{\text{inl}}}{389} \right) \tag{h}$$

To calculate the exergy b of the ball emission, formula (6.8) can be applied:

$$b = \sigma \left(T^4 - \frac{4}{3} T_0 T^3 + \frac{1}{3} T_0^4 \right) \tag{i}$$

Expressing the instantaneous ball temperature T from formula (b) the total exergy emitted by the ball in the considered time t is:

$$\int_0^t b \, dt = \sigma \left(\int_0^t T_{\text{inl}}^4 \, e^{4A_A t} dt - \frac{4}{3} T_0 \int_0^t T_{\text{inl}}^4 \, e^{3A_A t} dt + \frac{1}{3} T_0^4 \, dt \right)$$

$$= \sigma \left[\frac{T_{\text{inl}}^4}{4A_A} (e^{4A_A t} - 1) - \frac{4}{3} \frac{T_0 T_{\text{inl}}^3}{3A_A} (e^{3A_A t} - 1) + \frac{1}{3} T_0^4 t \right] \tag{j}$$

Based on the calculated terms of equation (g) the drop of the internal exergy of the ball (assumed as basic 100%) equal $B_{\text{ball},A} = 1065.5$ kJ is spent for the exergy of the ball emission (−63.72%) and on the exergy loss (−36.28%) due to irreversibility, as shown in column 3 of Table 6.4.

The same value 36.28% on the exergy loss can also be obtained based on the Gouy–Stodola law, equation (2.60), according to which the overall entropy growth Π is multiplied by T_0. The entropy growth Π consists of entropy calculated according to formula (5.24) for the entropy at the environment temperature:

$$s_0 = \frac{4}{3}\sigma T_0^3 t \tag{k}$$

for the ball emission entropy:

$$s = \frac{4}{3}\sigma \int_0^t T^3 dt = \frac{4}{3}\sigma \int_0^t T_{\text{inl}}^3 e^{3A_A t} dt = \frac{4}{3}\frac{\sigma T_{\text{inl}}^3}{3A_A}(e^{3A_A t} - 1) \tag{l}$$

and for the ball material entropy drop, calculated based on equation (2.38), neglecting the pressure term:

$$s_{\text{ball}} = -mc \, \ln \frac{T_{\text{inl}}}{T} \tag{m}$$

Thus:

$$\delta b = T_0 \, \Pi = T_0 \, (-s_0 + s_{\text{ball}} + s) \tag{n}$$

(B) Environment temperature T_0 is varying in time t:

Assume now the environment temperature growing in time, e.g., as follows:

$$T_0 = T_{0,\text{inl}} \, e^{A_t t} \tag{o}$$

where $A_t = 2.5 \times 10^{-5}\,1/\text{s}$ is the assumed constant and $T_{0,\text{inl}} = 280\,\text{K}$ is the initial value of the environment temperature which, for better comparison, is equal to the environment temperature assumed to be constant in case A: $T_{0,\text{inl}} = (T_0)_A = 280\,\text{K}$.

Assume the solution for T in the form analogous to (b):

$$T = T_{\text{inl}} e^{A_B t} \tag{p}$$

where A_B is a constant for case B.

Equation (a) has to be modified accordingly to the considered case (B) for $T_0 \neq \text{const}$. Dividing equation (a) by $m \times c \times dt$, expressing e_0 and e by respective temperatures; T_0 from equations (o) and T, and calculating the derivative $dT/d\tau$ based on equation (p), equation (a) becomes as follows:

$$T_{\text{inl}} A_B e^{A_B \, t} = \frac{\sigma}{mc}\left(T_{0,\text{inl}}^4 e^{4, A_t t} - T^4\right) \tag{r}$$

Substitute the condition that for $t = 0$, the ball temperature has to be $T = T_{\text{inl}}$. Then equation (r), with constant A_B instead of A_A, becomes exactly the same as condition (d). Thus, $A_B = A_A = 8.2561 \times 10^{-6}\,1/\text{s}$. Obviously, the determined value A_B is valid only for m, c, and $(T_0)_A = (T_{0,\text{inl}})_B$ assumed to be like case A.

For the considered case B, equation (a) can be now used again; with the emission e_0 and e expressed by the respective temperatures from equations (o) and (p). Integrating from 0 to t:

$$\int_0^t \sigma \left(T_{0,\text{inl}}\, e^{A_t t}\right)^4 dt = mc\,(T - T_{\text{inl}}) + \int_0^t \sigma \left(T_{\text{inl}} e^{A_B t}\right)^4 dt$$

and after calculations of integrals and rearranging:

$$T = T_{\text{inl}} + \frac{\sigma}{4\,mc}\left[\frac{T_{0,\text{inl}}^4}{A_t}\left(e^{4,A_t t} - 1\right) - \frac{T_{\text{inl}}^4}{A_B}\left(e^{4,A_B t} - 1\right)\right] \tag{s}$$

For example, when substituting $t = 3600$ s in formula (s), the obtained temperature is $T = 389.77$ K which is a little larger in comparison to values 389.00 K obtained in case A.

Further, consider the cooling ball during time period $t = 3600$ s from $T_{\text{inl}} = 400$ K to $T_{\text{fin}, B} = 389.77$ K. Based on the calculated terms of equation (a), interpreted for case B, the drop of the internal energy of the ball (assumed as 100%) equal to $E_{\text{ball}, B} = 3416$ kJ and the absorbed environment emission (44.20%) are spent for the ball emission (−144.20%), as shown in column 4 of Table 6.4.

For the considered case B ($T_0 \neq$ constant), the exergy balance equation for an infinitely short time period dt, is:

$$\int_0^t b_0\, dt = 0 = -b_{\text{ball}} + \int_0^t b\, dt + \delta b + \delta b_e \tag{t}$$

The drop of ball exergy:

$$b_{\text{ball}} = mc\left(T_{\text{fin}, B} - T_{0,\text{fin}} - T_{0,\text{fin}} \ln \frac{T_{\text{fin}, B}}{T_{0,\text{fin}}} - T_{\text{inl}, B} + T_{0,\text{inl}} + T_{0,\text{inl}} \ln \frac{T_{\text{inl}, B}}{T_{0,\text{inl}}}\right) \tag{u}$$

Based on equation (i) the exergy of the ball emission:

$$\int_0^t b\, dt = \sigma\left\{\frac{T_{\text{inl}}^4}{4A_B}\left(e^{4A_B t} - 1\right) - \frac{4}{3}\frac{T_{0,\text{inl}} T_{\text{inl}}^3}{A_t + 3A_B}\left[e^{(A_t + 3A_B)t} - 1\right] + \frac{1}{3}\frac{T_{0,\text{inl}}^4}{4A_t}\left(e^{4A_t t} - 1\right)\right\} \tag{w}$$

The exergy loss δb due to irreversibility can be presented by a formula with three members corresponding to three entropy changes: negative (disappearing) entropy of environment emission, negative (cooled ball) entropy change of the ball, and positive (appearing) entropy of the ball emission:

$$\delta b = -\int_0^t T_0 s_0\, dt + \int_0^t T_0 s\, dt - \int_0^t T_0 s_{\text{ball}}\, dt = -\frac{4}{3}\frac{\sigma\, T_{0,\text{inl}}^4}{4A_t}\left(e^{4A_t t} - 1\right)$$

$$+ \frac{4}{3}\frac{T_{0,\text{inl}} T_{\text{in}}^3}{A_t + 3A_B}\left[e^{(A_t + 3A_B)t} - 1\right] - mc\,T_{0\,\text{inl}} \ln \frac{T_{\text{inl}}}{T_{\text{fin}, B}}\left(e^{A_t t} - 1\right) \tag{x}$$

The exergy loss δb_e due to increasing environment temperature is calculated from equation (t). Based on the calculated terms of equation (t) the drop of the internal exergy of the ball (assumed as basic 100%) equal $B_{\text{ball},B} = 3{,}504.1$ kJ is spent for the exergy of the ball emission (−15.72%), for the exergy loss (−75.25%) due to irreversibility, and for the exergy loss (−9.03%) due to increased environment temperature, as shown in column 5 of Table 6.4.

On the other hand, from equation (t) the following formula on δb_e for the considered example can be derived:

$$\delta b_e = \sigma\left[\frac{T_{0,\text{inl}}^4}{4A_t}\left(e^{4A_t t}-1\right) - \frac{T_{\text{inl},B}^4}{4A_B}\left(e^{4A_B t}-1\right)\right] + mc[(T_{0,\text{inl}}-T_{0,fin})-(T_{\text{inl},B}-T_{\text{fin},B})]$$

$$+ mc\left\{T_{0,\text{inl}}\left[\left(e^{A_t t}-1\right)\ln\frac{T_{\text{inl},B}}{T_{\text{fin},B}} + \ln\frac{T_{\text{inl},B}}{T_{0,\text{inl}}}\right] - T_{0,\text{fin}}\ln\frac{T_{\text{fin},B}}{T_{0,\text{fin}}}\right\} \qquad \text{(y)}$$

Formula (y) shows that in the considered case B the value of δb_e with changing environment temperature is not simple. The effect of the varying environment temperature can be studied based on the exemplary numerical results shown in Table 6.4.

6.10 Radiation of Surface of Nonuniform Temperature

6.10.1 Emission Exergy at Continuous Surface Temperature Distribution

In practical engineering calculations, isothermal surfaces for which the temperature distribution is uniform and expressed with a constant value for the surface temperature at every point of the surfaces are usually considered. In real situations the surfaces usually are not isothermal and the surface temperature varies from point to point. The conductive heat transfer within a body renders the temperature of the body surface to be changing most continuously (smooth temperature distribution). For an elemental part dA of the considered surface A, the emission exergy dB can be determined based on formula (6.10) as follows:

$$dB = \varepsilon\,\frac{\sigma}{3}\left(3T^4 + T_0^4 - 4T_0 T^3\right)dA \qquad (6.71)$$

where the environment temperature $T_0 = \text{const}$, the local surface temperature T depends on location on surface; e.g., $T = T(x, y)$, and the emissivity ε can depend on the local temperature; $\varepsilon = \varepsilon(T) = \varepsilon[T(x, y)]$. The element of area A is $dA = dx \times dy$. The integration of equation (6.71):

$$B = \int_A \varepsilon\,\frac{\sigma}{3}\left(3T^4 + T_0^4 - 4T_0 T^3\right)dA \qquad (6.72)$$

can be difficult and the numerical interpretation of (6.72) would be as follows:

$$B = \sum_i \left[\varepsilon \, \frac{\sigma}{3} \left(3T^4 + T_0^4 - 4T_o T^3 \right) \right]_i \Delta A_i \qquad (6.73)$$

where B (kW) is the total emission exergy radiating from the considered surface of area A.

Example 6.5 A flat ceramic surface has dimensions 10×9 m. In the considered case, for rectangular coordinates x and y, $(0 \le x \le 10)$ and $(0 \le y \le 9)$ the surface temperature $T(x, y)$:

$$T = 400 + 2x^{1.6} + 0.5y^2 \qquad (a)$$

and the surface emissivity $\varepsilon(T)$:

$$\varepsilon = 0.86 + 0.153 \left(1 - \frac{T}{400} \right) \qquad (b)$$

To calculate the emission exergy, equation (6.73) is applied as follows:

$$B = \sum_{i=0}^{i=10} \sum_{j=0}^{j=9} \varepsilon(T_{i,j}) \left(3T_{i,j}^4 + T_0^4 - 4T_0 T_{i,j}^3 \right)(\Delta x \; \Delta y)_{i,j} \qquad (c)$$

If the constant increments $\Delta x = \Delta y = 1$ m are assumed, then $(\Delta x \times \Delta y)_{i,j} = 1$ and $x_i = i$ and $y_j = j$. The temperature and emissivity are then calculated based on equation (a) and (b), respectively, as follows:

$$T_{i,j} = 400 + 2i^{1.6} + 0.5j^2 \qquad (d)$$

$$\varepsilon_{i,j} = 0.86 + 0.153 \left(1 - \frac{T_{i,j}}{400} \right) \qquad (e)$$

For example, based on equation (d), Figure 6.13 shows the temperature distribution over the considered surface.

Figure 6.14 presents two different distributions (for $T_0 = 303$ K and $T_0 = 243$ K) of the emission exergy b radiated from the considered ceramic surface. Exergy values of the unchanged surface temperatures differs significantly and for high temperature $T_0 = 303$ K (30 °C), are smaller than those for $T_0 = 243$ K (–30 °C). The total emission exergy at $T_0 = 303$ K is $B_{303} = 44.211$ kW, whereas at $T_0 = 243$ K is $B_{243} = 78.719$ kW.

6.10.2   Effective Temperature of a Nonisothermal Surface

In some situations the considered surface, although nonisothermal, has its temperature not significantly diversified. If high exactness is not required in practical calculations, then, for better convenience and simplification, the *effective temperature* of the surface can be introduced

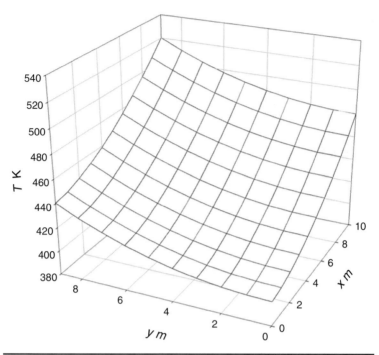

Figure 6.13 Temperature distribution for the considered surface.

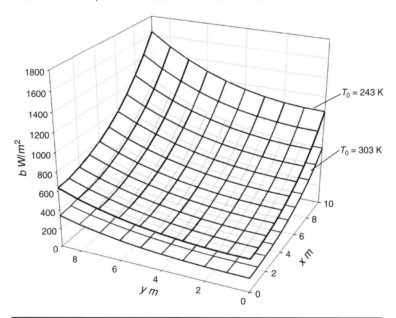

Figure 6.14 Emission exergy distribution for the considered surface for two different environment temperatures ($T_0 = 303$ K and $T_0 = 243$ K).

into consideration. Three different cases of energy analysis are possible. First, if only the convective heat transfer is considered, then the energetic convective effective temperature $T_{\text{eff},E,h}$ can be used. Second, if only the radiation is analyzed, then the energetic radiative effective temperature $T_{\text{eff},E,r}$ can be applied. In the third case the convection together with radiation is taken into account and then the energetic effective temperature $T_{\text{eff},E}$ can be considered. The definitions of these temperatures take into account the equivalence based on the averaged potential, which is the heat transfer coefficient h and surface temperature (for the convection) and the surface temperature to the fourth power (for the radiation).

For these three temperatures the definition equations for any surface A, together with numerical interpretation, are:

Energetic Effective Temperature

Pure convection:

$$Ah T_{\text{eff},E,h} = \int_A hT \, dA \qquad (6.74a)$$

or numerically:

$$Ah T_{\text{eff},E,h} = \sum_i h_i T_i \, \Delta A_i \qquad (6.74b)$$

pure radiation:

$$A\sigma T_{\text{eff},E,r}^4 = \sigma \int_A T^4 dA \qquad (6.75a)$$

or numerically:

$$A\sigma T_{\text{eff},E,r}^4 = \sigma \sum_i T_i^4 \Delta A_i \qquad (6.75b)$$

combined convection with radiation:

$$Ah T_{\text{eff},E} + A\sigma T_{\text{eff},E}^4 = \int_A hT \, dA + \sigma \int_A T^4 dA \qquad (6.76a)$$

or numerically:

$$A\left(h T_{\text{eff},E} + \sigma T_{\text{eff},E}^4\right) = \sum_i h_i T_i \Delta A_i + \sigma \sum_i T_i^4 \Delta A_i \qquad (6.76b)$$

Analogically for exergetic consideration the effective temperatures for any surface A can be defined as follows:

Exergetic Effective Temperature:

Pure convection:

$$AhT_{eff,B,h}\left(1 - \frac{T_0}{T_{eff,B,h}}\right) = \int_A hT\left(1 - \frac{T_0}{T}\right)dA \qquad (6.77a)$$

numerically:

$$AhT_{eff,B,h}\left(1 - \frac{T_0}{T_{eff,B,h}}\right) = \sum_i h_i T_i\left(1 - \frac{T_0}{T_i}\right)\Delta A_i \qquad (6.77b)$$

only radiation:

$$A\frac{\sigma}{3}\left(3T_{eff,B,r}^4 + T_0^4 - T_0 T_{eff,B,r}^3\right) = \frac{\sigma}{3}\int_A \left(3T^3 + T_0^4 - T_0 T^3\right)dA$$

$$(6.78a)$$

numerically:

$$A\frac{\sigma}{3}\left(3T_{eff,B,r}^4 + T_0^4 - T_0 T_{eff,B,r}^3\right) = \frac{\sigma}{3}\sum_i \left(3T_i^4 + T_0^4 - T_0 T_i^3\right)\Delta A_i$$

$$(6.78b)$$

including convection and radiation:

$$AhT_{eff,B}\left(1 - \frac{T_0}{T_{eff,B}}\right) + A\frac{\sigma}{3}\left(3T_{eff,B}^4 + T_0^4 - T_0 T_{eff,B}^3\right)$$

$$= \int_A hT\left(1 - \frac{T_0}{T}\right)dA + \frac{\sigma}{3}\int_A \left(3T^4 + T_0^4 - T_0 T^3\right)dA$$

$$(6.79a)$$

numerically:

$$AhT_{eff,B}\left(1 - \frac{T_0}{T_{eff,B}}\right) + A\frac{\sigma}{3}\left(3T_{eff,B}^4 + T_0^4 - T_0 T_{eff,B}^3\right)$$

$$(6.79b)$$

$$= \sum_i h_i T_i\left(1 - \frac{T_0}{T_i}\right)\Delta A_i + \frac{\sigma}{3}\sum\left(3T_i^4 + T_0^4 - T_0 T_i^3\right)\Delta A_i$$

Example 6.6 The accuracy of the effective temperature of a surface can be evaluated as follows. A black surface has a linear temperature distribution that can be approximated with three equal increments of a surface with three respective surface temperatures T_i, $(i = 1, 2, 3)$. It is assumed that a constant value of convection heat transfer coefficient $h = 5$ W/(m^2 K) and an environmental temperature $T_0 = 288.16$ K. Table 6.5 demonstrates numerically calculated values of the effective temperatures from respective equations. As a result, for a moderate growth of surface temperature (from 320 to 340 K) all the effective temperatures vary insignificantly. However, for the larger growth of surface temperatures, e.g., from 320 to 720 K, except $T_{eff,h}$ all the other effective temperatures change significantly. As might be expected, the effective temperature for radiation is higher than the effective temperature for radiation together with convection.

$T_{l=1}$	$T_{l=2}$	$T_{l=3}$	$T_{eff,E,h}$	$T_{eff,E,r}$	$T_{eff,E}$	$T_{eff,B,h}$	$T_{eff,B,r}$	$T_{eff,B}$
320	330	340	330	330.30	330.19	330	330.98	330.17
320	420	520	420	442.21	437.64	420	454.12	440.15
320	520	720	520	585.41	578.83	520	601.84	588.03
Applied equation			(6.74b)	(6.75b)	(6.76b)	(6.77b)	(6.78b)	(6.79b)

TABLE 6.5 Calculated Values of the Effective Temperatures of a Black Surface

Nomenclature for Chapter 6

A	surface area, m^2
A	constant value, 1/s
A, B, C, D, E	points in Figure 6.2
c	specific heat, J/(kg K)
C_b	$= 5.6693$ W/(m^2 K), constant for black radiation
c_0	speed of light in a vacuum, m/s
D	diameter, m
e	emission density, W/m^2
b	exergy of emission density, W/m^2
h	convective heat transfer coefficient, W/(m^2 K)
i, j	successive numbers
m	mass, kg
q	heat, W/m^2
s	entropy of emission density, W/(m^2 K)
t	time, s
T	absolute temperature, K
V	volume, m^3
W	work, J
x	temperature ratio T/T_0
x	coordinate, m
y	coordinate, m

Greek

δb	exergy loss due to irreversibility, W/m^2
δB	exergy loss due to irreversibility, W
Δ	increase
ε	emissivity of surface
φ	view factor
Π	overall entropy growth, W/(m^2 K), or W/K
ρ	density, kg/m^3
σ	$= 5.6693 \times 10^{-8}$ W/(m^2 K^4), Boltzmann constant for black radiation
ψ	ratio of emission exergy to emission energy

Subscripts

A, B	different cases
A, B, C, D, E	points in Figure 6.2
a	absorption
b	black
B	exergy
ball	ball
c	cold
E	energy
e	emission
ET	empty tank
eff	effective
h	convection
inl	inlet
0	environment
q	heat
sky	sky
S	system
S	solar
wv	water vapor
x	denotation
y	denotation
1, 2	denotation

## CHAPTER 7

Radiation Flux

7.1 Energy of Radiation Flux

In Section 5.7, the energy u, J/m^3, of trapped radiation within a space was discussed. In practice, radiation can also be considered as a flux, J/s, which originates from a surface of known properties. For example, emission energy can be calculated for the black or gray surfaces with equations (3.21) or (3.22), respectively. However, generally, the radiation flux propagating in space can consist of many emissions from unknown surfaces and with unknown temperatures. Such radiation flux can be categorized as the radiosity of an arbitrary spectrum. The radiosity can be calculated if the spectrum is determined, e.g., from measurement.

The elemental radiation energy flux, J/s, expressed as an elemental radiosity $d^3 J$ (the elemental order is selected based on the number of component elementals) propagating within a bundle of rays within a solid angle $d\omega$ and passing through an elemental surface dA is calculated as:

$$d^3 J = i_0 \cos \beta \, dA \, d\omega \qquad (7.1)$$

where β is the angle between the normal to the elementary surface dA and the direction of the considered solid angle $d\omega$. The magnitude i_0, W/(m^2 sr), is the normal radiation intensity and, as mentioned in Chapter 3, expresses energy passing within a unitary solid angle, in unit time and through a unitary surface area perpendicular to the direction of propagation. Generally, when radiation is polarized:

$$i_0 = \int\limits_{\nu=0}^{\nu=\infty} (i_{0,\nu,\min} + i_{0,\nu,\max}) \, d\nu \qquad (7.2)$$

The quantities $i_{0,\nu,\min}$ and $i_{0,\nu,\max}$ depend on frequency ν, 1/s, and are the principal (smallest and largest) mutually independent (incoherent), polarized at right angles to each other, values of the monochromatic component of radiation intensity, J/(m^2 sr).

However, for nonpolarized radiation $i_{0,\nu,\min} = i_{0,\nu,\max} = i_{0,\nu}$ and from (7.2):

$$i_0 = 2 \int_{\nu=0}^{\nu=\infty} i_{0,\nu} \, d\nu \tag{7.3}$$

or:

$$i_0 = 2 \int_{\lambda=0}^{\lambda=\infty} i_{0,\lambda} \, d\lambda \tag{7.4}$$

where $i_{0,\nu}$ and $i_{0,\lambda}$ are the normal monochromatic radiation intensity of the linearly polarized radiation depending on frequency, J/(m^2 sr), and wavelength, W/(m^3 sr), respectively.

Equations (7.3) and (7.4) express the same amount of energy. As mentioned already in Section 3.1, in experimental physics the spectrum of this energy is considered to be a function of wavelength, whereas for theoretical analysis it is more convenient using the spectrum as a function of the frequency, which does not change during propagation of radiation through different media. The relation between the wavelength and frequency ($\nu = c_0/\lambda$) is shown by equation (3.1).

If the algebraic sign is assumed the same for the considered intervals $d\nu$ and $d\lambda$, then, from (3.1), for propagation of radiation in vacuum is:

$$d\nu = c_0 \frac{d\lambda}{\lambda^2} \tag{7.5}$$

Equating the right-hand sides of equations (7.3) and (7.4) and using (7.5), one obtains:

$$i_{0,\lambda} = \frac{c_0 i_{0,\nu}}{\lambda^2} \tag{7.6}$$

The relation between the normal radiation intensity $i_{b,0,\lambda}$, and the black monochromatic emission density $e_{b,\lambda}$ expressed by equation (3.13) is as follows:

$$e_{b,\lambda} = 2\pi i_{b,0,\lambda} \tag{7.7}$$

where factor 2 appears because the two components shown in equation (7.2) are taken into account, and the value π results from equation (3.28). Using the expressions for the constants $c_1 = 2 \times \pi \times h \times c_0^2$ and $c_2 = h \times c_0/k$, equation (3.13) can be applied in the following form:

$$i_{b,0,\lambda} = \frac{c_0^2 h}{\lambda^5} \frac{1}{\exp\left(\frac{c_0 h}{k\lambda T}\right) - 1} \tag{7.8}$$

Combining relation $\lambda \times \nu = c_0$ with equations (7.5) and (7.6), one obtains the formula for the normal monochromatic (given frequency ν) radiation intensity of linearly polarized black radiation at temperature T as a function of frequency:

$$i_{b,0,\nu} = \frac{h\nu^3}{c_0^2} \frac{1}{\exp\left(\frac{h\nu}{kT}\right) - 1} \tag{7.9}$$

Expression (4.14) can be interpreted for calculation of any radiation energy from a certain surface A' arriving in the considered surface A. Therefore, the radiosity density $j_{A'}$, W/m², is considered as follows:

$$j_{A'} = \int_{\omega=0}^{\omega=2\pi} i_0 \cos \beta \, d\omega \tag{7.10}$$

where ω is the solid angle under which surface A' is seen from element dA of surface A as shown in Figure 7.1. The relation between distances shown in this Figure 7.1 is $r = R \times \sin \beta$; thus the elemental solid angle $d\omega$ is:

$$d\omega \equiv \frac{dA'}{R^2} = \frac{(R \, d\beta) \, (r \, d\varphi)}{R^2} = \sin \beta \, d\beta \, d\varphi \tag{7.11}$$

The flat angles of β (called *declination*) and φ (called *azimuth*) are defined in Figure 7.1. For a polarized radiation, using equation (7.2) and (7.11) in equation (7.10):

$$j_{A'} = \int_{\beta=0}^{\beta=\frac{\pi}{2}} \int_{\varphi=0}^{\varphi=2\pi} \int_{\nu=0}^{\nu=\infty} (i_{0,\nu,min} + i_{0,\nu,max}) \cos \beta \, \sin \beta \, d\beta \, d\varphi \, d\nu \tag{7.12}$$

FIGURE 7.1
Scheme for calculation of radiation flux (from Petela, 1962).

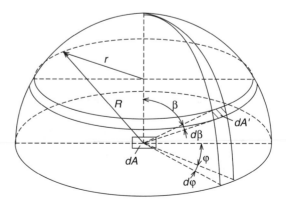

The monochromatic components $i_{0,\nu,\min}$ and $i_{0,\nu,\max}$ in equation (7.12) depend on angles β and φ (Figure 7.1), frequency ν, and the direction of polarization.

For nonpolarized radiation the components are equal, $i_{0,\nu,\min} = i_{0,\nu,\max} = i_{0,\nu}$ and from (7.12) is:

$$j_{A'} = 2 \int_{\beta=0}^{\beta=\frac{\pi}{2}} \int_{\varphi=0}^{\varphi=2\pi} \int_{\nu=0}^{\nu=\infty} i_{0,\nu} \cos\beta \, \sin\beta \, d\beta \, d\varphi \, d\nu \qquad (7.13)$$

To calculate the integral of equation (7.13), for each point P of surface A one has to know how the monochromatic radiation intensity depends on the angles β and φ (Figure 7.1) and on the frequency ν:

$$i_{0,\nu,P} = i_{0,\nu,P} \, (\beta, \varphi, \nu) \qquad (7.14)$$

When radiation is nonpolarized and propagates uniformly in all directions, then the monochromatic radiation intensity depends only on frequency, $i_{0,\nu}(\nu)$, and the density of radiosity of such radiation is:

$$j_{A'} = 2 \left(\iint_{\beta \ \varphi} \cos\beta \, \sin\beta \, d\beta \, d\varphi \right) \int_{\nu} i_{0,\nu} \, d\nu \qquad (7.15)$$

Additionally, if from any point of surface A the surface A' is seen within the solid angle 2π, then the double integral in equation (7.15) has value π and the density of radiosity is:

$$j = 2\pi \int_{\nu} i_{0,\nu} \, d\nu \qquad (7.16)$$

and, in such a case, the density of radiosity is the same at any point of the space between surfaces A and A'.

Another case to be considered is uniform radiation propagating in an arbitrary solid angle. Then, using equation (7.3) in (7.15) one obtains:

$$j_{A'} = i_0 \iint_{\beta \ \varphi} \cos\beta \, \sin\beta \, d\beta \, d\varphi \qquad (7.17)$$

or, dividing side by side equations (7.15) and (7.16):

$$j_{A'} = \frac{j}{\pi} \iint_{\beta \ \varphi} \cos\beta \, \sin\beta \, d\beta \, d\varphi \qquad (7.18)$$

For the black radiation of given temperature T, the energy spectrum is determined by formula (7.9). After substituting equation (7.9) into (7.16) and after integrating, we obtain:

$$j_{A'} \equiv j_b = \frac{ac_0}{4}T^4 \equiv e_b \qquad (7.19)$$

Comparison of formulae (7.19) and (3.21) confirms that for black radiation the radiosity is equal to the emission.

In conclusion, the formulae for radiosity presented in this section can be applied to many possible cases of radiation.

7.2 Entropy of Radiation Flux

7.2.1 Entropy of the Monochromatic Intensity of Radiation

Formula (5.23) for the entropy density of a photon gas residing in a system in an equilibrium state was derived in Section 5.5. However, the radiation entropy in the general case can be discussed by the following historically first derivations by Planck. The derivations are still recognized as leading to a sufficient approximation.

In analogy to equation (7.1) for radiosity, the elemental entropy of radiation, W/K, expressed as the radiosity entropy d^3S propagating within a bundle of rays in a solid angle $d\omega$ and passing through the elemental surface dA, is:

$$d^3S = L_0 \cos\beta \; dA \, d\omega \qquad (7.20)$$

where β is the angle between the normal to the elementary surface dA and the direction of the considered solid angle $d\omega$. The symbol "L" is assumed after Planck (1914); however, in the present consideration the subscript "0" is added to emphasize the meaning of the symbol; $L \equiv L_0$, W/(K m^2 sr). The magnitude L_0 is the entropy of the directional normal radiation intensity, which is the entropy passing within a unitary solid angle, in unit time, and through a unitary surface area perpendicular to the direction of propagation. Generally, when radiation is polarized:

$$L_0 = \int\limits_{\nu=0}^{\nu=\infty} (L_{0,\nu,\min} + L_{0,\nu,\max}) \, d\nu \qquad (7.21)$$

where $L_{0,\nu,\min}$ and $L_{0,\nu,\max}$ depend on frequency ν, 1/s, and are the principal (i.e., smallest and largest) mutually independent (incoherent), polarized at right angles to each other, values of the monochromatic component of the entropy of radiation intensity, J/(K m^2 sr).

For nonpolarized radiation, $L_{0,\nu,\min} = L_{0,\nu,\max} = L_{0,\nu}$ and from (7.21) is:

$$L_0 = 2 \int\limits_{\nu=0}^{\nu=\infty} L_{0,\nu} \, d\nu \qquad (7.22)$$

or

$$L_0 = 2 \int_{\lambda=0}^{\lambda=\infty} L_{0,\lambda} d\lambda \qquad (7.23)$$

where $L_{0,\nu}$ and $L_{0,\lambda}$ are the entropies of monochromatic intensity of linearly polarized radiation dependent, respectively, on frequency, $J/(m^2 \text{ K sr})$, and on wavelength, $W/(m^3 \text{ K sr})$. Equations (7.22) and (7.23) express the same amount of entropy.

Now, the pivotal formulae derived by Planck for the entropy considerations of black radiation are:

$$L_{b,0,\nu} = \frac{k\nu^2}{c_0^2} [(1+X)\ln(1+X) - X\ln X] \text{ where } X \equiv \frac{c_0^2 i_{b,0,\nu}}{\nu^3 h}$$

$$(7.24)$$

or

$$L_{b,0,\lambda} = \frac{c_0 k}{\lambda^4} [(1+Y)\ln(1+Y) - Y\ln Y] \text{ where } Y \equiv \frac{\lambda^5 i_{b,0,\lambda}}{c_0^2 h}$$

$$(7.25)$$

where $i_{b,0,\nu}$ and $i_{b,0,\lambda}$ are the monochromatic normal directional intensity for black radiation.

7.2.2   Entropy of Emission from a Black Surface

The entropy s, $W/(K \text{ m}^2)$, expresses the entropy density of radiation emitted by the unit surface area of a body in all the directions of the front hemisphere in unit time:

$$s = \int_{\omega=0}^{\omega=2\pi} L_0 \cos\beta \, d\omega \qquad (7.26)$$

where L_0, $W/(m^2 \text{ K sr})$, is the entropy of the directional normal radiation intensity of the emitting surface. If the emission of the surface is the same in all directions, then L_0 is constant and after using equation (7.11) in equation (7.26), is:

$$s = L_0 \int_{\varphi=0}^{\varphi=2\pi} d\varphi \int_{\beta=0}^{\beta=\pi/2} \cos\beta \, \sin\beta \, d\beta = \pi L_0 \qquad (7.27)$$

If at any point P the surface element dA emits energy only within a solid angle $\omega \leq 2\pi$, and if the emission within this solid angle is uniform ($L_0 = $ constant), then the entropy of such emission from point P is:

$$s_\omega = L_0 \int_{\beta=o}^{\beta=\pi/2} \int_{\varphi=0}^{\varphi=2\pi} \cos\beta \, \sin\beta \, d\beta \, d\varphi \qquad (7.28)$$

Another form of formula (7.28) can be obtained by division of equations (7.27) and (7.28) side by side:

$$s_\omega = \frac{s}{\pi} \int_{\beta=0}^{\beta=\pi/2} \int_{\varphi=0}^{\varphi=2\pi} \cos\beta \, \sin\beta \, d\beta \, d\varphi \qquad (7.29)$$

Equation (7.9) for the black emission of entropy can be inserted into (7.24) and the other Planck's formula is obtained:

$$L_{b,0,\nu} = \frac{h\nu^3}{c_0^2 T} \frac{1}{\exp\frac{h\nu}{kT} - 1} - \frac{k\nu^2}{c_0^2} \ln\left(1 - \exp\frac{-h\nu}{kT}\right) \qquad (7.30)$$

Now, the formula for entropy of the uniform emission s_b from a black surface at temperature T will be obtained. For a black surface the entropy L_0 of radiation intensity is denoted as $L_{b,0}$, ($L_0 = L_{b,0}$). Inserting equation (7.22) and (7.30) into (7.27), and integrating, as shown, e.g., by Petela (1961a), the following relations are obtained:

$$s_b = \frac{8\,\pi^5 k^4}{45\,c_0^2 h^3} T^3 = \frac{a c_0}{3} T^3 = \frac{4}{3}\sigma\, T^3 \qquad (7.31)$$

Similar to equation (3.22) for the emission density, the entropy density of emission from a perfectly gray surface can be determined as follows:

$$s = \varepsilon s_b = \varepsilon \frac{a c_0}{3} T^3 = \varepsilon \frac{4}{3}\sigma T^3 \qquad (7.32)$$

7.2.3   Entropy of Arbitrary Radiosity

Now, let us introduce the entropy flux, W/(K m^2), which is the entropy of radiosity density $s_{j,A'}$ passing the unit control surface area A' in a space, in unit time, and falling on the element dA of the considered surface A. Obviously, in certain particular case which is excluded for the time being, such entropy can be also interpreted as the entropy of emission in unit time and from the surface area of a body. The entropy $s_{j,A'}$ is:

$$s_{j,A'} = \int_\omega L_0 \cos\beta \, d\omega \qquad (7.33)$$

where ω is the solid angle under which the surface A' is seen from the element dA of the surface A.

Further presentation of entropy formulae is analogous to the energy formulae discussed in Section 7.1. For polarized radiation, after using (7.21) in (7.33), we obtain:

$$s_{j,A'} = \int_{\beta=0}^{\beta=\pi/2} \int_{\varphi=0}^{\varphi=2\pi} \int_{\nu=0}^{\nu=\infty} (L_{0,\nu,min} + L_{0,\nu,max}) \cos\beta \, \sin\beta \, d\beta \, d\varphi \, d\nu$$

(7.34)

The monochromatic components $L_{0,\nu,min}$ and $L_{0,\nu,max}$ in equation (7.34) depend on angles β and φ (Figure 7.1), frequency ν, and on the manner of polarization.

For nonpolarized radiation, the components are equal, $L_{0,\nu,max} = L_{0,\nu,max} = L_{0,\nu}$ and from (7.34):

$$s_{j,A'} = 2 \int_{\beta=0}^{\beta=\pi/2} \int_{\varphi=0}^{\varphi=2\pi} \int_{\nu=0}^{\nu=\infty} L_{0,\nu} \cos\beta \, \sin\beta \, d\beta \, d\varphi \, d\nu$$

(7.35)

For calculation of the integral in equation (7.35) the function (7.14) has to be known and used in (7.24).

If the entropy of radiosity density of surface A' is the same for all its points and in all directions, then the entropy of the radiosity density coming from surface A' to surface A is:

$$s_{j,A'} = 2 \left(\int_{\beta} \int_{\varphi} \cos\beta \, \sin\beta \, d\beta \, d\varphi \right) \int_{\nu} L_{0,\nu} \, d\nu$$

(7.36)

The entropy $L_{0,\nu}$ depends only on the frequency ν and for calculation of this entropy from equation (7.24) the function $i_{0,\nu}(\nu)$ has to be known.

For a radiation falling upon surface A within solid angle 2π the double integral in equation (7.36) has the value π and the entropy of radiosity density is:

$$s_j = 2\pi \int_{\nu} L_{0,\nu} \, d\nu$$

(7.37)

and in such case the radiosity density s_j is the same at any point of the space between surfaces A and A'.

For uniform radiation arriving at a point P of surface A within a solid angle ω the entropy of radiosity density is obtained by using equation (7.22) in (7.35):

$$s_{j,\omega} = L_0 \int_{\beta} \int_{\varphi} \cos\beta \, \sin\beta \, d\beta \, d\varphi$$

(7.38)

or by division side by side of equations (7.35) and (7.22):

$$s_{j,\omega} = \frac{s_j}{\pi} \int_{\beta} \int_{\varphi} \cos\beta \, \sin\beta \, d\beta \, d\varphi$$

(7.39)

where β and φ are the coordinates of directions within the solid angle ω.

For uniform black radiation of given temperature T the energy spectrum is determined by formula (7.9). After substituting equation (7.24) and (7.9) into (7.36) and after integrating, we obtain:

$$s_{j,b} = \frac{ac_0}{3}T^3 \equiv s_b \qquad (7.40)$$

Comparison of formulae (7.31) and (7.40) confirms that for uniform black radiation the entropy of radiosity is equal to the entropy of emission.

In conclusion, the formulae presented in this section concerning the entropy of radiosity can be applied in many possible cases of radiation.

7.3 Exergy of Radiation Flux

7.3.1 Arbitrary Radiation

Often the temperature and properties (e.g., emissivity) of the radiation source from which the considered radiation arrives are unknown. Such radiation is categorized as *arbitrary radiation*, which is the radiation of any irregular spectrum that is not expressible, e.g., by the ideal black or gray model. The arbitrary radiation can be specified by the radiosity calculated from the results of spectral measurement determining the solid angle of the arriving radiation and its components of monochromatic intensity as a function of wavelength. The following formulae for the calculation of exergy in various cases of arbitrary radiation flux were derived for the first time by Petela (1961) and then modified by Petela (1962, 1964).

As was explained in Section 6.1, the exergy of radiation is related to black radiation (i.e., photon gas) at the environment temperature T_0 and does not depend on environmental emissivity; it also does not depend on the emissivity or temperature of any other surrounding surfaces that are not components of the environment. Thus, when considering the exergy of arbitrary radiation arriving at a certain surface, the properties of this surface do not affect the final results of calculating the exergy. These properties can be assumed appropriately to make the calculation simplest. Thus, it is further assumed that a certain considered surface absorbing arbitrary radiation is the environment surface for which within the solid angle ω emissivity is $\varepsilon = 1$ and beyond the angle ω the emissivity is $\varepsilon = 0$.

In order to calculate the exergy $b_{A'}$ of arbitrary radiation, at its radiosity $j_{A'}$, originating from certain unknown surface A' and arriving in the point P of surface A, one can consider the element dA at

Figure 7.2 Scheme
of energy and
exergy balances
for arbitrary
radiation (from
Petela, 1962).

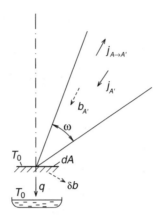

point P of the surface A (Figure 7.2). The elemental surface dA, due to connection with a heat source, maintains a steady environment temperature T_0. Surface A is black (i.e., radiosity is equal black emission) and able to emit and absorb radiation within the solid angle ω. The energy fluxes (solid arrows) and exergy fluxes (dashed arrows) shown in Figure 7.2 are referred to 1 m^2 surface. The energy balance equation for the elemental surface dA is:

$$j_{A'} = q + j_{A \to A'} \tag{a}$$

where q is heat transferred from the considered elemental surface dA to the heat source at environment temperature T_0. Quantity $j_{A \to A'}$ is the energy radiating from the element dA within the solid angle ω and can be determined by the following interpretation of formula (7.18):

$$j_{A \to A'} = \frac{(e_b)_{T=T_0}}{\pi} \int_\beta \int_\varphi \cos \beta \; \sin \beta \; d\beta \; d\varphi \tag{b}$$

where e_b in formula (b) is the emission energy of the black surface at temperature T_0, determined by formula (3.21). Substituting (b) into (a) the heat q can be determined as:

$$q = j_{A'} - \frac{(e_b)_{T=T_0}}{\pi} \int_\beta \int_\varphi \cos \beta \; \sin \beta \; d\beta \; d\varphi \tag{c}$$

According to the definition of exergy, the change of exergy of the heat source of the environmental temperatures is zero, and also zero is the exergy of emission at the environmental temperature. Therefore, the exergy $b_{A'}$ of unknown arbitrary radiation can be calculated from the following exergy balance equation (for 1 m^2 surface):

$$b_{A'} = \delta b \tag{d}$$

where δb is the exergy loss caused by the irreversibility of simultaneous emission and the absorption of radiation occurring at the considered surface dA. The exergy loss δb is calculated from the Gouy–Stodola law (2.60):

$$\delta b = T_0 \Pi \tag{e}$$

in which the overall entropy growth Π, is:

$$\Pi = \frac{q}{T_0} - s_{j,A'} + s_{j,A \to A'} \tag{f}$$

where $s_{j,A'}$ is the entropy of radiosity of radiation arriving at the elemental surface dA. Quantity $s_{j,A \to A'}$ is the entropy of emission of surface A within a solid angle ω, determined from formula (7.39) as follows:

$$s_{j,A \to A'} = \frac{(s_b)_{T=T_0}}{\pi} \int_\beta \int_\varphi \cos\beta \, \sin\beta \, d\beta \, d\varphi \tag{g}$$

where s_b in formula (g) is the entropy of the emission of the black surface at temperature T_0, which can be determined based on formula (7.32).

After using (f) and (g) in (e):

$$\delta b = q - T_0 \left(s_{j,A'} - \frac{(s_b)_{T=T_0}}{\pi} \int_\beta \int_\varphi \cos\beta \, \sin\beta \, d\beta \, d\varphi \right) \tag{h}$$

Substituting now (h) and (c) into (d) and expressing entropy s_b based on equation (7.32), and after rearranging, one obtains:

$$b_{A'} = j_{A'} - T_0 s_{j,A'} + \frac{\sigma \, T_0^4}{3 \, \pi} \int_\beta \int_\varphi \cos\beta \, \sin\beta \, d\beta \, d\varphi \tag{7.41}$$

Formula (7.41) can be used for any categorized case of radiation for which radiosity $j_{A'}$ and entropy $s_{j,A'}$ have to be determined appropriately. Total exergy $B_{A'-A}$, W, of arbitrary radiation arriving at surface A from the unknown surface A' can be determined as:

$$B_{A' \to A} = \int_A b_{A'} \, dA \tag{7.42}$$

Further, some typical formulae for categorized radiation cases are developed based on formula (7.41).

7.3.2 Polarized Radiation

Exergy of Arbitrary Polarized Radiation

The exergy $b_{A'}$, W/m^2, of the arbitrary polarized radiation originating from an unknown surface A' and arriving at point P of the considered surface A per unit time and unit absorbing surface area, can be calculated from the formula derived by substituting (7.12) and (7.34) into (7.41):

$$b_{A'} = \int_\beta \int_\varphi \int_\nu (i_{0,\nu,min} + i_{0,\nu,max}) \cos\beta \, \sin\beta \, d\beta \, d\varphi \, d\nu$$

$$- \int_\beta \int_\varphi \int_\nu (L_{0,\nu,min} + L_{0,\nu,max}) \cos\beta \, \sin\beta \, d\beta \, d\varphi \, d\nu$$

$$+ \frac{\sigma \, T_0^4}{3 \, \pi} \int_\beta \int_\varphi \cos\beta \, \sin\beta \, d\beta \, d\varphi \tag{7.43}$$

In order to utilize formula (7.43) one has to determine the solid angle ω within which the surface A' is seen from point P on surface A, and to make measurements for determination of $i_{0,\nu,min}$ and $i_{0,\nu,max}$ as a function of frequency ν and direction defined by β and φ. Dependence (7.6) between $i_{0,\nu,min}$ and $i_{0,\nu,max}$, and respective $i_{0,\lambda,min}$ and $i_{0,\lambda,max}$ can be useful. The respective entropy components $L_{0,\nu,min}$ and $L_{0,\nu,max}$ are determined based on formula (7.30). The total exergy of the considered arbitrary radiation arriving at all the points of surface A is calculated from formula (7.42).

7.3.3 Nonpolarized Radiation

Exergy of Arbitrary Nonpolarized Radiation

The formula for such radiation is obtained after substituting (7.13) and (7.35) into (7.41):

$$b_{A'} = \int_\beta \int_\varphi \int_\nu i_{0,\nu} \cos\beta \, \sin\beta \, d\beta \, d\varphi \, d\nu$$

$$- \int_\beta \int_\varphi \int_\nu L_{0,\nu} \cos\beta \, \sin\beta \, d\beta \, d\varphi \, d\nu$$

$$+ \frac{\sigma \, T_0^4}{3 \, \pi} \int_\beta \int_\varphi \cos\beta \, \sin\beta \, d\beta \, d\varphi \tag{7.44}$$

In order to utilize formula (7.44) one has to determine the solid angle ω within which the surface A' is seen from point P on surface

A, and based on measurements, the $i_{0,\nu}$ as a function of frequency ν and direction defined by β and φ, has to be determined. The formulae (7.6), (7.30), and (7.42) can be also useful.

7.3.4 Nonpolarized and Uniform Radiation

Exergy of Arbitrary, Nonpolarized, and Uniform Radiation
The formula for such radiation is derived by substituting equations (7.15) and (7.36) into (7.41):

$$
b_{A'} = \left(2 \int_\nu i_{0,\nu}\, d\nu - 2\, T_0 \int_\nu L_{0,\nu}\, d\nu + \frac{\sigma T_0^4}{3\,\pi} \right) \int_\beta \int_\varphi \cos\beta\ \sin\beta\ d\beta\ d\varphi
$$

$$(7.45)$$

Before utilizing formula (7.45) the solid angle ω within which the surface A' is seen from point P on surface A, has to be determined and based on measurements the radiation spectrum as function of frequency, $i_{0,\nu}(\nu)$, has to be given. Again, the formulae (7.6), (7.30), and (7.42) can be useful.

7.3.5 Nonpolarized, Uniform Radiation in a Solid Angle 2π

Exergy of Arbitrary, Nonpolarized, and Uniform Radiation Propagating Within a Solid Angle 2π
The formula for such radiation is derived by substituting equations (7.16) and (7.37) into (7.41):

$$
b = 2\pi \int_\nu i_{0,\nu}\, d\nu - 2\,\pi T_0 \int_\nu L_{0,\nu}\, d\nu + \frac{\sigma}{3} T_0^4 \qquad (7.46)
$$

To utilize formula (7.46) the function $i_{0,\nu}(\nu)$, based on measurements, is required. Formulae (7.6) and (7.30) can be useful. The total exergy of the considered radiation arriving to all the points of the surface A is calculated as follows:

$$
B = bA \qquad (7.47)
$$

Example 7.1 Figure 7.3 shows the measured monochromatic normal radiation intensity $i_{0,\lambda}$ (solid line) of radiation, as a function of wavelength λ, for the water vapor layer of the equivalent thickness 1.04 m at temperature 200°C according to Jacob (1957). The product of the thickness and the partial pressure for the vapor is 10.4 m kPa. The monochromatic normal intensity $i_{b,0,\lambda}$ for black radiation, calculated from equation (7.8), is also shown for comparison (dashed line). For approximate calculation, instead of the surface area under a solid line, the area of seven rectangles (dotted line) is taken into account as the integral energy emitted by the vapor upon the hemispherical enclosure. The areas of these rectangles can be recognized as the absorption bands of width $\Delta\lambda$ spread symmetrically on both sides of wavelength λ, the values of which are given in Table 7.1.

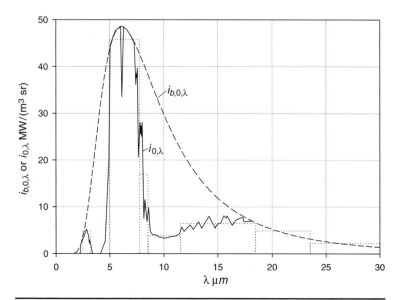

FIGURE 7.3 Radiation of water vapor layer of thickness 1.04 m at temperature 473.15 K and pressure 0.1 MPa (from Jacob, 1957).

Exergy of radiation arriving in 1 m^2 of the enclosing hemispherical wall can be calculated from formula (7.46), in which the frequency should be eliminated by wavelength. Each integral in formula (7.46) can be replaced by the sum of appropriate products:

$$b = \frac{\sigma}{3} T_0^4 + 2\pi \sum i_{0,\lambda}\, \Delta\lambda - 2\pi\, T_0 \sum L_{0,\lambda} \Delta\lambda \qquad (7.48)$$

#	λ	$\Delta\lambda$	$i_{0,\lambda} \times 10^{-6}$	$i_{0,\lambda} \times \Delta\lambda$	$L_{0,\lambda} \times 10^{-4}$	$L_{0,\lambda} \times \Delta\lambda$
	μ m		$\dfrac{W}{m^3\ sr}$		$\dfrac{W}{m^2\ K\ sr}$	
1	2.69	0.66	5.0	3.3	1.07	0.0071
2	6.15	2.8	45.7	128.0	11.72	0.3282
3	7.95	0.8	17.2	13.8	5.41	0.0433
4	9.8	2.9	3.7	10.7	1.56	0.0452
5	14.8	7.1	6.4	45.4	2.38	0.1690
6	21.0	5.3	5.1	27.0	1.83	0.0970
7	26.8	6.3	2.2	13.9	0.78	0.0491
Total				242.1	—	0.7389

TABLE 7.1 Radiation of Water Vapor Layer of Thickness 1.04 m at Temperature 473.15 K and Pressure 0.1 MPa

For the assumed temperature $T_0 = 300$ K formula (7.48) yields:

$$b = \frac{5.6693 \times 10^{-8}}{3} 300^4 + 2\pi \times 0.2421 - 2\pi \times 300 \times 0.7389 \times 10^{-3}$$

$$= 0.153 + 1.521 - 1.393 = 0.281 \, \text{kW/m}^2$$

The ratio of the exergy of radiation of the vapor to its energy emission is $b/e = 0.281/1.521 = 0.185$. More details of the considered example are discussed by Petela (1961a).

7.3.6 Nonpolarized, Black, Uniform Radiation in a Solid Angle 2π

Exergy of Arbitrary, Nonpolarized, Black, and Uniform Radiation, Propagating Within a Solid Angle 2π

The formula for such radiation is derived by substituting equations (7.19) and (7.40) into (7.41):

$$b_b = \frac{\sigma}{3} \left(3T^4 + T_0^4 - 4T_0 T^3 \right) \qquad (7.49)$$

To utilize formula (7.49), only the temperature of the black radiation is required. The total exergy arriving at surface A can be calculated from formula (7.47).

It is noteworthy that equation (7.49) is identical to equation (6.8) derived for the black emission. Formula (7.49) is also as formula (6.30), at $\varepsilon = 1$, derived from the exergy balance of a gray surface. This sameness is a confirmation that the exergy of the black radiation is equal to the exergy of the emission of the black surface, consequently to this that the radiosity of the blackbody is equal to its emission.

7.3.7 Nonpolarized, Black, Uniform Radiation Within a Solid Angle ω

Exergy of Nonpolarized, Uniform, Black Radiation Propagating Within a Solid Angle ω

The formula for such radiation can be established analogously to formula (7.18) for radiosity:

$$b_{b,\omega} = \frac{b_b}{\pi} \int_{\beta} \int_{\varphi} \cos\beta \, \sin\beta \, d\beta \, d\varphi \qquad (7.50)$$

where the solid angle ω has to be determined by the appropriate ranges of variation of the both flat angles β (declination) and φ (azimuth).

7.4 Propagation of Radiation

7.4.1 Propagation in a Vacuum

The radiation exergy formulae derived in Section 7.3 are for any radiation arriving at certain surface. The surface can be recognized either as a body surface, i.e., a certain cross section situated infinitely near to this surface (abutted on it), or as any arbitrary plane imagined in space. On its way, the propagating radiation flux can experience different energetic and optical adventures that can affect the radiation energy, entropy, and exergy.

In the simplest case the radiation propagates in a vacuum or within a neutral (nonradiating and nonabsorbing) and homogeneous (constant density) medium. For example, such a medium can practically be a monatomic gas (e.g., He, Ar, etc.) or a diatomic gas (e.g., pure air, CO, etc.), or their mixture. However, even these gases in the vicinity of any condensed body surface can affect the radiation process by generating the convective heat transfer that influences the surface temperature and thus the surface emission. Thus, only the vacuum is a perfectly nonaffecting propagation of radiation. In a vacuum, different radiation fluxes can propagate independently, unless they are trapped in the enclosed space lined up with the mirrorlike walls. Then the equilibrium state of the black photon gas appears at a uniform temperature. The trapped radiation was discussed in Section 5.2.

Although the intensity of propagating radiation decreases with the distance to the second power, the directional intensity in the neutral medium remains unchanged. The directional intensity has been considered, e.g., by Petela (1984), in the situation shown in Figure 7.4. A surface element dA is the source of the normal directional radiation intensity i_0, W/(m² sr), which arrives within the cone of the solid angle $d\omega$ to the element dA' of another surface. The consideration was carried out under the assumption that the surface dA represented a blackbody. Thus, the eventual radiosity of element dA is equal to the emission of radiation of the determined spectrum corresponding to the surface temperature. Consideration of radiosity would be more

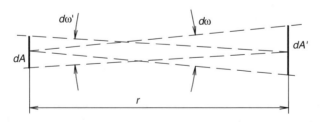

FIGURE 7.4   Irradiated surface elements.

complex because it would generally require inclusion of a number of fluxes of different temperatures and respective spectra.

The surface dA is irradiated by a certain normal directional intensity i'_0, which arrives within the solid angle $d\omega'$. The comparison of the energy that leaves the element dA for dA' with the energy arriving in the element dA' from dA leads to the following equation:

$$i_0 \, d\omega \, dA = i'_0 \, d\omega' \, dA' \qquad (7.51)$$

After determining the solid angles: $d\omega = dA'/r^2$ and $d\omega' = dA/r^2$, where r is the distance between dA and dA', it yields $i_0 = i'_0$. This means that the normal directional radiation intensity does not depend on the distance from the source. It also means that temperature (determined by spectrum) does not change; however, the amount of energy radiated within the solid angle is changed. The temperature measurement from a distance is based on such a rule.

A similar reasoning can be carried out with the entropy or exergy of radiation and, respectively, the same result can be obtained. In the considered propagation of radiation there is no irreversible process, such as, e.g., absorption, and the distributed energy of radiation can be theoretically reversibly concentrated by a perfectly (nonabsorbing) optical device. Therefore, radiation propagation in a vacuum or in a neutral medium is a reversible phenomenon and occurs without any loss of exergy.

The aforementioned attenuation of radiation in space during propagation can be considered as presented, e.g., by Petela (1983). Again, for simplicity, instead of radiosity j, only the emission e of the black surface element dA_1 shown in Figure 7.5, is considered. The element dA_1 radiates into the front vacuumed hemisphere, but only a part of the radiation arrives in an arbitrarily situated other surface element dA_2. The distance between the two elements is r.

The flat angles β_1 and β_2 are between the straight line linking the surface elements, and the respective perpendicular straight lines (normal) n_1 and n_2 to these elements (dA_1 and dA_2). From element dA_1 the element dA_2 is seen within a solid angle $d\omega$, and the element dA_2 can be replaced for the considerations by dA'_2 determined as follows:

$$dA'_2 = dA_2 \, \cos \beta_2 \qquad (7.52)$$

The element dA'_2 is a projection of element dA_2 on the surface of hemisphere of radius r. Therefore, the solid angle $d\omega$ can be determined as follows:

$$d\omega = \frac{dA_2 \, \cos \beta_2}{r^2} \qquad (7.53)$$

The directional radiosity density expressed by formula (3.29) can be applied to the emission density ($j_\beta = e_\beta$) of element dA_1 in the

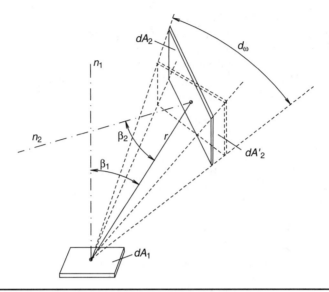

Figure 7.5 Surface elements.

direction determined by angle β_1:

$$e_{1,\beta_1} = \frac{e_1}{\pi}\cos\beta_1 \tag{7.54}$$

and the total energy emitted from dA_1 and arriving in dA_2 within $d\omega$ is:

$$E_{dA_1 \to dA_2} \equiv d^2 E_{1-2} = e_{1,\beta_1}\, dA_1\, d\omega \tag{7.55}$$

Using (7.53) and (7.54) in (7.55):

$$d^2 E_{1-2} = \frac{e_1}{\pi}\cos\beta_1\, dA_1\, \frac{dA_2\cos\beta_2}{r^2} \tag{7.56}$$

and after integration:

$$E_{1-2} = \frac{e_1}{\pi}\int\limits_{A_1}\int\limits_{A_2}\frac{\cos\beta_1\ \cos\beta_2}{r^2}dA_1\, dA_2 \tag{7.57}$$

Formula (7.57) determines this part E_{1-2} of the emission E_1 propagating from the whole surface A_1, which arrives at the surface A_2. There is the characteristic inverse square of distance (r^2), which determines the energy attenuation of the radiation propagating in a neutral medium. Based on the consideration in Section 7.5.3, by replacing emission energy e_1 with the exergy b_1 of this emission in equation (7.57), the calculation of the respective amount of exchanged exergy B_{1-2} is possible.

7.4.2 Some Remarks on Propagation in a Real Medium

It was shown in the previous section that, except for attenuation due to the geometric divergence of flux, nothing happens to the radiating beam if the beam travels through a vacuum or eventually through a nonradiative (i.e., perfectly transmitting medium) gas. However, in general, the radiation flux can fall partly, or totally, upon bodies, e.g., a cloud of absorbing gas (CO_2, H_2O CH_4, etc,), condensed particles (droplets, dust, etc.) dispersed in the gaseous medium through which the beam is traveling. The radiation can weaken, which can then result in a reduction of energy and a change in the radiation spectrum.

The dispersed bodies can to some extent absorb, reflect, or transmit the beam. An absorbing body emits its own radiation, at its own temperature. Reflection can be specular or diffuse (see Section 3.2). Both reflected and emitted portions can then fall again upon other bodies, and thus the original radiation can be transformed in multiple processes into radiation with less energy, a changed spectrum, and a changed solid angle of propagation. In addition, e.g., in the case of a solar radiation beam, some other beams from the sun initially not aimed at the considered surface on the earth can be redirected to the considered surface. Therefore, the surface at which received radiation is considered can also receive emitted and reflected radiation redirected from beyond the solid angle within which the radiation source is seen from the considered surface.

Perfect transmission and specular reflection of radiation are reversible and do not change the exergy of radiation. Besides absorption and emission, which are irreversible (due to temperature difference as discussed in Section 6.6), radiation can also be subjected to diffuse reflection, dispersion, refraction, and other simple or combined phenomena that belong to the area of optics. To date, the energy, entropy, and exergy analyses of these optical phenomena have not been developed much—see, e.g., Candau (2003)—and are open for future analyses not considered in the present book.

In comparison to energy, the calculation of exergy propagation in a real medium is even more difficult because the newly emitted radiation of a suspended body generally occurs at a different temperature from the emitting body, and each absorption and emission causes losses in exergy. The global exergy loss cannot be estimated by comparison of the exergy from the radiation source aimed at the considered surface when entering the medium, to the exergy estimated based on the measured spectrum of exergy of the source radiation arriving at the considered surface. This is because, as mentioned, some undetermined exergy from the radiation source, initially not aiming at the considered surface, can be redirected to the considered surface.

As an alternative to the approximate calculations of the real radiation energy or exergy arriving at the considered surface is the method of measuring the radiation spectrum and geometry directly at the considered surface.

In summary, if the medium has the properties of a specular surface of emissivity $\varepsilon = 0$—i.e., the medium does not absorb the radiation, but at most is exposed only to the multiple reflections or refractions—the annihilation process of the exergy of radiation does not occur. However, there can appear some geometric consequences, i.e., to the given surface there can arrive the radiation of the directional distribution of energy, or exergy, which differs from the case when between the radiation source and given surface there is a vacuum or perfectly neutral medium. In other words, the radiation can come within the changed solid angle.

However, if the medium has even a small ability to absorb ($\varepsilon > 0$) and so to emit, and the medium temperature differs from that of the source surface, then radiation traveling through such a medium partly loses its exergy and, due to reflections or refractions, its geometric parameters in respect to the given surface also change as in the previous case.

An example of radiation traveling through a real medium is solar radiation arriving at the earth through the atmosphere. This process is particularly important for life on earth. The total radiation received by the surface of the earth can consist of the direct solar radiation arriving within the solid angle under which the sun is seen and of the diluted radiation arriving usually under the solid angle of the hemisphere. Radiation traveling through a real medium is considered, e.g., by Landsberg and Tonge (1979). Recently, using the solar spectral radiation databank developed by Gueymard (2008) and based on the radiation exergy interpretation by Candau (2003), some new data on the extraterrestrial and terrestrial solar radiation exergy were obtained by Chu and Liu (2009).

With a clear sky, the surface of the earth receives direct solar radiation at a high temperature and at high energy at a high exergetic value within a small solid angle. With an overcast sky, the energy received can be even larger than from a clear sky because the radiation arrives within a large solid angle; however, such diluted radiation can have a lower exergetic value.

Exact description of the fate of radiation penetrating a real medium and the heat transfer through such a complex medium are both very difficult. In the specialist books in this area, the formulated various mathematical models, involving a number of simplifying assumptions, allow for calculation of energy propagation. However, the obtained results are usually only a rough approximation. To date, the exergetic interpretation of such real propagation of radiation has not been considered.

7.5 Radiation Exergy Exchange Between Surfaces

7.5.1 View Factor

Propagation of radiation from different surfaces results in energy exchange between the surfaces. The effect of such exchange is described in textbooks on heat transfer. The exchange depends on the properties of surfaces and on the view factor defined in Section 3.3. The same energy exchange process can be also interpreted by the respective exergy exchange which, beside the surfaces, properties, and the view factor, depends also on the environment temperature.

Generally, for any two surfaces 1 and 2, the view factor φ_{1-2} for surface 1 is defined as the ratio of radiosity J_{1-2} arriving from surface 1 to surface 2 to the total (in all directions) radiosity J_1 of surface 1:

$$\varphi_{1-2} = \frac{J_{1-2}}{J_1} \tag{7.58}$$

Interpreting formula (7.57) for the radiosity ($E = J$) and taking into account that $J_1 = A_1 \times j_1$, the average view factor results from definition (7.58):

$$\varphi_{1-2} = \frac{1}{\pi A_1} \int_{A_1} \int_{A_2} \frac{\cos \beta_1 \cos \beta_2}{r^2} \, dA_1 \, dA_2 \tag{7.59}$$

However, because the different elements dA_1 and dA_2 of surfaces A_1 and A_2 can be differently situated, the local view factor $\varphi_{d,1-2}$ of any element dA_1 relating to the total surface A_2 may be appropriately defined:

$$\varphi_{d,1-2} = \frac{d^2 J_{1-2}}{dJ_1} \tag{7.60}$$

Assuming $dJ_1 = j_1 \times dA_1$ and introducing (7.56) to (7.60):

$$\varphi_{d,1-2} = \frac{1}{\pi} \int_{A_2} \frac{\cos \beta_1 \cos \beta_2}{r^2} \, dA_2 \tag{7.61}$$

The significance of the local view factor is illustrated in Figure 7.6. The spherical surface 1 is surrounded by the spherical surface 2 with two elemental surfaces dA_2 and dA_2'. Due to different solid angles ($\omega' > \omega$), there are significantly different respective view factors, $\varphi_{d,2-1}'$ > $\varphi_{d,2-1}$. The uniform values of the view factor $\varphi_{d,2-1} = \varphi_{2-1} = \text{const}$ would be for the case of concentric surfaces 1 and 2.

To apply numerical interpretation of formula (7.61) the surface A_1 can be divided into k finite elements ΔA_i in such a fashion that for each surface element the constant value $\varphi_{\Delta,1-2} = \text{const}$ can be assumed. To

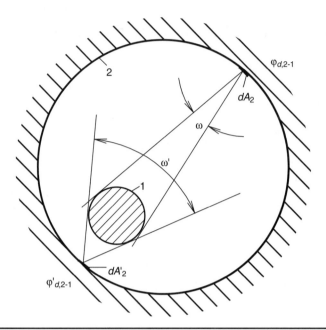

FIGURE 7.6 Two different local view factors.

determine the average value of the view factor, instead of formula (7.59), the following formula can be used:

$$\varphi_{1-2} = \frac{\sum\limits_{i=1}^{i=k} (\varphi_{\Delta,1-2})_i \, \Delta A_i}{\sum\limits_{i=1}^{i=k} \Delta A_i} \tag{7.62}$$

where i is the successive number $(i = 1, \ldots, k)$ of the finite elements of the surface and of the respective local view factor, whereas

$$\sum_{i=1}^{i=k} \Delta A_i = A_1 \tag{7.63}$$

A simple application of the view factors in calculation of exchanged energy or exergy between surfaces will be demonstrated later in Example 7.2. Calculation of the view factor from formula (7.59) for the complex configuration of considered surfaces can sometimes be very difficult, and application of various graphical or optical methods described in textbooks on heat transfer can often be useful.

As a result from formula (7.58), radiation energy J_{1-2} propagating from any surface 1 to surface 2 can be determined if we have the radiosity J_1 of surface 1 and the respective view factor φ_{1-2}. In order to determine radiation energy exchanged between many different surfaces enclosing a certain space, not only the radiosities of these surfaces but also the view factors for the given surfaces, configuration have to be known. The real situation often is considered under the assumptions that the enclosed space is a vacuum or filled up with a neutral medium (with neglected eventual convection of heat), and additionally, the surfaces are perfectly gray. In such a case the view factors represent only the geometry of the considered system of surfaces embracing the space.

To calculate the set of required view factors for the system some rules can be applied. One of them is the *reciprocity rule* applied for any two surfaces. To derive the rule, the formula (7.59) can be interpreted also for propagating radiation from surface 2 to surface 1 as follows:

$$\varphi_{2-1} = \frac{1}{\pi A_2} \int\limits_{A_1} \int\limits_{A_2} \frac{\cos \beta_1 \cos \beta_2}{r^2} \, dA_1 \, dA_2 \qquad (7.64)$$

Dividing equations (7.59) and (7.64) side by side one obtains:

$$\varphi_{1-2} A_1 = \varphi_{2-1} A_2$$

or for any ith and jth surfaces:

$$\varphi_{i-j} A_i = \varphi_{j-i} A_j \qquad (7.65)$$

which is the general reciprocity rule.

The *complacency rule* results from the energy conservation equation applied for any of n surfaces, numbered from 1 to n, enclosing the system space. The radiosity J_1 of surface 1 distributes as follows:

$$J_1 = J_{1-1} + J_{1-2} + J_{1-3} + \cdots + J_{1-n} \qquad (7.66)$$

Dividing equation (7.66) by J_1 and applying the general definition (7.58) of the view factor, equation (7.66) changes as:

$$1 = \varphi_{1-1} + \varphi_{1-2} + \varphi_{1-3} + \cdots + \varphi_{1-n} \qquad (7.67)$$

Formula (7.67) expresses the complacency rule. The view factor φ_{1-1} represents the possibility of a concave curved surface that may radiate on itself.

In certain situations, the *Polak's rule* of a "crossed-string," described, e.g., by Gray and Müller (1974), can be applied. The rule can be applied to the two considered surfaces (i and j) when they are the sides of parallel and infinitely long cylinders, not necessarily circular but not concave. In practice, the rule can be applied also for the finite but sufficiently long cylinders. The method introduces the

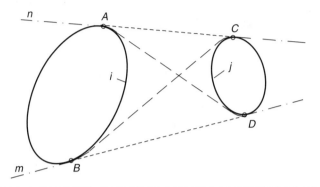

FIGURE 7.7 The profiles of cylindrical surfaces i and j.

two lengths of imagined strings, shown in Figure 7.7, which cross (L_c) or do not cross (L_n) when they gird the surfaces:

$$\varphi_{i-j} = \frac{L_c + L_n}{L_i} \tag{7.68}$$

Figure 7.7 shows the cross-sections of the two considered cylinders (i and j). The two tangents (n and m) to the surface profiles determine the four tangency points A, B, C and D. The total length L_n of the noncrossed string is measured from A to C and from B to D. The total length of the crossed string is from A to D and from B to C. The length L_i is measured from A to B over the profile of surface i.

Sometimes Polak's rule can be applied to certain two surfaces for which it is difficult at first glance to recognize the crossed and non-crossed lengths required for formula (7.68). In such a case, it is helpful to apply imaginary displacement of the two surfaces as far from each other until the lengths become easy to notice. Then, while displacing the surfaces back to their original position, it can be noticed how the lengths vary. For example, the system of a long circular cylindrical surface laid along a plate is shown in Figure 7.8. In the original situation (a), at first glance, it is not easy to identify the crossed and noncrossed

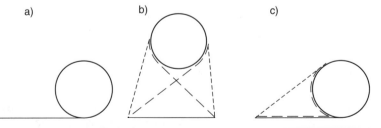

FIGURE 7.8 Example of the strings interpretation.

lengths. However, after replacement of the surface in situation (b), the strings appear clearly. By restoring the initial situation (c) it is clearly seen that, when approaching the surfaces, the right-hand side of the noncrossed string disappears, whereas the crossed string consists of the length of the cylinder profile and of the width of the plate.

In determination of the view factor the characteristic geometric feature of the considered surface system can be additionally used. For example, if any xth surface is flat, then $\varphi_{x-x} = 0$. If any xth surface is entirely unseen from any yth surface, then $\varphi_{x-y} = 0$. If any surfaces, e.g., surfaces 2, 4, and 6, are geometrically situated in the same way regarding surface 1, then $\varphi_{1-2} = \varphi_{1-4} = \varphi_{1-6}$.

In order to increase the radiating flux from any gray surface of emissivity ε the surface can be grooved, which increases the effective emissivity ε_g according to Surinow's formula:

$$\varepsilon_g = \frac{\varepsilon}{1 - \rho\,\varphi} \tag{7.69}$$

where ρ is the surface reflectivity and φ is the view factor expressing the depth of the grooves. If the grooves are very deep then the view factor is close to unity ($\varphi \approx 1$). Then, from formula (7.69), using formulae (3.5) and (3.31), the result is $\varepsilon_g \approx 1$, which means that the effect of the deep groove is similar to the model of a black surface shown in Figure 3.4.

Example 7.2 A space (e.g., a vacuum or a space filled with a neutral medium for which the convective effect is neglected) is enclosed with n different surfaces ($j = 1, \ldots, n$). Figure 7.9 shows the example in which $n = 3$ and the surfaces belong to the long and parallel cylinders. The surface system is in the steady state and each surface is at a constant and uniform temperature due to connection with the heat sources of the respective surface temperatures. Radiative heat is positive if delivered to the surface or negative when taken off the surface. The space contains radiation (photon gas), which is not in an equilibrium state because the radiating fluxes are traveling between the surfaces at different temperatures.

If the system, defined by the system boundary shown in Figure 7.9, is in thermal equilibrium, then the algebraic sum of all the heat fluxes is zero:

$$\sum_{j=1}^{j=n} Q_j = 0 \tag{7.70}$$

Referring to any ith surface selected from all n surfaces, the energy balance can be considered for the subsystem which is a thin layer of the ith surface. The input to such subsystem is heat Q_i and the absorbed part of radiosity arriving from all other surfaces to the considered ith surface. The part of radiosities reflected from the ith surface may not be considered in the balance because it cancels out. The subsystem output is the emission of the ith surface. The energy conservation equation takes the form:

$$Q_i + \alpha_i \sum_{j=1}^{j=n} \varphi_{j-i} J_j = E_i \tag{7.71}$$

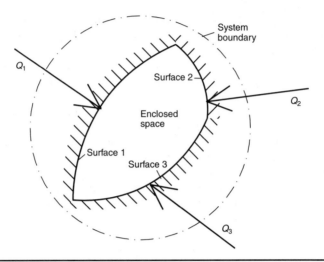

Figure 7.9 Space enclosed with three nonconvex surfaces 1, 2, and 3.

where $i = 1, \ldots, n$ is the successive number of considered surfaces selected from n surfaces and α_i is the absorptivity of the ith surface. The variables J_j appearing in formula (7.71) can be determined in a way as shown, e.g., for surface $i = 1$ (at $n = 3$):

$$J_1 = E_1 + \rho_1 \left(\varphi_{1-1} J_1 + \varphi_{2-1} J_2 + \varphi_{3-1} J_3 \right) \tag{7.72}$$

In the general case of a multisurface system, formula (7.72) for the ith surface takes the form:

$$J_i = E_i - \rho_i \sum_{j=1}^{j=n} \varphi_{j-i} J_j \tag{7.73}$$

The surface emissions can be determined from equation (3.22), at valid relation (3.5). If additionally the reflectivities and temperatures of the surfaces as well as all the view factors are given, then equation (7.73) represents the set of n equations with n unknowns, which are the radiosities of all the surfaces.

For numerical illustration there is further consideration of three surfaces ($n = 3$). There are nine different view factors determining the configuration of radiating surfaces; thus the nine equations are needed for calculations of the factors. The three equations are obtained by assuming the surfaces to be flat; $\varphi_{1-1} = \varphi_{2-2} = \varphi_{3-3} = 0$. The other six equations follow from the complacency rule (7.67):

$$1 = \varphi_{1-1} + \varphi_{1-2} + \varphi_{1-3} \tag{a}$$

$$1 = \varphi_{2-1} + \varphi_{2-2} + \varphi_{2-3} \tag{b}$$

$$1 = \varphi_{3-1} + \varphi_{3-2} + \varphi_{3-3} \tag{c}$$

and the reciprocity rule (7.65):

$$\varphi_{1-2} A_1 = \varphi_{2-1} A_2 \tag{d}$$

$$\varphi_{2-3} A_2 = \varphi_{3-2} A_3 \tag{e}$$

$$\varphi_{1-3} A_1 = \varphi_{3-1} A_3 \tag{f}$$

Assuming $L_1 = 10$ m, $L_2 = 7$ m and $L_3 = 8$ m the view factors calculated from relations (a) to (f) are $\varphi_{1-2} = 0.450$, $\varphi_{1-3} = 0.550$, $\varphi_{2-1} = 0.643$, $\varphi_{2-3} = 0.357$, $\varphi_{3-1} = 0.688$, and $\varphi_{3-2} = 0.312$. The surfaces are very long and their emissions E_i are calculated in relation to 1 m of the length. The surface areas are expressed by the respective lengths L_i of the surface profiles. All the surfaces are perfectly gray with reflectivities at $\rho_1 = 0.1$, $\rho2 = 0.2$, and $\rho3 = 0.3$ and from (3.5) the absorptivities are $\alpha_1 = \varepsilon_1 = 0.9$, $\alpha_2 = \varepsilon_2 = 0.8$ and $\alpha_3 = \varepsilon_3 = 0.7$.

$$E_i = L_i \varepsilon_i \sigma T_i^4 \tag{g}$$

For $T_1 = 400$ K, $T_2 = 600$ K, and $T_3 = 1000$ K the calculated from (g) emissions are: $E_1 = 13.06$ kW/m, $E_2 = 41.15$ kW/m, and $E_3 = 317.48$ kW/m.

According to (7.73) the radiosities, also related to 1 m of surface length, are calculated from the following set of three equations:

$$J_1 = E_1 + \rho_1 \left(\varphi_{1-1} J_1 + \varphi_{2-1} J_2 + \varphi_{3-1} J_3 \right) \tag{h}$$

$$J_2 = E_2 + \rho_2 \left(\varphi_{1-2} J_1 + \varphi_{2-2} J_2 + \varphi_{3-2} J_3 \right) \tag{i}$$

$$J_3 = E_3 + \rho_3 \left(\varphi_{1-3} J_1 + \varphi_{2-3} J_2 + \varphi_{3-3} J_3 \right) \tag{j}$$

Calculated values of radiosities, higher than respective emissions, are $J_1 = 40.03$ kW/m, $J_2 = 65.44$ kW/m, and $J_3 = 331.10$ kW/m. Heat fluxes, related to 1 m of surface length, result from (7.71):

$$Q_1 = E_1 - \alpha_1 \left(\varphi_{1-1} J_1 + \varphi_{2-1} J_2 + \varphi_{3-1} J_3 \right) \tag{k}$$

$$Q_2 = E_2 - \alpha_2 \left(\varphi_{1-2} J_1 + \varphi_{2-2} J_2 + \varphi_{3-2} J_3 \right) \tag{l}$$

$$Q_3 = E_3 - \alpha_3 \left(\varphi_{1-3} J_1 + \varphi_{2-3} J_2 + \varphi_{3-3} J_3 \right) \tag{m}$$

and their values are $Q_1 = -229.67$ kW/m, $Q_2 = -56.04$ kW/m, and $Q_3 = 285.71$ kW/m. According to (7.70) the sum of heat fluxes is zero: $-229.67 - 56.04 + 285.71 = 0$.

The following calculations of exergy are simplified. All the considered surfaces are now assumed to be black ($\rho_i = 0$, $\alpha_i = 1$). For the same temperatures of surfaces ($T_1 = 400$ K, $T_2 = 600$ K, and $T_3 = 1000$ K) and for unchanged configuration of the surfaces (i.e., the same values of all the view factors), the calculated emissions, equal to respective radiosities ($E_i = J_i$), are respectively smaller: $E_1 = 14.51$ kW/m, $E_2 = 51.43$ kW/m, and $E_3 = 453.54$ kW/m. The heat fluxes are significantly changed ($Q_1 = -330.36$ kW/m, $Q_2 = -96.83$ kW/m. and $Q_3 = 427.19$ kW/m) and their sum is zero.

Exergy B_i radiating from the surface can be calculated based on formulas (7.47) and (7.49) adjusted as follows:

$$B_i = L_i \frac{\sigma}{3} \left(3 \, T_i^4 + T_0^4 - 4 \, T_0 T_i^3 \right) \tag{n}$$

Using equation (n) at $T_0 = 300$ K yields: $B_1 = 1.53$ kW/m, $B_2 = 18.22$ kW/m, and $B_3 = 273.35$ kW/m. These calculated values can be used in the exergy balance equations for a particular surface. The exergy input considered for any ith surface consists of radiation exergy arriving from other surfaces and of exergy of heat exchanged with the considered ith surface. On the output side of the equation is exergy radiating from the considered ith surface and the irreversible exergy loss due to absorption and emission occurring at the ith surface. Exergy balance equations for the consecutive surfaces 1, 2, and 3 are:

$$\varphi_{1-1}B_1 + \varphi_{2-1}B_2 + \varphi_{3-1}B_3 + Q_1\left(1 - \frac{T_0}{T_1}\right) = B_1 + \delta B_1 \tag{o}$$

$$\varphi_{3-2}B_3 + \varphi_{2-2}B_2 + \varphi_{1-2}B_1 + Q_2\left(1 - \frac{T_0}{T_2}\right) = B_2 + \delta B_2 \tag{p}$$

$$\varphi_{1-3}B_1 + \varphi_{2-3}B_2 + \varphi_{3-3}B_3 + Q_3\left(1 - \frac{T_0}{T_3}\right) = B_3 + \delta B_3 \tag{q}$$

The sum of the exergy loss terms from equations (o) to (q) expresses the exergy loss of the whole surface system: $\delta B = \delta B1 + \delta B2 + \delta B3 = 115.52 + 19.48 + 33.03 = 168.03$ kW/m. The same value of the global exergy loss δB can be also determined from the Guoy–Stodola law presented by equation (2.60):

$$\delta B = T_0\Pi = T_0\left(\frac{Q_1}{T_1} + \frac{Q_2}{T_2} + \frac{Q_3}{T_3}\right) \tag{r}$$

Obviously, the same value of the exergy loss for a particular surface can be also determined from formula (2.60) which, e.g., for surface 1 can be used as follows:

$$\delta B_1 = T_0\left(L_1\frac{4}{3}\sigma\,T_1^3 - L_1\varphi_{1-1}\frac{4}{3}\sigma T_1^3 - L_2\varphi_{2-1}\frac{4}{3}\sigma T_2^3 - L_3\varphi_{3-1}\frac{4}{3}\sigma T_3^3 - \frac{Q_1}{T_1}\right) \tag{s}$$

7.5.2 Emission Exergy Exchange Between Two Black Surfaces

The radiative exergy exchange based on formula (7.49) can be analyzed with more details. The simple example for considerations is assumed. The model of two parallel surfaces 1 and 2, as determined in Figure 6.5, is applied. The surfaces parameters are denoted with subscript 1 and 2. The surfaces are black ($\varepsilon_1 = \varepsilon_2 = 1$) and flat ($\varphi_{1-1} = \varphi_{2-2} = 0$ and $\varphi_{1-2} = \varphi_{2-1} = 1$). Temperature T_1 is constant whereas temperature T_2 varies from 0 to $T_2 = T_1$.

The exchanged exergy b_{1-2} per 1 m^2 of surface is equal to the difference b_1 and b_2 which are the exergy of emission densities of the surfaces 1 and 2:

$$b_{1-2} = b_1 - b_2 \tag{7.74}$$

where b_1 and b_2 are determined from formula (7.49). Heat q_{1-2} exchanged by radiation is equal to the difference e_1 and e_2 of emission densities of the surfaces:

$$q_{1-2} = e_1 - e_2 \tag{7.75}$$

where e_1 and e_2 are determined from formula (3.21). The changes of exergy b_{q1} and b_{q2} for both the heat sources, according to

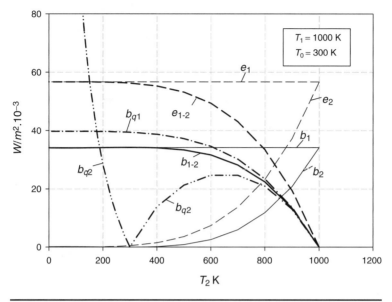

FIGURE 7.10 Exchanged heat and exergy between black surfaces at temperatures T_1 and T_2.

formula (2.61) are:

$$b_{q1} = q_{1-2}\left(1 - \frac{T_0}{T_1}\right) \quad \text{and} \quad b_{q2} = \left|q_{1-2}\left(1 - \frac{T_0}{T_2}\right)\right| \qquad (7.76)$$

Figure 7.10 shows the emission densities, e_1 and e_2, exergy of emission densities, b_1 and b_2, and the exergy change of the heat sources, b_{q1} and b_{q2} as function of temperature T_2 at constant temperatures $T_1 = 1000$ K and $T_0 = 300$ K. With the growing temperature T_2 there grow the emission density e_2 and the exergy of emission density b_2, whereas e_1 and b_1 remain constant due to the constant temperature T_1. The exchanged exergy b_{1-2} decreases from 34,016 W/m^2 to zero and the exchanged heat q_{1-2} also decreases to zero but from 56,693 W/m^2. The exergy change b_{q1} of the heat source at temperature T_1 decreases from 39,685 W/m^2 to zero (because q_{1-2} reaches zero). However, the exergy change b_{q2} of the heat source at temperature T_2 decreases from infinity to zero at $T_2 = T_0$ and then it grows reaching the maximum ($\sim$24,673 W/m^2) for $T_2 = \sim$600 K, and then diminishes to zero. The appearing maximum is the effect of the growing value of the exergy of heat due to growing its temperature and, on the other hand, of the decreasing amount of this heat.

7.5.3 Exergy Exchange Between Two Gray Surfaces

7.5.3.1 Significance and Description of the Problem

One of the important problems of radiation is the practical calculation of the exchanged radiation energy or exergy between different surfaces. Consideration of systems composed of black surfaces is relatively easy because each black surface can only emit and absorb radiation without any reflection. However, a system of gray surfaces is more complex and the calculation accounts for the successive reflections of each emitted radiation. Simplification of the problem by neglecting reflections to calculate the approximate value of the exchanged energy (heat) is eventually acceptable only in the case of the surfaces with emissivities not much different than 1.

The calculation methods for exchanged radiation energy are relatively well described in the heat transfer textbooks, e.g., Holman (2009). However, calculation of exchanged exergy is not well discussed and is more complex. Generally, each flux of radiation exergy, arriving at a nontransmitting surface, can be partly absorbed, partly reflected, and the remaining part is lost due to irreversibility. The exergetic considerations have to account for many multiprocesses of absorption, emission, reflection, and losses.

The mechanism principle of the exchange of exergy between two gray surfaces can be sufficiently analyzed based on the simple system of a vacuumed space (no heat transfer by conduction or convection) enclosed only with the two surfaces, which are infinitely large, facing each other, parallel, and flat as shown in Figure 7.11. Temperatures of the surfaces are constant in time and at every place on the respective surfaces (i.e., isothermal surfaces). The properties of surface 1, denoted

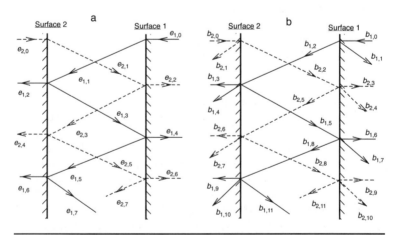

Figure 7.11 The scheme of exchange of radiation energy (a) and exergy (b) between surfaces 1 and 2.

with subscript 1, are temperature T_1, emissivity ε_1, absorptivity α_1, and reflectivity ρ_1. Analogically, the same properties (T_2, ε_2, α_2, ρ_2) of surface 2 are denoted with subscript 2.

7.5.3.2   Transfer of Radiation Energy

As mentioned, consideration of the net heat exchanged between the surfaces is commonly known and here is briefly recalled only to give a comparison background for consideration of exergy exchange. Figure 7.11(a) presents the scheme of the radiation emission from both surfaces. The energy stream e, related to a 1 m^2 surface area, is labeled with the two number subscripts (n and m); the first number (n) determines the surface (1 or 2), whereas the second number (m) determines the stage of the stream after successive happening.

For example, considering surface 1, energy stream $e_{1,0}$ represents heat transferred from the heat source at temperature T_1 to surface 1. The same energy is then emitted from surface 1 to surface 2 ($e_{1,0} = e_{1,1}$). Energy $e_{1,1}$ arrives in surface 2 and is partly ($e_{1,2}$) absorbed by surface 2 and partly ($e_{1,3}$) reflected back to surface 1. The portion $e_{1,3}$ arriving in surface 1 is again partly ($e_{1,4}$) absorbed by surface 1 and partly ($e_{1,5}$) reflected into surface 2. In the same way the successive reflections and absorptions continue, and the energy streams are gradually diminishing.

Analogously, the energy $e_{2,0}$ of heat transferred from the heat source at temperature T_2 takes part in the successive happenings. The values of successive streams of both surfaces are developed as follows:

$$e_{1,0} = \varepsilon_1 e_{b,1}$$
$$e_{1,1} = \varepsilon_1 e_{b,1}$$
$$e_{1,2} = \alpha_2 \varepsilon_1 e_{b,1}$$
$$e_{1,3} = \rho_2 \varepsilon_1 e_{b,1}$$
$$e_{1,4} = \rho_1 \rho_2 \varepsilon_1 e_{b,1}$$
$$e_{1,5} = \rho_1 \rho_2 \varepsilon_1 e_{b,1}$$
$$e_{1,6} = \alpha_2 \rho_1 \rho_2 \varepsilon_1 e_{b,1}$$
$$e_{1,7} = \rho_1 \rho_2^2 \varepsilon_1 e_{b,1}$$

etc.

where, the density of emission energy $e_{b,1}$ of a black surface at temperature T_1 is:

$$e_{b,1} = \sigma T_1^4 \tag{a}$$

and for the energy streams of surface 2:

$$e_{2,0} = \varepsilon_2 e_{b,2}$$
$$e_{2,1} = \varepsilon_2 e_{b,2}$$
$$e_{2,2} = \alpha_1 \varepsilon_2 e_{b,2}$$

$$e_{2,3} = \rho_1 \varepsilon_2 e_{b,2}$$
$$e_{2,4} = \rho_2 \rho_1 \varepsilon_2 e_{b,2}$$
$$e_{2,5} = \rho_2 \rho_1 \varepsilon_2 e_{b,2}$$
$$e_{2,6} = \alpha_1 \rho_2 \rho_1 \varepsilon_2 e_{b,2}$$
$$e_{2,7} = \rho_2 \rho_1^2 \varepsilon_2 e_{b,2}$$

etc.

where, the density of emission energy $e_{b,2}$ of a black surface at temperature T_2 is:

$$e_{b,2} = \sigma T_2^4 \tag{b}$$

Surface 2 absorbs the sum e_1 of the energy portions ($e_{1,2}, e_{1,6}$, etc.) from surface 1 as follows:

$$e_1 = \alpha_2 \varepsilon_1 e_{b,1} \left(1 + \rho_1 \rho_2 + \rho_1^2 \rho_2^2 + \cdots\right) = \alpha_2 \varepsilon_1 e_{b,1} \frac{1}{1 - \rho_1 \rho_2} \tag{c}$$

The expression in the brackets is replaced by the sum of the terms of the infinite geometric progression with the common ratio $\rho_1 \times \rho_2$. Absorbed energy e_1 is transferred to the interior (heat source at temperature T_2) to maintain the steady state of radiating surfaces.

Analogously, surface 1 absorbs the sum e_2 of the energy portions ($e_{2,2}, e_{2,6}$, etc.) from surface 2 as follows:

$$e_2 = \alpha_1 \varepsilon_2 e_{b,2} \left(1 + \rho_2 \rho_1 + \rho_2^2 \rho_1^2 + \cdots\right) = \alpha_1 \varepsilon_2 e_{b,2} \frac{1}{1 - \rho_1 \rho_2} \tag{d}$$

Also, the absorbed energy e_2 is transferred to the interior (heat source at temperature T_1) to maintain the steady state of radiating surfaces. It is to emphasize that the processes occurring at both surfaces occur simultaneously; only for calculation purposes are they considered separately.

The effective heat e_{1-2} transferred from surface 1 to surface 2 is:

$$e_{1-2} = e_1 - e_2 \tag{e}$$

Taking into account that $\alpha + \rho = 1$, $\alpha = \varepsilon$, and substituting equations (a)–(d) to (e) the known formula is obtained:

$$e_{1-2} = \varepsilon_{1-2} \sigma \left(T_1^4 - T_2^4\right) \tag{f}$$

where

$$e_{1-2} = \frac{1}{\frac{1}{\varepsilon_1} + \frac{1}{\varepsilon_2} - 1} \tag{g}$$

Formula (g) is used for calculation of the net heat exchanged between surfaces 1 and 2.

7.5.3.3 Transfer of Radiation Exergy

Analogously to the radiation energy exchange, the exchange of radiation exergy streams between the same considered surfaces [Figure 7.11(b)] can be analyzed.

The exergy stream b, related to a 1 m² surface area, is also denoted with the two number subscripts (n and m); the first number (n) determines the surface (1 or 2), whereas the second number (m) determines the stage of the stream after successive happening. For example, considering surface 1, stream $b_{1,0}$ represents the exergy of heat transferred from the heat source at temperature T_1 to the surface 1. The surface 1 supplied by the exergy stream $b_{1,0}$ is able to emit exergy $b_{1,2}$ although at the irreversible exergy loss $b_{1,1}$.

Exergy $b_{1,2}$ arrives in surface 2 and is partly ($b_{1,3}$) absorbed by surface 2, partly ($b_{1,5}$) reflected back to surface 1 and the rest ($b_{1,4}$) is lost due to irreversibility occurring at surface 2. The exergy portion $b_{1,5}$ arriving in surface 1 is again partly ($b_{1,6}$) absorbed by surface 1, partly ($b_{1,8}$) reflected into surface 2 and the rest of exergy ($b_{1,7}$) is lost due to irreversibility occurring at surface 1. In the same way the successive reflections and absorptions are continuing, and gradually diminishing the exergy streams.

Analogically, the exergy $b_{2,0}$ of heat transferred from the heat source at temperature T_2 takes part in the successive happenings. The values of successive streams of both surfaces are developed as follows:

$$b_{1,0} = \varepsilon_1 e_{b,1} \left(1 - \frac{T_0}{T_1} \right)$$

$$b_{1,1} = b_{1,0} - b_{1,2}$$

$$b_{1,2} = \varepsilon_1 b_{b,1}$$

$$b_{1,3} = \alpha_2 \varepsilon_1 e_{b,1} \left(1 - \frac{T_0}{T_2} \right)$$

$$b_{1,4} = b_{1,2} - b_{1,3} - b_{1,5}$$

$$b_{1,5} = \rho_2 \varepsilon_1 b_{b,1}$$

$$b_{1,6} = \alpha_1 \rho_2 \varepsilon_1 e_{b,1} \left(1 - \frac{T_0}{T_1} \right)$$

$$b_{1,7} = b_{1,5} - b_{1,6} - b_{1,8}$$

$$b_{1,8} = \rho_1 \rho_2 \varepsilon_1 b_{b,1}$$

$$b_{1,9} = \alpha_2 \rho_1 \rho_2 \varepsilon_1 e_{b,1} \left(1 - \frac{T_0}{T_2} \right)$$

$$b_{1,10} = b_{1,8} - b_{1,9} - b_{1,11}$$

$$b_{1,11} = \rho_1 \rho_2^2 \varepsilon_1 b_{b,1}$$

$$b_{1,12} = \alpha_1 \rho_1 \rho_2^2 \, \varepsilon_1 e_{b,1} \left(1 - \frac{T_0}{T_1} \right)$$

$$b_{1,13} = b_{1,11} - b_{1,12} - b_{1,14}$$
$$b_{1,14} = \rho_1^2 \rho_2^2 \varepsilon_1 b_{b,1}$$
etc.

where the emission exergy $b_{b,1}$ of a black surface at temperature T_1 is:

$$b_{b,1} = \frac{\sigma}{3} \left(3\,T_1^4 + T_0^4 - 4\,T_0 T_1^3\right) \tag{h}$$

and for the exergy radiation of surface 2:

$$b_{2,0} = \varepsilon_2 e_{b,2} \left(1 - \frac{T_0}{T_2}\right)$$
$$b_{2,1} = b_{2,0} - b_{2,2}$$
$$b_{2,2} = \varepsilon_2 b_{b,2}$$
$$b_{2,3} = \alpha_1 \varepsilon_2 e_{b,2} \left(1 - \frac{T_0}{T_1}\right)$$
$$b_{2,4} = b_{2,2} - b_{2,3} - b_{2,5}$$
$$b_{2,5} = \rho_1 \varepsilon_2 b_{b,2}$$
$$b_{2,6} = \alpha_2 \rho_1 \varepsilon_2 e_{b,2} \left(1 - \frac{T_0}{T_2}\right)$$
$$b_{2,7} = b_{2,5} - b_{2,6} - b_{2,8}$$
$$b_{2,8} = \rho_2 \rho_1 \varepsilon_2 b_{b,2}$$
$$b_{2,9} = \alpha_1 \rho_2 \rho_1 \varepsilon_2 e_{b,2} \left(1 - \frac{T_0}{T_1}\right)$$
$$b_{2,10} = b_{2,8} - b_{2,9} - b_{2,11}$$
$$b_{2,11} = \rho_1^2 \rho_2 \varepsilon_2 b_{b,2}$$
etc.

where the emission exergy $b_{b,2}$ of a black surface at temperature T_2 is:

$$b_{b,2} = \frac{\sigma}{3} \left(3\,T_2^4 + T_0^4 - 4\,T_0 T_2^3\right) \tag{i}$$

The portions of the radiation exergy of surface 1 delivered to surface 2 are:

$$b_1 = (b_{1,2} - b_{1,5}) + (b_{1,8} - b_{1,11}) + \cdots$$
$$= \alpha_2 \varepsilon_1 b_{b,1} \left(1 + \rho_1 \rho_2 + \rho_1^2 \rho_2^2 + \cdots\right) = \frac{\alpha_2 \varepsilon_1 b_{b,1}}{1 - \rho_1 \rho_2} \tag{j}$$

The portions of the radiation exergy of surface 2 delivered to surface 1 are:

$$b_2 = (b_{2,2} - b_{2,5}) + (b_{2,8} - b_{2,11}) + \cdots$$
$$= \alpha_1 \varepsilon_2 b_{b,2} \left(1 + \rho_1 \rho_2 + \rho_1^2 \rho_2^2 + \cdots\right) = \frac{\alpha_1 \varepsilon_2 b_{b,2}}{1 - \rho_1 \rho_2} \tag{k}$$

The net radiation exergy b_{1-2} transferred from surface 1 to surface 2 is:

$$b_{1-2} = b_1 - b_2 \tag{l}$$

After taking into account that $\alpha + \rho = 1$, $\alpha = \varepsilon$, and substituting equations (h)–(k) into (l) the net radiation exergy transferred from surface 1 to surface 2 is:

$$b_{1-2} = \varepsilon_{1-2}\left(b_{b,1} - b_{b,2}\right) = \varepsilon_{1-2}\sigma\left[T_1^4 - T_2^4 - \frac{4}{3}T_0\left(T_1^3 - T_2^3\right)\right] \tag{7.77}$$

where ε_{1-2} is determined by formula (g) and in derivation of formula (7.77), the formulae (h) and (i) were used.

7.5.3.4 Exergy of Heat Sources

Analysis of the streams of exergy can take different forms. For example, following the history of the exergy streams in Figure 7.11(b), the exergy decrease $b_{q,1,1}$ of the heat source 1, caused by radiation of surface 1, can be determined as:

$$b_{q,1,1} = b_{1,0} - b_{1,6} - b_{1,12} - \cdots$$

thus:

$$b_{q,1,1} = \varepsilon_1 e_{b,1}\left(1 - \frac{T_0}{T_1}\right)\left[1 - \alpha_1\rho_2\left(1 + \rho_1\rho_2 + \rho_1^2\rho_2^2 + \cdots\right)\right]$$

$$= \varepsilon_1 e_{b,1}\left(1 - \frac{T_0}{T_1}\right)\left(1 - \frac{\alpha_1\rho_2}{1 - \rho_1\rho_2}\right) \tag{m}$$

The exergy increase $b_{q,1,2}$ of the heat source 1, caused by radiation of surface 2 is:

$$b_{q,1,2} = b_{2,3} + b_{2,9} + b_{2,15} + \cdots = \varepsilon_2 e_{b,2}\frac{\alpha_1}{1 - \rho_1\rho_2}\left(1 - \frac{T_0}{T_1}\right) \tag{n}$$

Analogously, the exergy decrease $b_{q,2,2}$ of the heat source 2, caused by radiation of surface 2, can be determined as:

$$b_{q,2,2} = b_{2,0} - b_{2,6} - b_{2,12} - \cdots = \varepsilon_2 e_{b,2}\left(1 - \frac{T_0}{T_2}\right)\left(1 - \frac{\alpha_2\rho_1}{1 - \rho_1\rho_2}\right) \tag{o}$$

The exergy increase $b_{q,2,1}$ of the heat source 2, caused by radiation of surface 1 is:

$$b_{q,2,1} = b_{1,3} + b_{1,9} + b_{1,15} + \cdots = \varepsilon_1 e_{b,1}\frac{\alpha_2}{1 - \rho_1\rho_2}\left(1 - \frac{T_0}{T_2}\right) \tag{p}$$

The total exergy decrease $b_{q,1}$ of the heat source 1 is:

$$b_{q,1} = b_{q,1,1} - b_{q,1,2} \tag{q}$$

and the total exergy increase $b_{q,2}$ of the heat source 2 is:

$$b_{q,2} = b_{q,2,1} - b_{q,2,2} \tag{r}$$

7.5.3.5 Exergy Losses

Another aspect of the exergy exchange mechanism shown in Figure 7.11(b) can be the analysis of exergy losses.

It is worth emphasizing that any partial exergy balance, e.g., equations expressing exergy streams $b_{1,1}$, $b_{1,7}$, $b_{1,13}$, etc., are used only for simplicity of calculations, and the losses determined by such equations can produce unrealistic negative values. However, real positive values of the losses are always obtained if the partial exergy balances are taken together appropriately for the processes occurring simultaneously. Thus, the applied simplification for calculation does not violate the second law of thermodynamics according to which the overall entropy growth should be nonnegative even in the elemental step of the process.

After such an explanation, let us consider the calculative partial exergy loss $\delta b_{1,1}$ caused only by emission of surface 1 and absorption of the portions of this emission reflected from surface 2 and absorbed on surface 1. Based on the partial exergy balances:

$$\begin{aligned}
\delta b_{1,1} &= b_{1,1} + b_{1,7} + b_{1,13} + \cdots \\
&= (b_{1,0} - b_{1,2}) + (b_{1,5} - b_{1,6} - b_{1,8}) \\
&\quad + (b_{1,11} - b_{1,12} - b_{1,14}) + \cdots
\end{aligned}$$

the following formula can be derived:

$$\delta b_{1,1} = \varepsilon_1 e_{b,1} \left(1 - \frac{T_0}{T_1}\right) \left(1 - \frac{\varepsilon_1 \rho_2}{1 - \rho_1 \rho_2}\right) - \frac{\varepsilon_2 \varepsilon_1 b_{b,1}}{1 - \rho_1 \rho_2} \tag{s}$$

The calculative partial exergy loss $\delta b_{1,2}$ caused only by emission of surface 2 and absorption of the portions of this emission reflected from surface 2 and absorbed on surface 1, can be defined as follows:

$$\begin{aligned}
\delta b_{1,2} &= b_{2,4} + b_{2,10} + b_{2,16} + \cdots \\
&= (b_{2,2} - b_{2,3} - b_{2,5}) + (b_{2,8} - b_{2,9} - b_{2,11}) \\
&\quad + (b_{1,14} - b_{1,15} - b_{1,17}) + \cdots
\end{aligned}$$

from which:

$$\delta b_{1,2} = \frac{\varepsilon_1}{1 - \rho_1 \rho_2} \left[\varepsilon_2 b_{b,2} - \varepsilon_2 e_{b,2} \left(1 - \frac{T_0}{T_1}\right)\right] \tag{t}$$

The joint exergy loss δb_1 on surface 1 is:

$$\delta b_1 = \delta b_{1,1} + \delta b_{2,1} \tag{u}$$

By analogy the respective exergy losses can be derived for surface 2. The calculative partial exergy loss $\delta b_{2,1}$ caused only by emission of surface 2 and absorption of the portions of this emission reflected from surface 1 and absorbed on surface 2, is:

$$\delta b_{2,1} = \varepsilon_2 e_{b,2} \left(1 - \frac{T_0}{T_2}\right) \left(1 - \frac{\varepsilon_2 \rho_1}{1 - \rho_1 \rho_2}\right) - \frac{\varepsilon_1 \varepsilon_2 b_{b,2}}{1 - \rho_1 \rho_2} \qquad \text{(w)}$$

The calculative partial exergy loss $\delta b_{2,2}$ caused only by emission of surface 1 and absorption of the portions of this emission reflected from surface 1 and absorbed on surface 2, is:

$$\delta b_{2,2} = \frac{\varepsilon_2}{1 - \rho_1 \rho_2} \left[\varepsilon_1 b_{b,1} - \varepsilon_1 e_{b,1} \left(1 - \frac{T_0}{T_2}\right)\right] \qquad \text{(x)}$$

The joint exergy loss δb_2 on surface 2 is:

$$\delta b_2 = \delta b_{2,1} + \delta b_{2,2} \qquad \text{(y)}$$

The correctness of the all presented considerations can be confirmed by fulfillment of the exergy balance equation for the global process. The exergy decrease of the heat source 1 is spent for the exergy increase of the heat source 2 and for the irreversible exergy losses occurring on both surfaces (1 and 2):

$$e_{1-2} \left(1 - \frac{T_0}{T_1}\right) = e_{1-2} \left(1 - \frac{T_0}{T_2}\right) + \delta b_1 + \delta b_2 \qquad \text{(z)}$$

Interpretation of the radiative heat exchange between two gray surfaces was analyzed in terms of energy and exergy viewpoints. The significant differences in the viewpoints were disclosed, and Figure 7.11 particularly illustrated the difference in energetic and exergetic interpretations of occurring mechanisms.

The simplified configuration was considered to clearly demonstrate the method. In various complex situations the principle of the method can be also be applied—however, together with appropriate inclusion of the view factors.

Example 7.3 The presented considerations can be illustrated by the calculation example for the two surfaces, shown in Figure 7.11 and having temperature $T_1 = 1000$ K, $T_2 = 500$ K, and respective emissivities $\varepsilon_1 = 0.95$ and $\varepsilon_2 = 0.9$. Environment temperature is $T_0 = 300$ K.

The numerical values of the particular energy and exergy streams, respectively, $e_{n,m}$ and $b_{n,m}$, are calculated according to formulae given in Sections 7.5.3.2 and 7.5.3.3 and are shown in Table 7.2. The subscript denotation of energy or exergy stream is given in column 1. Columns 2–4 illustrate the streams of heat emitted, absorbed, and successively reflected between surfaces 1 and 2.

Subscript n,m	Energy streams $e_{n,m}$, W			Exergy streams $b_{n,m}$, W				
	Surface 2	Space	Surface 1	Surface 2		Space	Surface 1	
				Heat	Loss		Heat	Loss
1	2	3	4	5	6	7	8	9
Surface 1:								
1,0			−53,858				−37,700	
1,1		53,858						5240
1,2	48,473					32,460		
1,3		5385		19,389				
1,4			5116		9825			
1,5		269				3246		
1,6	242						3581	
1,7		27						−497
1,8			25.7			162		
1,9		1.3		97				
1,10	1.2				48.8			
1,11		0.12						
1,12						16.2	17.9	
1,13								−2.5
1,14						0.80		

Surface 2:						
2,0	-3189			-1275		
2,1		3189			499	
2,2		159	3030			776
2,3						2121	-1384
2,4	-143	16	15.2			
2,5						39
2,6		0.8		57.4		
2,7					-22.3	
2,8						3.9
2,9						10.6
2,10					0.19		-8.6
2,11						
Total	45,671	—	-45,671	18,269	10,352	-31,970	3349
Symbol	e_{1-2}	—	e_{1-2}	$b_{q,2}$	δb_2	$b_{q,1}$	δb_1

TABLE 7.2 The Calculated Streams of Energy ($e_{n,m}$) and Exergy ($b_{n,m}$) Shown in Figure 7.11

For example, from heat source 1 the amount $e_{1,0} = -53{,}858$ W/m^2 (column 4) is transferred to surface 1 and then the same amount $e_{1,1} = 53{,}858$ W/m^2 (column 3) is emitted to surface 2. However, surface 2 absorbs only 90% ($\varepsilon_2 = 0.9$), i.e., $e_{1,2} = 48{,}473$ W/m^2 (column 2) of the arrived emission, and the rest is reflected as $e_{1,3} = 5385$ W/m^2 (column 3).

Analogously to the above energy consideration, the fate of heat $e_{2,0}$ (column 2) transferred from heat source 2 to surface 2, and then, the further emissions and absorptions, can be tracked.

The process of successive absorptions and reflections progresses to infinity with gradually reduced values of the processed streams. The energetic effect in the form of the net rate of transferred heat e_{1-2} is determined from formulae (f) and (g):

$$e_{1-2} = \frac{5.669 \times 10^{-8}\left(1000^4 - 500^4\right)}{\dfrac{1}{0.95} + \dfrac{1}{0.9} - 1} = 45.67\text{kW/m}^2$$

Columns 5–9 illustrate the streams of exergy emitted, absorbed, and successively reflected between surfaces 1 and 2. Each step of emission or absorption is accompanied by the respective exergy loss due to irreversibility. For example, from heat source 1 the exergy amount $b_{1,0} = -37{,}700$ W/m^2 (column 8) is transferred to surface 1 and then the reduced amount of exergy $b_{1,2} = 32{,}460$ W/m^2 (column 7) is emitted to surface 2. The difference between $b_{1,0}$ and $b_{1,2}$ is the exergy loss $b_{1,1} = 5240$ W/m^2 (column 9). However, according to the emissivity $\varepsilon_2 = 0.9$, surface 2 absorbs only 90%, of the arrived emission, and the rest, at the reflectivity $\rho_2 = 0.1$, is reflected as $b_{1,5} = 3246$ W/m^2 (column 7). Meanwhile, the absorption at surface 2 causes the exergy growth $b_{1,3} = 19{,}389$ W/m^2 (column 5) of heat source 2 at the involved exergy loss $b_{1,4} = 9825$ W/m^2 (column 6).

Analogously, to the above exergy consideration, the rate of exergy decrease $b_{2,0} = -1275$ W/m^2 (column 5) of heat source 2, transferred to surface 2, and then the further emissions, absorptions, and losses can be tracked.

The process of successive absorptions, losses, and reflections is progressing to infinity with gradually reduced values of the processed streams. The exergetic effect in the form of the net rate of transferred exergy b_{1-2} is determined from formulae (7.77) and (g):

$$b_{1-2} = \frac{5.669 \cdot 10^{-8}\left[1000^4 - 500^4 - \dfrac{4}{3}300 \times \left(1000^3 - 500^3\right)\right]}{\dfrac{1}{0.9} + \dfrac{1}{0.95} - 1} = 28.62 \text{ kW/m}^2$$

The total exergy decrease $b_{q,1}$ of the heat source 1 can be determined from formula (q):

$$b_{q,1} = b_{q,1,1} - b_{q,1,2} = -34{,}101 + 2131 = -31{,}970 \quad \text{W/m}^2$$

and the total exergy increase $b_{q,2}$ of the heat source 2 from formula (r) is:

$$b_{q,2} = b_{q,2,1} - b_{q,2,2} = 19{,}486 - 1217 = 18{,}269 \quad \text{W/m}^2$$

The joint exergy loss δb_1 on surface 1 from formula (u) is:

$$\delta b_1 = \delta b_{1,1} + \delta b_{2,1} = 4740 - 1392 = 3349 \quad \text{W/m}^2$$

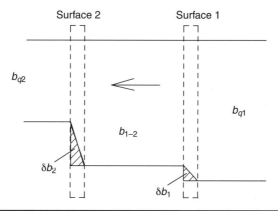

FIGURE 7.12 Scheme of the radiation exergy exchange considered in Example 7.3.

The joint exergy loss δb_2 on surface 2 from formula (x) is:

$$\delta b_2 = \delta b_{2,1} + \delta b_{2,2} = 447 + 9875 = 10352 \quad W/m^2$$

The correctness of the all presented calculations is confirmed by the complete exergy balance equation (y) for the global process:

$$e_{1-2}\left(1 - \frac{T_0}{T_1}\right) - e_{1-2}\left(1 - \frac{T_0}{T_2}\right) - \delta b_1 - \delta b_2 = 31970 - 18269 - 3349 - 10352 = 0$$

It is also worth noting that the change in exergy sources is

$$\left|b_{q,1}\right| - \left|b_{q,2}\right| = 31.97 - 18.269 = 13.701 \quad kW/m^2,$$

whereas the radiation exergy transferred from surface 1 to 2 is larger b_{1-2} = 28.43 kW/m². The difference is caused by the two exergy losses occurring on the surfaces 1 and 2, of which the loss on surface 2 is larger ($\delta b_2 > \delta b_1$). Figure 7.12 shows schematically the calculated values in the exergy balance of the considered surfaces 1 and 2.

The consideration addressed was the case of two parallel infinite surfaces facing each other as shown in Figure 7.11. In a real case, the considered surfaces can usually be finite, relatively small, and arbitrarily situated in regard to each other. In the calculations for such cases the view factors should be involved for determination of the multireflected fluxes between surfaces. This necessity seemingly make the consideration more complex; however, multiplying the considered fluxes by the view factors causes such quick weakening of the exchanged fluxes that for practical purposes it is usually sufficient to take into account only the first reflections or even to entirely ignore the reflections, especially in cases of low values for the view factors and surface reflectivities.

Real systems are composed mostly of more than two surfaces and the presented principles of determination of the radiation exergy exchange has to be appropriately applied taking into account all the possible combinations of the mutually radiative interaction between system surfaces.

7.6 Exergy of Solar Radiation

7.6.1 Significance of Solar Radiation

Solar energy is the most important renewable source of energy on the earth. Solar energy is a high-temperature source; however, harvesting occurs inefficiently due to extensive degradation of the energy. The degradation of solar energy is well demonstrated by consideration of exergy. Therefore the engineering thermodynamics of thermal radiation addresses mainly exergy analyses of diversified problems for the utilization of solar radiation. The potential for maximum work produced from radiation has been the subject of intensive research. For a better understanding of possible utilization, some basics of solar radiation are described in the following. More details are discussed by Duffie and Beckman (1991).

Solar energy, although rich, is poorly concentrated, and thus it requires a relatively large surface to harvest the sun's radiation. From this viewpoint, solar radiation is especially valuable for countries that have large unused areas (e.g., deserts). The small concentration of energy needs intensive theoretical studies in order to obtain acceptable efficiency of energy utilization. Effective method for such purpose is exergy analysis.

Solar radiation is the result of the fusion of atoms inside the sun. Part of the fusion energy delivers heat to the outer layer of the sun (the chromosphere), which is much cooler than the sun's interior. Thus, the solar radiation incident on earth is the chromosphere radiation, not much different from the radiation of any surface at about 5800 K.

Extraterrestrial solar radiation is about 47% in the visible wavelengths (380–780 nm), about 46% in the infrared wavelengths (greater than 780 nm), and about 7% in the ultraviolet wavelengths (below 380 nm). A large portion of the ultraviolet radiation is absorbed and scattered by the atmosphere. For example, air molecules scatter the shorter-wavelength radiation more strongly than the longer wavelengths, i.e., they scatter out more blue light, making the sky appear blue.

Generally, solar radiation passing through the atmosphere is absorbed, scattered, and reflected not only by air molecules but also by water vapor, clouds, dust, pollutants, smoke from forest fires and

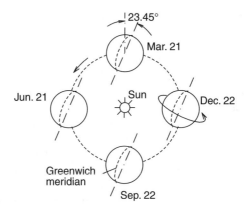

FIGURE 7.13
Orientation of earth
to sun.

volcanoes, etc. These factors cause diffusion (called also *dilution*) of solar radiation. The portion of solar radiation that reaches the earth's surface without being diffused is called *direct beam solar radiation*. Thus, global solar radiation (global irradiance) consists of the diffuse and direct solar radiation. For example, during cloudy days the atmosphere reduces direct beam radiation to zero.

If the considered surface is tilted with respect to the horizontal, the global irradiance consists of the incident diffuse (dilute) radiation of the normal irradiance projected onto the tilted surface and of the ground-reflected irradiance that is incident on the considered surface. The amount of direct radiation on an arbitrarily oriented surface can be calculated from Lambert's law and is based on direct normal irradiance.

The solar radiation incident outside the earth's atmosphere is called *extraterrestrial radiation* and its average value is 1367 W/m². This value varies ±3% as the earth revolves around the sun in an elliptical orbit; the earth's closest distance to the sun is on around January 4, and it is the furthest from the sun on around July 5. The orientation of the earth relative to the sun is schematically shown in Figure 7.13. The earth's axis always points in the same direction as it orbits around the sun. The 23.45° tilt in the earth's axis of revolution results in longer days in the northern hemisphere from March 21 to September 22 and longer days in the southern hemisphere during the other six months. The solar angle varies at a given spot on earth throughout the year, causing the year's seasons and determining the length of daylight each day.

At noon on a cloudless day about 25% of the extraterrestrial solar radiation is scattered and absorbed by the atmosphere and only about 1000 W/m² reaches the earth's surface as direct normal irradiance (*beam irradiance*).

To describe the sun's path across the sky or to determine the instant position of the sun in the sky as seen from the earth, one needs the values of two parameters, which are *declination* and *azimuth* shown, e.g., in Figure 7.1 as β and φ, respectively. Declination β is a flat angle measured from the north–south axis and azimuth φ is measured from the 0 meridian (passing at Greenwich, England).

Because the earth is round, the sun's rays arrive at the earth's surface at different angles ranging from the 0° declination (just above the horizon) to the 90° declination (directly overhead). The vertical rays supply the most possible radiation energy. The more slanted are the rays, the longer are their path through the atmosphere, and the sunlight is therefore more scattered and diluted.

Only a part of scattered sunlight reaches the earth because some sunlight is scattered back into space. Also some radiation of the earth, together with sunlight scattered off the earth's surface, is re-scattered into the atmosphere. This effect can be significant, e.g., when the earth's surface is covered with snow.

Solar radiation is difficult to calculate because, as discussed, the radiation energy reaching the surface of the earth is composed of direct and diluted radiation components, and depends on geographic location, time of day, season of year, local weather, and even on local landscape. One relatively effective method of determining solar radiation is by spectral measurement and application of the obtained results in the formulae derived in Section 7.3.

Example 7.4 Regarding the solar radiation as nonpolarized, black, uniform, and propagating within a solid angle ω, the exergy of the extraterrestrial solar radiation may be approximately calculated by means of equation (7.50). The required exergy b_b of emission density can be calculated from (7.49) for the sun surface temperature $T = 6000$ K and for the environment temperature $T_0 = 300$ K as follows:

$$b_b = \frac{5.6693 \times 10^{-8}}{3} \left(3 \times 6000^4 + 300^4 - 4 \times 300 \times 6000^3\right) = 68.5 \quad \text{MW/m}^2$$

(a)

Approximately, the radius of the sun is $R_S = 695{,}500$ km and the mean distance from the sun to the earth is $L_S = 149{,}500{,}000$ km. The integral in formula (7.50) expresses the solid angle ω and is equal the area of circle of radius R_S divided by the square distance L_S, thus :

$$\iint_{\beta \ \varphi} \cos \beta \ \sin \beta \ d\beta \ d\varphi = \frac{R_S^2 \ \pi}{L_S^2} = 2.16 \times 10^{-5} \pi \quad \text{sr}$$

(7.78)

By substitution of (a) and (7.78) into (7.50):

$$b_{b,\ \omega} = \frac{68{,}500}{\pi} 2.16 \times 10^{-5} \pi = 1.48 \quad \text{kW/m}^2$$

(b)

From formulae (7.50) and (7.18), in which radiosity can be interpreted as emission ($j = e$), the ratio of exergy to energy of emission is $b_{b,\omega}/e_{b,\omega} = 0.9333$.

Example 7.5 More exact computations of the exergy of solar radiation were carried out by Petela (1961a) based on the extraterrestrial radiation spectrum determined experimentally by Kondratiew (1954). Calculations are based on equation (7.45) for nonpolarized and uniform radiation. Table 7.3 presents some exemplary Kondratiew's data on the intensity of radiation $i_{0,\lambda}$ (column 2) as a function of wavelength λ (column 1). The part of the spectrum is shown in Figure 7.14 together with three spectra (dashed lines), for comparison, for black radiation at temperatures 6000, 5800, and 5600. The $i_{0,\lambda}$ values in Table 7.3 are assumed to be constant for the respective ranges of wavelengths $\Delta\lambda$ (column 5). Corresponding ranges of frequency $\Delta\nu$ calculated based on equation (7.5), for $c_0 = 2.9979 \times 10^8$ m/s, are shown in column 6, whereas equation (7.6) was used to determine $i_{0,\nu}$ in column 4. The ν values in column 3 were determined from equation (3.1). The $L_{0,\nu}$ values of column 8 are calculated from equation (7.30). Columns 7 and 9 are calculated as respective products of columns 4 and 6; $(i_{0,\nu} \times \Delta\nu)$, and 6 and 8; $(L_{0,\nu} \times \Delta\nu)$.

Formula (7.45) is applied in the following numerical form:

$$b_{A'} = \left(2\sum i_{0,\nu}\Delta\nu - 2\,T_0 \sum L_{0,\nu}\Delta\nu + \frac{\sigma T_0^4}{3\,\pi}\right)\int\limits_{\beta}\int\limits_{\varphi} \cos\beta\,\sin\beta\,d\beta\,d\varphi \qquad (7.79)$$

Assuming the environment temperature $T_0 = 300$ K, substituting formula (7.78) into (7.79) and using data from Table 7.3:

$$b_{A'} = \left(2 \times 10,079,300 - 2 \times 300 \times 2263.3 + \frac{5.6693 \times 10^{-8} \times 300^4}{3\,\pi}\right)$$

$$\times\,\pi \times 2.16 \times 10^{-5}$$

$$= 1367.9 - 92.151 + 0.0033 = 1275.8 \text{ W/m}^2$$

The obtained result 1275.8 W/m^2 is the exergy of the extraterrestrial solar radiation arriving at the 1 m^2 surface, which is perpendicular to the direction of the sun. The obtained ratio of radiosity to exergy is 1275.8/1367.9 = 0.9326.

7.6.2 Possibility of Concentration of Solar Radiation

The possibility of concentrated radiation can be illustrated with use of a simple model of two surfaces shown in Figure 7.15. The imagined surface of area A_S represents the black ($\varepsilon_S = 1$) solar irradiance IR at constant temperature T_S. The other surface of area A is gray at emissivity ε, and its temperature T is controlled by the cooling heat Q. The vacuum space between the two surfaces is enclosed by a cone-shaped surface that is mirrorlike ($\varepsilon_0 = 0$). The surface areas ratio $a_S = A_S/A$. The energy balance of the cooled surface A is:

$$a_S\varepsilon\,IR = \varepsilon\sigma T^4 + k\,(T - T_0) \qquad (7.80)$$

where k is the heat transfer coefficient at which heat Q is extracted from surface A. The heat rate

$$q = k\,(T - T_0) \qquad (7.81)$$

$\lambda \times 10^{10}$ m 1	$i_{0,\lambda} \times 10^{-10}$ $\dfrac{W}{m^3 sr}$ 2	$\nu \times 10^{-11}$ $\dfrac{1}{s}$ 3	$i_{0,\nu} \times 10^{12}$ $\dfrac{J}{m^2 sr}$ 4	$\Delta\lambda \times 10^{10}$ m 5	$\nu \times 10^{-12}$ $\dfrac{1}{s}$ 6	$i_{0,\nu} \times \Delta\nu$ $\dfrac{W}{m^2 sr}$ 7	$L_{0,\nu} \times 10^{13}$ $\dfrac{J}{m^2 s\, sr}$ 8	$L_{0,\nu} \times \Delta\nu$ $\dfrac{W}{m^2 K\, sr}$ 9
2200	10	13,627	15	100	62.0	960	0.03	0.205
2300	26	13,035	47	100	56.7	2650	0.10	0.540
2400	31	12,492	59	100	52.1	3090	0.12	0.639
.	.	.	.	.	.	.	.	.
.	.	.	.	.	.	.	.	.
.	.	.	.	.	.	.	.	.
60,000	1	500	1765	10,000	0.714	1260	7.19	0.514
70,000	1	428	1201	10,000	0.535	640	5.48	0.293
Total						10,079,300	—	2263.306

Table 7.3 Spectrum of the Extraterrestrial Solar Radiation (from Petela, 1962)

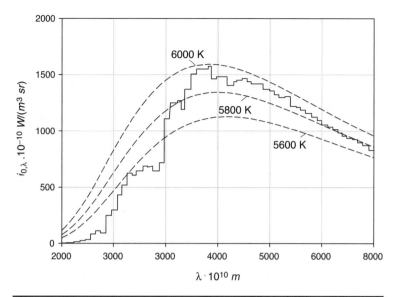

FIGURE 7.14 Spectrum of extraterrestrial solar radiation (from Petela, 1962).

can be used to express the total heat Q absorbed by surface A:

$$Q = \frac{A_S}{a_S} q \qquad (7.82)$$

The energy efficiency η_E of concentration of solar radiation can be measured as the ratio of absorbed heat Q and the solar irradiance IR:

$$\eta_E = \frac{Q}{IR} \qquad (7.83)$$

For comparison also exergetic efficiency η_B can be considered based on the following definition:

$$\eta_B = \frac{B_Q}{IR \, \psi} \qquad (7.84)$$

FIGURE 7.15
Scheme of
concentrated
radiation.

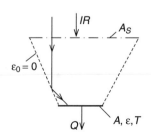

where ψ is the exergy energy ratio discussed in Section 6.4.1 and the exergy B_Q of heat absorbed by surface A is:

$$B_Q = Aq \left(1 - \frac{T_0}{T} \right) \tag{7.85}$$

The reality of the discussed effect of concentration of solar radiation can be evaluated by the calculated value of the overall entropy growth Π, which consists of the positive entropies of heat Q, of the emission of surface A, and of the negative entropy of absorbed solar radiation:

$$\Pi = \frac{Q}{T} + A\varepsilon\, \frac{4}{3}\sigma T^3 - A_S\varepsilon\ SR \tag{7.86}$$

where SR is the entropy of irradiance IR. The magnitude SR can be evaluated from the assumed ratio SR/IR to be equal the ratio s/e of the black emission entropy and emission energy, $SR/IR = s/e$. With the use of formulae (3.21) and (5.24) the following relation can be derived:

$$SR = \frac{4}{3} \frac{IR}{T_S} \tag{7.87}$$

The overall entropy growth determined from equation (7.86) should be positive ($\Pi > 0$). If the overall entropy growth is negative ($\Pi \leq 0$), then the concentration of solar radiation is impossible because it is against the second law of thermodynamics.

Example 7.6   The concentration of solar radiation can be considered, e.g., at $IR = 800\ \text{W/m}^2$ arriving at the imagined surface of area $A_S = 1\ \text{m}^2$, as shown in Figure 7.15 (thin solid line). Assuming also that $k = 3\ \text{W/(m}^2\text{K)}$ and the environment temperature $T_0 = 300\ \text{K}$, equation (7.80) allows for determining temperature T of surface A as a function of the surface ratio a_S. With the growing a_S the temperature T grows; also, the heat rate q grows, determined by formula (7.81), as is shown in Figure 7.16 by a long-dash line.

However, according to formula (7.82), with growing a_S the total heat Q is varying (short-dash line) with a maximum of about 134 W at about $a_S \approx 2$. The maximum appears because with growing a_S its effect becomes stronger than the effect of the growing heat rate q.

The energy efficiency η_E of the concentration of solar radiation, based on definition (7.83), is varying as shown with the thick-dashed line in Figure 7.16. The efficiency η_E has a maximum of about 16.8% appearing also at about $a_S \approx 2$, correspondingly to the maximum of Q.

Exergy B_Q of absorbed heat is determined by (7.85) and is shown in Figure 7.16 by a dotted line. The exergy B_Q varies and has a maximum of about 45.8 W, which appears at the surface area ratio about $a_S \approx 6$. The maximum is a result of two factors varying with growing a_S: one is the growing exergy of heat due to growing temperature T; the other is due to a decrease of the absorbed heat Q.

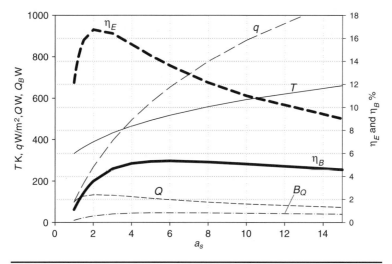

FIGURE 7.16 Exemplary effects of the concentration of solar radiation.

Assuming $\psi = 0.933$, as for black radiation at temperature $T_S = 6000$ K the efficiency η_B can be determined from formula (7.84) and shown in Figure 7.16 (thick solid line). The efficiency η_B has a maximum of about 5.34%, which also corresponds to a value of $a_S \approx 6$.

The overall entropy growth determined from equation (7.86) for the data used in the example is always positive ($\Pi > 0$) and with growing a_S diminishes to zero ($\Pi = 0$) for $a_S = 91{,}843$ corresponding to temperature $T = 6000$ K. For further growing of a_S the overall entropy growth becomes negative ($\Pi < 0$), i.e., the further concentration of solar radiation is impossible.

Based on the calculations, the process of "deconcentration" of solar radiation, which would correspond to reducing a_S below 1, is irreversible and can occur but heat absorbed by the surface A is negative which means that the surface should be heated.

The data used in the present example were also used for the computation of the results shown in Table 7.4. These data illustrate the trends of the output data in response to changes in some input parameters. The values in column 3 of Table 7.4 are considered to be the reference values for studying the influence of varying input parameters in the output. Therefore, each of the next columns (4–6) corresponds to the case in which the input is changed only by the values shown in a particular column, whereas the other input parameters remain at the reference level.

For example, column 4 corresponds to a change in the emissivity ε, which increases from 0.9 to 1. The 10% ε increase causes the increase of temperature T from 517.8 to 519.9 K, q from 653.3 to 659.5 W/m^2, Q from 108.9 to 109.9 W, η_E from 13.61% to 13.74%, B_Q from 45.79 to 46.49 W, and η_B from 5.34% to 5.42%.

Columns 5 and 6 can be similarly interpreted. For example, increasing the heat transfer coefficient k from 3 to 5 W/(m^2 K) causes an increase in the exergetic efficiency from 5.34% to 8.04%, which is the result of increased heat rate q from 653.3 to 1021 W/m^2.

Quantity	Units	Reference value	Mono-variant changes of input parameters and resulting outputs		
1	2	3	4	5	6
Input:					
ε		0.9	1		
k	W/(m² K)	3		5	
T_0	K	300			270
Output:					
T	K	517.8	519.9	504.2	514.8
q	W/m²	653.3	659.6	1021	734.6
Q	W	108.9	109.9	170.2	122.4
B_Q	W	45.79	46.49	68.94	58.23
η_E	%	13.61	13.74	21.28	15.30
η_B	%	5.34	5.42	8.04	6.79

TABLE 7.4 Responsive Trends of Output to Change of Some Input Parameters (for $IR = 800$ W/m², $T_S = 6000$ K, $\psi = 0.933$, $A_S = 1$ m²)

Nomenclature for Chapter 7

A	surface area, m²
A'	comparable surface area, m²
a	universal radiation constant, $a = 7.764 \times 10^{-16}$ J/(m³ K⁴)
a_S	surface area ratio
B	exergy flux, W
b	exergy emission density (rate), W/m²
c_0	speed of propagation of radiation in vacuum $c_0 = 2.9979 \times 10^8$ m/s
c_1	the first Planck's constant, $c_1 = 3.74 \times 10^{-16}$ W m²
c_2	the second Planck's constant, $c_2 = 1.4388 \times 10^{-2}$ m K
E	energy emission, W
e	energy emission density, W/m²
h	Planck's constant, $h = 6.625 \times 10^{-34}$J s.
IR	solar irradiance, W/m²
i	directional radiation density, W/(m² sr)
i	successive number
J	radiosity, W
j	radiosity density, W/m²
j	successive number
k	number of elements
k	Boltzmann constant, $k = 1.3805 \times 10^{-23}$ J/K
k	heat transfer coefficient, W/(m² K)

L_0 normal entropy radiation intensity, W/(K m^2 sr)
L_c length of crossed strings, m
L_i length of profile of surface i, m
L_n length of not crossed strings, m
L_0 normal entropy radiation intensity, W/(m^2 K)
m successive number of happening
n successive number of the considered surface
n number of elements
P point on a surface
Q heat, W
q heat rate, W/m^2
R radius, m
r   distance, m
S entropy of radiosity or emission, W/K
SR entropy of solar irradiance, W/(m^2 K)
s entropy of emission density, W/(m^2 K)
s_j entropy of radiosity density, W/(m^2 K)
T absolute temperature, K
X expression in formula (7.24)
Y expression in formula (7.25)

Greek

α absorptivity
β flat angle, (declination), deg
Δ increment
δb irreversible loss of exergy, W/m^2
δB irreversible loss of exergy, W
ε emissivity of surface
φ view factor
φ flat angle, (declination), deg
η efficiency of solar radiation concentration
λ wavelength, m
ν oscillation frequency, 1/s
Π overall entropy growth, W/(m^2 K)
ρ reflectivity
σ Boltzmann constant for black radiation,
     $\sigma = 5.6693 \times 10^{-8}$ W/(m^2 K^4)
ω solid angle, sr

Subscripts

A, A' comparable surfaces
B exergetic
b black
d local
E energetic
j radiosity

max	maximum
min	minimum
q	heat
S	solar
Δ	local
λ	wavelength
ν	frequency
ω	solid angle
0	for $\beta = 0$
0	environment
1, 2	denotation

CHAPTER 8

Radiation Spectra of a Surface

8.1 Introductory Remarks

In the present chapter the analysis of spectra and emissivities of surfaces is developed according to Petela (2010).

The surface emission spectrum depends on the temperature and properties of the emitting surface. Radiation reflected from a surface affects neither the temperature nor the properties of the considered surface. However, the radiation absorbed by the surface can affect the surface temperature and thus can also affect the surface spectra emission. A spectrum of surface radiosity is the effect of many different spectra including the emission from the considered surface and all the radiation fluxes reflected. The untainted radiation spectrum of a body can be measured only if the body is not irradiated from other radiation sources. In practice, such a pure spectrum of a body does not occur, and consideration of such a spectrum has a rather theoretical meaning, allowing for better understanding of radiation processes.

The following example illustrates the problem of a spectrum. Snow that is strongly irradiated by the sun remains at a low temperature because the solar radiation is mainly reflected, not absorbed. The radiosity of the snow consists of the reflected solar radiation, the reflected radiosities from other surfaces, and the emission of the snow. Thus, the radiosity spectrum of snow consists of different spectra including the spectrum of the snow's emission, the spectrum of the sun, and the spectra of other surfaces contributing to the snow's radiosity. On the other hand, because the emission of snow at a temperature close to 0°C is relatively small, and the emissions of other surfaces are usually small also, the solar radiation dominates in the snow's radiosity. Thus, any measurement of the snow's radiosity spectrum will be near the sun's emission spectrum.

A body always emits black radiation, although the rate of this black emission depends on the properties of the body. The emission ability is generally different for every wavelength λ (range $d\lambda$) or respective frequency ν (range $d\nu$). The possible maximum value of any monochromatic emission appears in the theoretical model of a black surface. The monochromatic emission of a real surface is always smaller, and the departure from the value for monochromatic black emission is determined by the monochromatic emissivity (ε_λ) of the real surface. For the model of a perfectly gray surface, it is assumed that the monochromatic emissivity of the surface is smaller for all wavelengths, by the same ratio as the respective monochromatic emissivity of a black surface.

In order to emphasize the role of this property in further considerations of emissivity ε, which determines the rate of black emission from any nonblack surface, it can also be called the *energetic emissivity* $\varepsilon \equiv \varepsilon_E$. Any real surface, e.g., a gray surface, in spite of its nonblack emission rate, emits a photon gas that is black radiation. Correspondingly, any real surface, e.g., a gray surface, emits black emission entropy and emits the black emission exergy in spite of the nonblack properties of the surface. The amount of the entropy or exergy of the black energy emission, coming from a nonblack surface, is determined also by the energetic emissivity ε.

The spectrum of an emitting surface should be distinguished from the spectrum of the emitted photon gas, which is always black. The concept of emissivity has application only for the surface.

Badescu (1988) proposed a formula for the exergy spectrum component per volume unit, which is a function of the unclear reference state determined by both the environment temperature and the atmospheric pressure. The black photon gas has the exergy reference state sufficiently defined only by temperature or pressure, since both these are related. In spite of this, Moreno et al. (2003) developed the formula alteration according to the reduction and splitting of the photon quantum states.

The radiation exergy for an arbitrary energy spectrum has also been considered, e.g., by Karlsson (1982) and Wright et al. (2002). The formula for monochromatic radiation exergy was introduced by Candau (2003); however, diagrams for spectra have not been considered.

Exergy is an interpretive concept that can be proposed for describing the properties of any matter. Thus, beside the rates of black emission energy and its entropy and exergy, the spectra of the surface radiating the energy, entropy, or exergy can also be separately analyzed. Consequently, since we have considered the energetic emissivity we also will consider the respective entropic and exergetic emissivities.

8.2 Energy Radiation Spectrum of a Surface

The formula (7.8) for the monochromatic normal directional intensity $i_{b,0,\lambda}$, for linearly polarized black radiation propagating within a unit solid angle and dependent on wavelength λ, was established by Planck (1914):

$$i_{b,0,\lambda} = \frac{c_0^2 h}{\lambda^5} \frac{1}{\exp\left(\dfrac{c_0 h}{k\lambda T}\right) - 1} \tag{8.1}$$

where $c_0 = 2.9979 \times 10^8$ m/s is the speed of propagation of radiation in vacuum, $h = 6.625 \times 10^{-34}$J s is the Planck's constant, $k = 1.3805 \times 10^{-23}$ J/K is the Boltzmann constant, λ is the wavelength, and T is the absolute temperature.

The energetic emissivity ε_E can be applied to the radiation intensity and is defined as the ratio of the black radiation intensity emitted by the gray surface to the black radiation intensity emitted by the black surface at the same temperature. Therefore, the intensity represented by formula (8.1) can be used as follows:

$$\varepsilon_E = \left(\frac{\int_\lambda \varepsilon_{E,\lambda}\, i_{b,0,\lambda} d\lambda}{\int_\lambda i_{b,0,\lambda} d\lambda}\right)_T \equiv \varepsilon \tag{8.2}$$

where $\varepsilon_{E,\lambda}$ is the monochromatic energetic emissivity of a surface.

For example, the spectrum of

$$i_{0,\lambda} = \varepsilon i_{b,0,\lambda} \tag{8.3}$$

for the five values of emissivity ε and temperature $T = 6000$ K, is shown in Figure 8.1 for the gray surface. The presented radiation energy spectra are commonly known (and are shown here only for the convenience of comparison to other spectra discussed later), have smaller values the lower is the value of ε, and the spectrum maxima appear for the same wavelength.

8.3 Entropy Radiation Spectrum of a Surface

Considerations on entropy are based on equation (7.25) established by Planck (1914), which for the entropy of monochromatic directional normal radiation intensity and for linearly polarized black radiation propagating within a unit solid angle and dependent on wavelength λ, is:

$$L_{b,0,\lambda} = \frac{c_0 k}{\lambda^4} [(1 + Y)\ln(1 + Y) - Y \ln Y] \quad \text{where} \quad Y \equiv \frac{\lambda^5 i_{b,0,\lambda}}{c_0^2 h} \tag{8.4}$$

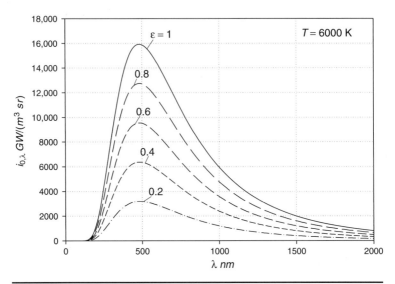

Figure 8.1 Examples of energetic spectra of surface at temperature 6000 K and for five different values of emissivity ε.

If replacement of the black radiation intensity $i_{b,0,\lambda}$ by the product $\varepsilon \times i_{b,0,\lambda}$ is justified, then equation (8.4) can be interpreted for a gray surface as follows:

$$L_{0,\lambda} = \frac{c_0 k}{\lambda^4} \left[(1 + Y_\varepsilon) \ln (1 + Y_\varepsilon) - Y_\varepsilon \ln Y_\varepsilon \right] \quad \text{where} \quad Y_\varepsilon \equiv \frac{\lambda^5 \varepsilon i_{b,0,\lambda}}{c_0^2 h} \tag{8.5}$$

Figure 8.2 shows examples of the entropic surface spectra calculated from equation (8.5) for a temperature of 6000 K and for five different values of energetic emissivity ε. It can be noticed that the ordinates of the spectra points are slightly larger than the values that would correspond to the values determined by the respective energetic emissivities. This means that for certain λ the monochromatic entropic emissivity $\varepsilon_{S,\lambda}$ would be larger than the monochromatic energetic emissivity ε_λ for the same temperature:

$$\varepsilon_{S,\lambda} \equiv \frac{L_{0,\lambda}}{L_{b,0,\lambda}} > \varepsilon_{E,\lambda} \equiv \varepsilon_\lambda \equiv \frac{i_{0,\lambda}}{i_{b,0,\lambda}} \tag{8.6}$$

The average entropic emissivity ε_S for the whole spectrum, in the wavelength range from 0 to ∞, can be determined as the ratio of the areas under the entropy spectrum curve for the considered gray surface—equation (8.5), and under the entropy spectrum curve for the

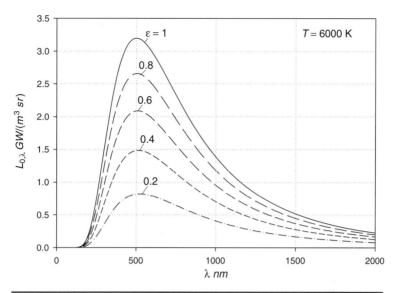

FIGURE 8.2   Examples of entropic spectra of a surface at temperature of 6000 K and for five different values of energetic emissivity ε (from Petela, 2010).

black surface—equation (8.4):

$$
\varepsilon_S = \frac{\left(\int\limits_0^\infty L_{0,\lambda}d\lambda \right)_{T,c}}{\left(\int\limits_0^\infty L_{b,0,\lambda}d\lambda \right)_T}
\tag{8.7}
$$

Both the integrals in formula (8.7) can be solved analytically or determined graphically (numerically) as the surface areas under the respective spectra curves. After using formula (8.7) in the numerical calculations of both integrals, the results are shown in Figure 8.3. The entropic emissivity ε_S is always larger than the energetic emissivity ε and differs more from ε with the decreasing value ε and with the growing surface temperature.

The discussion in the present section was inspired by Candau (2003). In his analysis of the entropy of a gray surface, he called attention to the fact that the entropic emissivity can be different from the energetic emissivity.

8.4   Radiation Exergy Derived from Exergy Definition

Considerations of the exergy spectrum are convenient when they are based on the radiation exergy formula derived in the shape resulting

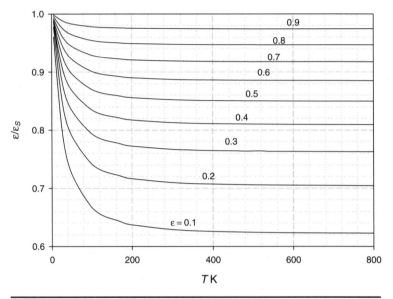

Figure 8.3 Ratio of the energetic emissivity ε to the entropic emissivity ε_S as a function of temperature T ($\varepsilon < \varepsilon_S$) (from Petela, 2010).

from direct interpretation of the general exergy definition for a substance. The methods of obtaining of such a formula will now be outlined.

There are several methods for derivation of exergy of radiation, some of them developed by Petela (2003). Historically, the formula for exergy b_b of a black surface emission density given by:

$$b_b = \frac{\sigma}{3}\left(3\,T^4 + T_0^4 - 4\,T_0 T^3\right) \qquad (8.8)$$

where T and T_0 are temperatures of the considered surface and environment, respectively, and $\sigma = 5.6693 \times 10^{-8}$ W/(m² K⁴) is the Boltzmann constant for black radiation, was derived for the first time by Petela (1961) based on consideration of the exergy balance of radiating surfaces. For an enclosed photon gas, independent derivation of a formula with the characteristic expression shown in the brackets of formula (8.8), was also shown by Petela (1964) through consideration of the useful work performed by isentropic expansion of the photon gas in the cylinder with a piston. Petela (1962) also derived for the first time the formula (7.43) for exergy b of the arbitrary and polarized radiation (as a function of frequency), which is equivalent to the following formula as a function of wavelength:

$$b = \int_\beta \int_\varphi \int_\lambda (i_{0,\lambda,\min} + i_{0,\lambda,\max}) \cos \beta \, \sin \beta \, d\beta \, d\varphi \, d\lambda$$

$$- \int_\beta \int_\varphi \int_\lambda (L_{0,\lambda,\min} + L_{0,\lambda,\max}) \cos \beta \, \sin \beta \, d\beta \, d\varphi \, d\lambda$$

$$+ \frac{\sigma T_0^4}{3 \pi} \int_\beta \int_\varphi \cos \beta \, \sin \beta \, d\beta \, d\varphi \qquad (8.9)$$

where:

β, φ -flat angle coordinates (declination and azimuth), deg;

$i_{0,\lambda,\max}, i_{0,\lambda,\min}$ - maximum and minimum monochromatic directional (normal) radiation intensity, $W/(m^2 \, m \, sr)$;

$L_{0,\lambda,\max}, L_{0,\lambda,\min}$ - maximum and minimum monochromatic entropy of directional radiation intensity, $W/(m^2 \, m \, sr \, K)$.

In all these methods mentioned above, and in many other later derivation methods, the mathematical definition (2.45) of exergy formulated for a substance was not applied for radiation. The interpretation of variables appearing in the definition was not obvious for radiation, nor was the version (components of exergy to be included—physical, chemical, kinetic, potential, etc.) of the substance formula to be selected for interpreting radiation. The problem of corresponding variables was already discussed in Section 5.9.

Now, when the shape of the formula of exergy of black radiation is already known without doubt, it is possible to discuss another derivation method, shown by Petela (1974), based on the analogy to the definition formula for the thermal exergy B of a substance:

$$B = H - H_0 - T_0 (S - S_0) \qquad (8.10)$$

where H, S, and H_0, S_0 are, respectively, the enthalpies and entropies of the considered substance (H, S) currently, and (H_0, S_0) in the case of equilibrium with an environment at temperature T_0. The successful interpretation of the analogy between the substance and the radiation discloses that the substance enthalpy corresponds to the radiosity j, and the substance entropy corresponds to the radiosity entropy s_j. The interpretation of B for the radiation exergy b ($B \rightarrow b$), is based on the following analogies:

$$H \to j = \int\limits_{\beta} \int\limits_{\varphi} \int\limits_{\lambda} (i_{0,\lambda,\max} + i_{0,\lambda,\min})_T \, d^2C \, d\lambda \tag{8.11}$$

$$H_0 \to j_0 = \int\limits_{\beta} \int\limits_{\varphi} \int\limits_{\lambda} (i_{0,\lambda,\max} + i_{0,\lambda,\min})_{T_0} \, d^2C \, d\lambda \tag{8.12}$$

$$S \to s_j = \int\limits_{\beta} \int\limits_{\varphi} \int\limits_{\lambda} (L_{0,\lambda,\max} + L_{0,\lambda,\min})_T \, d^2C \, d\lambda \tag{8.13}$$

$$S_0 \to s_{j,0} = \int\limits_{\beta} \int\limits_{\varphi} \int\limits_{\lambda} (L_{0,\lambda,\max} + L_{0,\lambda,\min})_{T_0} \, d^2C \, d\lambda \tag{8.14}$$

where the abbreviation:

$$d^2C \equiv \cos\beta \sin\beta \, d\beta \, d\varphi \tag{8.15}$$

and where:

j, j_0 - radiosity density of considered radiation and environment, W/m^2;

$s_j, s_{j,0}$ - entropy of radiosity density of considered radiation and the environment, W/(m^2 K);

T, T_0 - absolute temperature of the radiating surface and the environment, K.

For nonpolarized radiation $i_{0,\lambda,\max} = i_{0,\lambda,\min}$, thus $i_{0,\lambda,\max} + i_{0,\lambda,\min} = 2 \times i_{0,\lambda}$. Additionally, $L_{0,\lambda,\max} = L_{0,\lambda,\min}$, thus $L_{0,\lambda,\max} + L_{0,\lambda,\min} = 2 \times L_{0,\lambda}$.

For example d^2C used for the case of surface radiating to the forward hemisphere is:

$$\int\limits_{\beta} \int\limits_{\varphi} d^2C \equiv \int\limits_{\beta=0}^{\beta=\pi/2} \int\limits_{\varphi=0}^{\varphi=2\pi} \cos\beta \sin\beta \, d\beta \, d\varphi = \pi \tag{8.16}$$

Substituting in formula (8.10) the formulae (8.11)–(8.15) and the value (π) of the integral (8.16), the interpretation of the exergy b of a nonpolarized radiation is:

$$B \to b = 2\pi \left\{ \int\limits_{\lambda} (i_{0,\lambda})_T \, d\lambda - \int\limits_{\lambda} (i_{0,\lambda})_{T_0} \, d\lambda \right.$$

$$\left. - T_0 \left[\int\limits_{\lambda} (L_{0,\lambda})_T \, d\lambda - \int\limits_{\lambda} (L_{0,\lambda})_{T_0} \, d\lambda \right] \right\} \tag{8.17}$$

Formula (8.17) can be rearranged to a form identical to formula (8.9), obtained by consideration of the exergy balance of the elements of the radiating surface. Equation (8.17) confirms that the formulae for the exergy of radiation can be derived from the formulae for the exergy of a substance. Additionally, the equation shows how the analogy between the substance and the radiation allows for derivation of the verified formula for the exergy of arbitrary (i.e., with nonregular spectrum) radiation. Studies have pointed out an analogy between such variables as the specific energy of a substance (related to the unit of the substance) and the density of radiation related to the units of the surface area. However, such a derivation, derived after a long time from other methods, has mainly didactic significance because it confirms the general possibility of interpreting the variables of formula (8.10) for eventual application to matter other than substance or radiation.

The general formula (8.17) for exergy of arbitrary radiation is especially convenient for analyzing the radiation spectrum. Because formula (8.17) can be applied to any arbitrary case of radiation, it can also be applied to different ranges of wavelengths. In particular, equation (8.17) can be used to determine the monochromatic exergy, which was discussed by Candau (2003).

8.5 Exergy Radiation Spectrum of a Surface

8.5.1 Spectrum of a Black Surface

Formula (8.17) is convenient to determine the exergy of the monochromatic radiation intensity. Any black radiation at the environment temperature T_0 is in thermodynamic equilibrium with the environment, regardless of the diversified values of emissivities of the environment surfaces, which all have temperature T_0. Therefore, the reference state for determination of the exergy of radiation is the black radiation at the environment temperature T_0.

Practical observations confirm that beside the fluctuation of the environment emissivity, the variation of the solid angle or the wavelength range of propagating radiation of the environment radiation at temperature T_0, can never be utilized in practice for obtaining useful work, which is the measure of the exergy value. Such statements determine a freedom of assumptions about all other parameters of the environment except the environment temperature T_0. This is in accordance with the fact that the exergy is one of the thermodynamic functions of the instant state, and such a state can be defined only by the instant thermodynamic parameters, without the need of using any geometric or other nonthermodynamic properties.

Based on formula (8.17) the exergy $b_{b,\omega,\lambda}$ of black radiation propagating within an elemental solid angle $d\omega$ and within a wavelength range $d\lambda$ (monochromatic) can be derived. For this purpose, the exergy b of equation (8.17), taken for black radiation, $b = b_b$, can be related to the unit solid angle of a hemisphere ($b_{b,\omega} = b_b/2\pi$), and the exergy $b_{b,\omega}$ can be interpreted for the elemental wavelength range $d\lambda$. Thus the exergy spectral component $b_{b,\omega,\lambda}$ is:

$$b_{b,\omega,\lambda} = \frac{\partial b_{b,\omega}}{d\lambda} \tag{8.18}$$

and from equation (8.17):

$$b_{b,\omega,\lambda} = (i_{b,0,\lambda})_T - (i_{b,0,\lambda})_{T_0} - T_0\{[L_{b,0,\lambda}\,(i_{b,0,\lambda})]_T - [L_{b,0,\lambda}\,(i_{b,0,\lambda})]_{T_0}\} \tag{8.19}$$

where $i_{b,\omega,\lambda}$ is the directional intensity of black monochromatic emission and $L_{b,0,\lambda}$ is the respective entropy of monochromatic radiation intensity. The formula for monochromatic radiation exergy was introduced by Candau (2003), and applied later by Chu and Liu (2009); however, the spectrum diagrams have not been considered for perfectly gray surfaces.

The black radiation exergy/energy ratio $\psi = b_b/e_b$, expressed by formula (6.22), is:

$$\psi = 1 + \frac{1}{3}\left(\frac{T_0}{T}\right)^4 - \frac{4}{3}\frac{T_0}{T} \tag{8.20}$$

and can be analogously used for the monochromatic exergy/energy ratio:

$$\psi_\lambda = \frac{b_{b,\omega,\lambda}}{i_{b,0,\lambda}} \tag{8.21}$$

For comparison, Figure 8.4 shows the calculation results of the radiation spectra of energy (dashed line) and exergy (solid line) for the temperature $T = 1000$ K. In comparison to the energy spectrum, the spectrum curve for exergy has a similar shape; however, it represents smaller values. The ratio ψ_λ (double dotted line) determined from formula (8.21) is monotonically decreasing with the growing wavelength. Figure 8.4 shows also the constant value of ratio ψ (dotted line) defined by formula (8.20).

The wavelength $\lambda_{max,B}$, which corresponds to the maximum value of $b_{b,\omega,\lambda,max}$, is smaller than the respective wavelength $\lambda_{max,E}$ for which the maximum energy value $i_{b,\omega,\lambda,max}$ appears, as presented in Figure 8.5, e.g., for three different temperatures (1100, 1700, and 2000 K). In comparison to the energetic spectra, the maxima of the exergetic

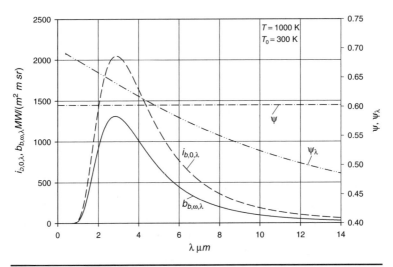

FIGURE 8.4 Comparison of energy emission, exergy emission, and exergy/energy ratios for a black surface at $T = 1000$ K and $T_0 = 300$ K (from Petela, 2010).

spectra at the same temperature, although smaller, appear displaced toward the smaller wavelength. This could mean that based on the exergy interpretation, the radiation at smaller wavelengths (or larger frequencies) is more valuable.

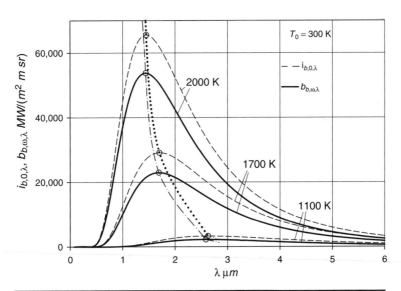

FIGURE 8.5 Different maxima of energy and exergy black spectra (from Petela, 2010).

In contrast to Wien's displacement law (3.18):

$$\lambda_{max,E} T = 2.8976 \times 10^{-3} \quad \text{m K} \tag{8.22}$$

the product of the temperature T and the wavelength $\lambda_{B,max}$ is not constant; it depends on temperature and is smaller than the respective product for the energy spectrum:

$$T\lambda_{max,B} < T\lambda_{max,E} \tag{8.23}$$

The ratio $\lambda_{max,B}/\lambda_{max,E}$ grows with temperature T and tends to 1 (for $T \to \infty$), for which the energy and exergy spectra overlap each other. For example, Figure 8.6 shows the ratio values for the temperature range from 600 to 6000 K. However, Figure 8.7 for the temperature range $T < 350$ K, shows that the ratio has the singular point of zero for $T = T_0 = 300$ K because for $T = T_0$, regardless of the wavelength, the exergy spectrum is always zero.

In comparison to Figure 8.4, Figures 8.8 and 8.9 present two examples of high (1500 K) and low (350 K) temperatures T, respectively. With growing temperature T (Figure 8.8) the energy and exergy spectra tend to overlap each other. However, with decreasing temperature T and approaching T_0 (Figure 8.9), the exergy spectrum gradually disappears.

The energy and exergy spectra of radiation for $T < T_0$ are shown in Figures 8.10 and 8.11. They show the comparison of diagrams for

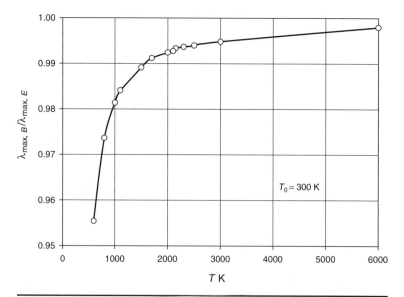

FIGURE 8.6 Ratio $\lambda_{max,B}/\lambda_{max,E}$, as a function of temperature T in the range from 600 to 6000 K (from Petela, 2010).

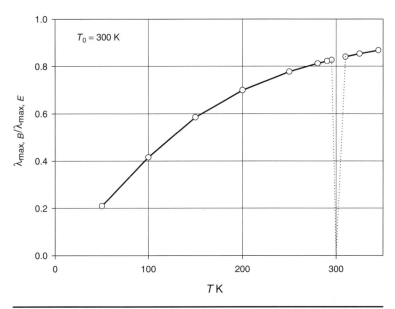

FIGURE 8.7 Ratio $\lambda_{max,B}/\lambda_{max,E}$, as a function of temperature T smaller than 350 K (from Petela, 2010).

the two different temperatures T (200 K and 100 K, respectively). It can be observed that with decreasing T the components of the energy spectrum decrease, whereas on the contrary, the exergy spectrum components increase.

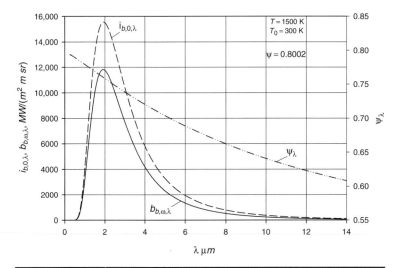

FIGURE 8.8 Comparison of energy emission, exergy emission, and exergy/energy ratios for a black surface at $T = 1500$ K and $T_0 = 300$ K (from Petela, 2010).

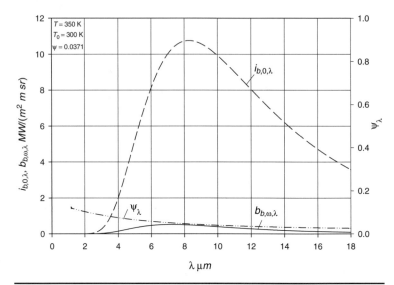

Figure 8.9 Comparison of emission, exergy of emission, and exergy/emission ratios for a black surface at $T = 350$ K and $T_0 = 300$K (from Petela, 2010).

Based on Figures 8.4 and 8.8–8.11, the comparison of the exergy/energy ratio ψ, determined by formula (8.20), with the monochromatic exergy/energy ratio ψ_λ, determined by formula (8.21), is possible. The ratio ψ is constant for a given temperature T; however, the

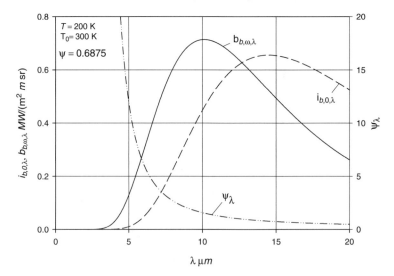

Figure 8.10 Comparison of energy emission, exergy emission, and exergy/energy ratios for a black surface at $T = 200$ K and $T_0 = 300$K (from Petela, 2010).

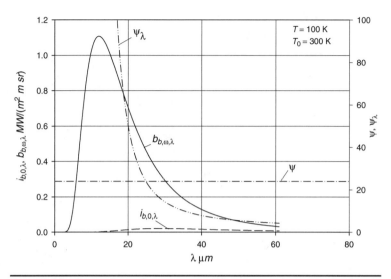

FIGURE 8.11 Comparison of energy emission, exergy emission, and exergy/energy ratios for a black surface at $T = 100$ K and $T_0 = 300$ K (from Petela, 2010).

monochromatic ratio ψ_λ varies significantly and decreases with the growing wavelength λ.

In comparison to the presented model spectra, the real curves of energy and exergy radiation spectra would not be so smooth and regular. For example, the presented spectra of the perfectly black and gray surfaces could be used as a theoretical comparative basis for analyses of the results obtained from the Simple Model of the Atmospheric Radiative Transfer of Sunshine (SMARTS), formulated and successively improved by Gueymard (2008) for calculation of the sky spectral irradiances. Chu and Liu (2009) applied SMARTS for calculation of the exergy spectra of terrestrial solar radiation.

8.5.2 Spectrum of a Gray Surface

Formula (8.19) can be applied for determination of the monochromatic exergy $b_{\omega,\lambda}$ of the gray surface emission. The energetic emissivity ε is used for determination of the energy and entropy, according to equation (8.3), and the appropriately interpreted formula (8.19) is:

$$b_{\omega,\lambda} = (\varepsilon i_{b,0,\lambda})_T - (i_{b,0,\lambda})_{T_0} - T_0 \left\{ [L_{b,0,\lambda} (\varepsilon i_{b,0,\lambda})]_T - [L_{b,0,\lambda} (i_{b,0,\lambda})]_{T_0} \right\}$$

$$(8.24)$$

The terms of the environment reference for energy and entropy (being a function only of T_0), in formulae (8.24) do not need any

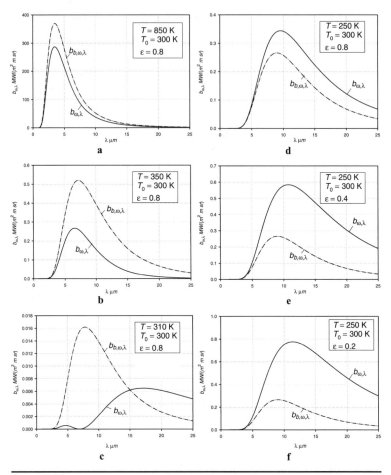

FIGURE 8.12 Exergy spectra for different T and ε (from Petela, 2010).

adjustment with ε, because, as mentioned before, the radiation exergy does not depend on the environment emissivity.

Some peculiarities disclosed by exergetic interpretation of the exergy emission spectra can be discussed. Based on the calculation with formula (8.24), the black (dotted line) and gray (solid line) exergy spectra are shown in Figures 8.12a–d for emissivity $\varepsilon = 0.8$ and for the four different surface temperatures T, diminishing from 800 through 350 and 310 to 250 K.

Figures 8.12b–d illustrate the exergetic spectra of a gray surface for temperature T, which is close to the environment temperature T_0 (T changes from 350 to 250 K). Figure 8.12c for temperature $T = 310$ K shows the two local maxima of the spectrum $b_{\omega,\lambda}$; one (left) for the wavelength around 5 μm and another (right) for the larger wavelength—about 15.5 μm. With increasing temperature T (350 K), Figure 8.12b, the right maximum gradually disappears and only one (left) remains and grows, whereas with decreasing temperature T (250 K), Figure 8.12d, the left maximum disappears and the right one grows even above the values for the black surface spectrum. As also results from Figures 8.12b–d, in contrast to the gray spectrum, the black spectrum $b_{b,\omega,\lambda}$ discloses always only one maximum.

For temperatures $T < T_0$ the gray surface spectrum occurs above the spectrum for the black surface at the respective temperature T.

Figures 8.12d–f illustrate the influence of decreasing surface emissivity ε (from 0.8 to 0.2) at the same surface temperature $T = 250$ K. In contrast to the unchanged spectrum of the black surface, the spectrum of the gray surface is larger with smaller values of emissivity.

It can be observed that the diminishing ε corresponds to the diminishing surface ability to emit radiation, which is similar to the case of extreme "cold" radiation or of an "empty tank" discussed in Section 6.3.

8.5.3 Exergetic Emissivity

As shown in Section 8.5.2, the exergetic spectrum of a gray surface differs significantly from the exergetic spectrum of a black surface. Analogously to the entropic emissivity, the monochromatic exergetic emissivity ε_B can be introduced:

$$\varepsilon_{B,\lambda} = \frac{b_{\omega,\lambda}}{b_{b,\omega,\lambda}} \qquad (8.25)$$

The average exergetic emissivity ε_B for the whole spectrum, in the wavelength range from 0 to ∞, can be calculated as the ratio of the areas under the exergetic spectrum curve for the considered gray surface, equation (8.24), and under the exergetic spectrum curve for the black surface, equation (8.19):

$$\varepsilon_B = \frac{\left(\int\limits_0^\infty b_{\omega,\lambda}d\lambda\right)_{T,\varepsilon}}{\left(\int\limits_0^\infty b_{b,\omega,\lambda}d\lambda\right)_T} \qquad (8.26)$$

Similarly to entropic emissivity, both integrals in formula (8.26) can be solved analytically or determined graphically as the surface areas under the respective spectra curves. With the use of formula (8.26) in the numerical calculations of both the integrals (for the wavelength range from 0.01 nm to 10 μm) the example of obtained results are shown in Figure 8.13. With the surface temperature T growing

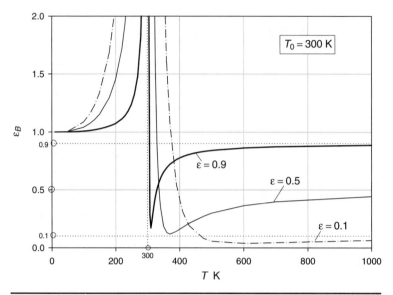

FIGURE 8.13   Some exemplary values of exergetic emissivity ε_B as a function of temperature T and energetic emissivity ε (from Petela, 2010).

from zero, the exergetic emissivity ε_B grows initially rapidly from value 1 to the values significantly larger from 1 (probably infinity) for $T = T_0$. However, for the temperature T growing further from values $T = T_0$, the exergetic emissivity decreases again rapidly, passes the minimum, and then approaches gradually the value of energetic emissivity ε.

Based on the value of the exergetic emissivity ε_B the exergetic spectrum $b_{\varepsilon B}$ of the considered surface can be determined:

$$b_{\varepsilon_B} = \varepsilon_B b_b \qquad (8.27)$$

In spite of a dramatic variation of exergetic emissivity ε_B, the exergetic spectrum of the radiating gray surface varies smoothly as shown in Figure 8.14. With growing temperature T the exergy $b_{\varepsilon B}$, from the finite value $\sim$153 W/m^2, decreases, passes the minimum, and then grows significantly. The minimum value for the gray surface occurs at certain temperature which is the larger the smaller is the energetic emissivity ε. It is noteworthy that the values of $b_{\varepsilon B}$ for $T = T_0$ are indefinite, because, as results from formula (8.26), $b_b = 0$, and as shown in Figure 8.13, ε_B is infinity.

Example 8.1   Application of exergy to the exergetic spectrum of the radiating surface can be illustrated by the following example. Emission of the element of the surface at temperature $T = 420$ K and emissivity $\varepsilon_\lambda = 0.6$ ($\varepsilon_\lambda = \varepsilon$) is considered

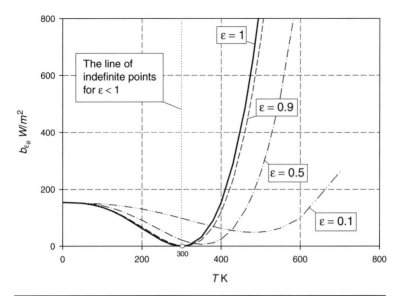

FIGURE 8.14 The gray surface exergy of radiation as a function of temperature T and energetic emissivity ε (from Petela, 2010).

within the wavelength range $d\lambda$ at $\lambda = 1\,\mu m$. Environment temperature $T_0 = 300$ K.

The comparable basis for the main results of calculation can be the monochromatic normal directional intensity $i_{b,0,\lambda}$ for linearly polarized black radiation of the surface, expressed by formula (8.1). Formula (8.1) is used for calculations of the intensities for T and T_0:

$$(i_{b,0,\lambda})_T \equiv i = \frac{(2.9979 \times 10^8)^2 \times 6.625 \times 10^{-34}}{(1 \times 10^{-6})^5} \frac{1}{\exp\left(\frac{2.9979 \times 10^8 \times 6.625 \times 10^{-34}}{1.3805 \times 10^{-23} \times 1 \times 10^{-6} \times 420}\right) - 1}$$

$$= 0.0791 \quad W/(m^3 sr)$$

$$(i_{b,0,\lambda})_{T_0} \equiv i_0 = 1.939 \times 10^{-4} \quad W/(m^3 sr)$$

According to formula (8.4) the entropy $L_{b,0,\lambda}$ of monochromatic intensity of linearly polarized radiation black radiation of the surface is:

$$(L_{b,0,\lambda})_T \equiv L = \frac{c_0 k}{\lambda^4}[(1 + Y)\ln(1 + Y) - Y \ln Y \tag{i}$$

where

$$Y = \frac{\lambda^5 i}{c_0^2 h} \tag{ii}$$

or for temperature T_0:

$$(L_{b,0,\lambda})_{T_0} \equiv L_0 = \frac{c_0 k}{\lambda^4}[(1 + Y_0)\ln(1 + Y_0) - Y_0 \ln Y_0] \tag{iii}$$

where:

$$Y_0 = \frac{\lambda^5 i_0}{c_0^2 h} \tag{iv}$$

Substituting in equation (ii):

$$Y = \frac{(1 \times 10^{-6})^5 \times 0.0791}{(2.9979 \times 10^8)^2 \times 6.625 \times 10^{-34}} = 1.328 \times 10^{-15} \; 1/\text{sr}$$

and substituting to equation (i):

$$L_{b,0,\lambda} \equiv L = \frac{2.9979 \times 10^8 \times 1.3805 \times 10^{-23}}{(1 \times 10^{-6})^4}[(1 + 1.328 \times 10^{-15})$$

$$\ln(1 + 1.328 \times 10^{-15}) - 1.328 \times 10^{-15} \ln 1.328 \times 10^{-15}] = 1.939 \times 10^{-4} \; \text{W/(m}^3\text{K sr)}$$

from equation (iv):

$$Y_0 = 1.4889 \times 10^{-21} \; 1/\text{sr}$$

and from equation (iii):

$$L_0 = 2.955 \times 10^{-10} \; \text{W/(m}^3\text{K sr)}$$

According to the discussion in Section 8.1, the gray surface emits black radiation exergy at the rate $b_{\varepsilon,0,\lambda}$ determined by emissivity ε:

$$b_{\varepsilon,0,\lambda} \equiv b = \varepsilon \left[i - i_0 - T_0 \left(L - L_0 \right) \right] \tag{8.28}$$

whereas the radiation exergy $b_{b,0,\lambda}$ of the considered surface, if the surface was black ($\varepsilon = 1$), is:

$$b_{b,0,\lambda} \equiv b_b = \frac{b}{\varepsilon} \tag{v}$$

Substituting to equation (8.28):

$$b = 0.6 \times [0.0791 - 8.865 \times 10^{-8} - 300 \times (1.939 \times 10^{-4} - 2.955 \times 10^{-10})]$$
$$= 0.0126 \; \text{W/(m}^3\text{sr)}$$

Substituting to equation (v):

$$b_b = \frac{0.0126}{0.6} = 0.0210 \; \text{W/(m}^3\text{sr)}$$

The obtained results can now be used for the main calculations. The monochromatic exergetic spectrum component for $\lambda = 1 \, \mu\text{m}$ of the considered surface is interpreted for the surface emission determined by ε and for the reference state. which is the black radiation at environment temperature T_0:

$$b_{SP,0,\lambda} \equiv b_{SP} = \varepsilon i - i_0 - T_0(L_{SP} - L_0) \qquad \text{(vi)}$$

where:

$$(L_{SP,0,\lambda})_{T_0} \equiv L_{SP} = \frac{c_0 k}{\lambda^4}[(1 + Y_{SP})\ln(1 + Y_{SP}) - Y_{SP}\ln Y_{SP}] \qquad \text{(vii)}$$

and

$$Y_{SP} = \frac{\lambda^5 \varepsilon i}{c_0^2 h} \qquad \text{(viii)}$$

From equations (vi)–(viii), respectively, we obtain:

$$Y_{SP} = 7.973 \times 10^{-16}\,1/\text{sr}$$
$$L_{SP} = 1.1839 \times 10^{-4}\,\text{W}/(\text{m}^3\text{K sr})$$
$$b_{SP} = 0.0120\,\text{W}/(\text{m}^3\,\text{sr})$$

The exergetic emissivity $\varepsilon_{B,0,\lambda}$ for the considered surface is:

$$\varepsilon_{B,0,\lambda} \equiv \varepsilon_B = \frac{b_{SP}}{b_b} \qquad \text{(ix)}$$

and substituting to equation (ix):

$$\varepsilon_B = \frac{0.012}{0.021} = 0.5706 \, < \, \varepsilon$$

Also for comparison the entropic emissivity $\varepsilon_{S,0,\lambda}$ of the considered surface can be calculated as:

$$\varepsilon_{S,0,\lambda} \equiv \varepsilon_S = \frac{L_{SP}}{L_{b,0,\lambda}} = \frac{1.1839 \times 10^{-4}}{1.939 \times 10^{-4}} = 0.611 > \varepsilon \qquad \text{(x)}$$

The calculation results show, e.g., that the considered surface, at $T = 420\,\text{K}$, at $T_0 = 300\text{K}\,(T > T_0)$, $\lambda = 1\,\mu\text{m}$ and $\varepsilon = 0.6$, emits the radiation exergy larger than the respective exergetic surface spectrum component ($b > b_{SP}$). In comparison to the energetic emissivity the respective entropic emissivity is larger ($\varepsilon_S > \varepsilon$), whereas the respective exergetic emissivity is smaller ($\varepsilon_B < \varepsilon$).

8.6 Application of Exergetic Spectra for Exergy Exchange Calculation

Application of surface exergy spectra, instead of surface radiating products, in the calculation of exchanged radiative exergy can be examined for comparison. The two surfaces exchanging exergy by radiation, analyzed in Section 7.5.3 based on Figure 7.11 with use of the energetic emissivities ε, can be considered now with application of the exergetic emissivities ε_B. Figure 7.11b can be used for the following interpretation:

For surface 1:

$$b_{1,0} = \varepsilon_1 e_{b,1} \left(1 - \frac{T_0}{T_1}\right)$$

$$b_{1,1} = b_{1,0} - b_{1,2}$$

$$b_{1,2} = \varepsilon_{B,1} b_{b,1}$$

$$b_{1,3} = \alpha_2 \varepsilon_1 e_{b,1} \left(1 - \frac{T_0}{T_2}\right)$$

$$b_{1,4} = b_{1,2} - b_{1,3} - b_{1,5}$$

$$b_{1,5} = \rho_2 \varepsilon_{B,1} b_{b,1}$$

$$b_{1,6} = \alpha_1 \rho_2 \varepsilon_1 e_{b,1} \left(1 - \frac{T_0}{T_1}\right)$$

$$b_{1,7} = b_{1,5} - b_{1,6} - b_{1,8}$$

$$b_{1,8} = \rho_1 \rho_2 \varepsilon_{B,1} b_{b,1}$$

$$b_{1,9} = \alpha_2 \rho_1 \rho_2 \varepsilon_1 e_{b,1} \left(1 - \frac{T_0}{T_2}\right)$$

$$b_{1,10} = b_{1,8} - b_{1,9} - b_{1,11}$$

$$b_{1,11} = \rho_1 \rho_2^2 \varepsilon_{B,1} b_{b,1}$$

$$b_{1,12} = \alpha_1 \rho_1 \rho_2^2 \varepsilon_1 e_{b,1} \left(1 - \frac{T_0}{T_1}\right)$$

$$b_{1,13} = b_{1,11} - b_{1,12} - b_{1,14}$$

$$b_{1,14} = \rho_1^2 \rho_2^2 \varepsilon_{B,1} b_{b,1}$$

etc., where $\varepsilon_{B,1}$ is the exergetic emissivity of surface 1, and according to formula (8.26):

$$\varepsilon_{B,1} = \left(\frac{\int_0^\infty b_{\omega,\lambda} d\lambda}{\int_0^\infty b_{b,\omega,\lambda} d\lambda}\right)_{T_1} \tag{a}$$

The emission exergy $b_{b,1}$ of a black surface at temperature T_1 is determined from formula (h) in Section 7.5.3.3.

For the exergy radiation of surface 2:

$$b_{2,0} = \varepsilon_2 e_{b,2} \left(1 - \frac{T_0}{T_2}\right)$$

$$b_{2,1} = b_{2,0} - b_{2,2}$$

$$b_{2,2} = \varepsilon_{B,2} b_{b,2}$$

$$b_{2,3} = \alpha_1 \varepsilon_2 e_{b,2} \left(1 - \frac{T_0}{T_1}\right)$$

$$b_{2,4} = b_{2,2} - b_{2,3} - b_{2,5}$$

$$b_{2,5} = \rho_1 \varepsilon_{B,2} b_{b,2}$$

$$b_{2,6} = \alpha_2 \rho_1 \varepsilon_2 e_{b,2} \left(1 - \frac{T_0}{T_2}\right)$$

$$b_{2,7} = b_{2,5} - b_{2,6} - b_{2,8}$$

$$b_{2,8} = \rho_2 \rho_1 \varepsilon_{B,2} b_{b,2}$$

$$b_{2,9} = \alpha_1 \rho_2 \rho_1 \varepsilon_2 e_{b,2} \left(1 - \frac{T_0}{T_1}\right)$$

$$b_{2,10} = b_{2,8} - b_{2,9} - b_{2,11}$$

$$b_{2,11} = \rho_1^2 \rho_2 \varepsilon_{B,2} b_{b,2}$$

etc., where the $\varepsilon_{B,2}$ is the exergetic emissivity of surface 2, and according to formula (8.26):

$$\varepsilon_{B,2} = \left(\frac{\int_0^\infty b_{\omega,\lambda} d\lambda}{\int_0^\infty b_{b,\omega,\lambda} d\lambda}\right)_{T_2} \tag{b}$$

The emission exergy $b_{b,2}$ of a black surface at temperature T_2 is determined from formula (i), Section 7.5.3.3.

The portions of the radiation exergy of surface 1 delivered to surface 2 are:

$$b_{SP,1} = (b_{1,2} - b_{1,5}) + (b_{1,8} - b_{1,11}) + \cdots$$

$$= \alpha_2 \, \varepsilon_{B,1} b_{b,1} \left(1 + \rho_1 \rho_2 + \rho_1^2 \rho_2^2 + \cdots\right) = \frac{\alpha_2 \varepsilon_{B,1} b_{b,1}}{1 - \rho_1 \rho_2} \tag{8.29}$$

The portions of the radiation exergy of surface 2 delivered to surface 1 are:

$$b_{SP,2} = (b_{2,2} - b_{2,5}) + (b_{2,8} - b_{2,11}) + \cdots$$

$$= \alpha_1 \, \varepsilon_{B,2} b_{b,2} \left(1 + \rho_1 \rho_2 + \rho_1^2 \rho_2^2 + \cdots\right) = \frac{\alpha_1 \varepsilon_{B,2} b_{b,2}}{1 - \rho_1 \rho_2} \tag{8.30}$$

The net radiation exergy $b_{SP,1-2}$ transferred from surface 1 to surface 2 is:

$$b_{SP,1-2} = b_{SP,1} - b_{SP,2} \tag{8.31}$$

Substituting to equation (8.31) equations (h) and (i) from Section 7.5.3.3, (8.29) and (8.30), as well as taking into account relations at

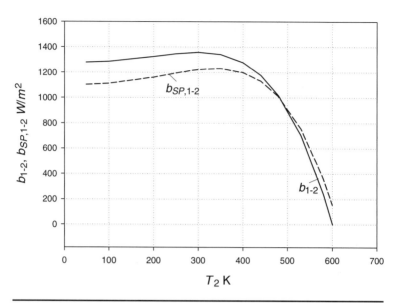

FIGURE 8.15 Comparison of exchanged radiation exergy for $T_1 = 600$ K, $T_0 = 300$ K, $\varepsilon_1 = 0.8$, $\varepsilon_2 = 0.6$, and $\varepsilon_{B,1} = 0.7272$, for varying values of temperature T_2 and appropriate emissivity $\varepsilon_{B,2}$.

$\alpha + \rho = 1$ and $\alpha = \rho$, the net radiation exergy $b_{SP,12}$ determined with exergetic emissivities is:

$$b_{SP,1-2} = \frac{\varepsilon_2\varepsilon_{B,1}\, b_{b,1} - \varepsilon_1\varepsilon_{B,2}b_{b,2}}{\varepsilon_1 + \varepsilon_2 - \varepsilon_1\varepsilon_2} \tag{8.32}$$

Obviously if $\varepsilon_{B,1} = \varepsilon_1$ and $\varepsilon_{B,2} = \varepsilon_2$, then equation (8.32) becomes like (7.77). But, e.g., if only $\varepsilon_{B,1} = 1$, then for $T >> T_0$:

$$b_{SP,1-2} = \varepsilon_2 b_{b,1} - \varepsilon_{B,2}b_{b,2} > \varepsilon_2\,(b_{b,1} - b_{b,2}) \tag{8.33}$$

The comparison of the radiative exergy exchange determined by exergetic emissivities, according to formula (8.33), and correctly determined by the energetic emissivities, according to formula (7.77), is illustrated in Figure 8.15. The following calculations were used $T_1 = 600$ K, $T_0 = 300$ K, $\varepsilon_1 = 0.8$, and $\varepsilon_2 = 0.6$. Determination of exergetic emissivity is based on the numerical calculation of the surface areas under the curves for the gray and black spectra. The exergetic emissivity $\varepsilon_{B,1} = 0.7272$ for surface 1 is determined based on formula (a) at $T_1 = 600$ K. The values of emissivities $\varepsilon_{B,2}$ are determined based on formula (b) for varying T_2. For the relatively small surface temperatures the exchanged exergy $b_{SP,1-2}$ can be slightly smaller or larger than exchanged exergy b_{1-2}.

However, as shown in Figure 8.16, for the larger surface temperatures, the difference between the values of $b_{SP,1-2}$ and b_{1-2} becomes negligible. In calculation for Figure 8.16, $T_2 = 500$ K, $T_0 = 300$ K,

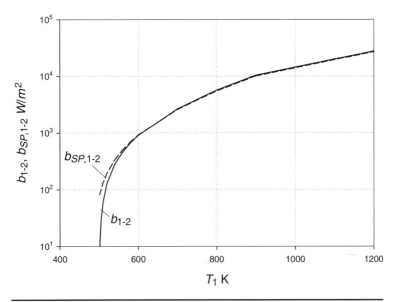

FIGURE 8.16 Comparison of exchanged radiation exergy for $T_2 = 500$ K, $T_0 = 300$ K, $\varepsilon_1 = 0.8$, $\varepsilon_2 = 0.6$, and $\varepsilon_{B,2} = 0.4149$, for varying values of temperature T_1 and appropriate emissivity $\varepsilon_{B,1}$.

$\varepsilon_1 = 0.8$, and $\varepsilon_2 = 0.6$. The exergetic emissivity $\varepsilon_{B,2} = 0.4149$ for surface 2 is determined based on formula (b) at $T_2 = 500$ K. The values of emissivities $\varepsilon_{B,1}$ are determined based on formula (a) for varying T_1.

8.7 Conclusion

The topics considered in Chapter 8 contribute to the theory of thermal radiation of a surface. In this chapter, the interpretative concept of exergy was applied to emphasizing and distinguishing the sensible exergetic spectrum of the emitting surface from the spectrum exergy of the emitted product (photon gas). Consequently, the entropic spectrum was also applied. The entropic and exergetic spectra were analyzed in comparison to the common energetic spectrum.

This method, based on the analogy between substance and radiation, was shown for the derivation of the radiation exergy formula, which is convenient for spectra analyses. In comparison to the energy spectra, it was found that the maxima of the exergetic spectra are smaller and displaced toward the smaller wavelengths, i.e., the larger frequencies.

Following discussion of the existing common notion of emissivity, referred to in this book as *energetic emissivity*, both entropic and exergetic emissivities were proposed. For example, it was shown that for surfaces at the same temperature T, the energetic emissivity is not larger than the entropic emissivity ($\varepsilon_S \geq \varepsilon$), whereas for temperature T larger than the environment temperature, the exergetic emissivity is not smaller than the energetic emissivity ($\varepsilon_B \leq \varepsilon$). In the vicinity of T_0, ($T \approx T_0$), some singularities occur and for $T < T_0$ is always $\varepsilon_B \geq \varepsilon$.

The radiation entropy production from a gray surface, determined based on the entropic emissivity ε_S is larger than the radiating entropy determined based on the energetic emissivity ε. The opposite effect is observed in the case of exergy. If excluded from consideration are the low-temperature ranges in the vicinity of the environment temperature and the zero absolute temperature, then for a gray surface the radiation exergy determined based on the exergetic emissivity ε_B is smaller than the exergy determined by using energetic emissivity ε.

It was also emphasized that all the emissivities can be addressed only at the considered surfaces, not at the emission product, which is always black. None of the discussed conclusions have significant meaning or practical applicability; however, they can contribute to better understanding of the surface radiation nature. The discussion of spectra of the perfectly black and gray surfaces can be also used, e.g., as a theoretical comparative basis for analyses of the real-sky spectral irradiances.

Nomenclature for Chapter 8

B	exergy of substance, J
C	abbreviation defined by formula (8.15), sr
c_0	speed of propagation of radiation in vacuum $c_0 = 2.9979 \times 10^8$ m/s
H	enthalpy of substance, J
h	Planck's constant, $h = 6.625 \times 10^{-34}$ J s.
$i_{b,0,\lambda}$	monochromatic normal directional intensity for linearly polarized black radiation propagating within a unit solid angle, dependent on λ, W/(m^2 m sr)
j	radiosity density, W/m^2
k	Boltzmann constant, k = 1.3805×10^{-23} J/K
$L_{b,0,\lambda}$	entropy of normal monochromatic directional intensity for linearly polarized black radiation propagating within unit solid angle and dependent on wavelength λ, W/(m^2 K sr)
S	entropy of substance, J/K
s_j	entropy of radiosity density, W/(m^2 K)
SMARTS	Simple Model of the Atmospheric Radiative Transfer of Sunshine
T	absolute temperature, K
Y	expression in formula (8.4)
Y_ε	expression in formula (8.5)

Greek

β	flat angle, (declination), deg
ε	emissivity of surface
φ	flat angle, (azimuth), deg
λ	wavelength, m
ν	oscillation frequency, 1/s
σ	Boltzmann constant for black radiation, $\sigma = 5.6693 \times 10^{-8}$ W/(m^2 K^4)
ψ	ratio of emission exergy to emission energy

Subscripts

B	exergetic
b	black
E	energetic
j	radiosity
ε	energetic emissivity
λ	wavelength
max	maximum
min	minimum
S	entropic
SP	spectrum
ω	solid angle
0	for $\beta = 0$
0	environment
1, 2	denotation

CHAPTER 9

Discussion of Radiation Exergy Formulae Proposed by Researchers

9.1 Polemic Addressees

In the present chapter we analyze different formulae about exergy proposed by many researchers, which have been selected based on the significance of their contribution, according to Petela (2003). In particular, we review the literature about radiation exergy. At the stage in the book, after the discussion in previous chapters, the reader will be sufficiently prepared to follow critically the presented viewpoints.

It was well observed by Bejan (1997) that any discussion of the efficiency or economics of solar radiation utilization should be based on understanding the potential of the radiation for maximum work performance, and such potential is expressed by the exergy of the thermal radiation. Researchers agree that thermal radiation received from the sun is rich in exergy; however, their quantitative determination of the radiation exergy often differs.

The formula discussed by various researchers is mainly that for the exergy of blackbody emission. This formula basically determines the general concept of radiation exergy, and any other versions for different cases of radiation follow as a result.

To date, researchers have focused mainly on three radiation exergy formulae—as derived by Petela (1961a, 1961b), Spanner (1964), and Jeter (1981). The following discussion is developed based on the approaches of Spanner and Jeter, as well as the discussions by Bejan (1987, 1997) and Wright et al. (2002). Also addressed are the aspects of radiation exergy raised by some researchers such as Boehm (1986),

Gribik and Osterle (1984, 1986), Landsberg and Tonge (1979), Fraser and Kay (2001), and Badescu (2008), all of whom quote the work of Petela (1964), as well as other researchers who do not, such as Spanner (1964), Press (1976), Parrott (1978, 1979), and Wall (1993).

9.2 What Work Represents Exergy?

The First Law of Thermodynamics, applied to any medium (e.g., a working fluid) undergoing the process within an enclosed system, leads to the equation of energy conservation for a certain time period between the beginning medium state, 1, and the end state, 2. The heat Q_{1-2} delivered to the medium from external sources is spent on raising the internal energy of the medium from U_1 to U_2, and on performing the absolute work W_{1-2}:

$$Q_{1-2} = U_2 - U_1 + W_{1-2} \tag{9.1}$$

The radiation matter can be assumed to be the processed medium. The absolute work W_{1-2} consists of the useful work W_u and the work W_e spent for the "compression of environment":

$$W_{1-2} = W_u + W_e \tag{9.2}$$

where:

$$W_u = \int_1^2 (p - p_0)\, dV \tag{9.3}$$

and

$$W_e = p_0 \, (V_1 - V_0) \tag{9.4}$$

where p is the current radiation pressure and p_0 is the radiation pressure at the environment temperature T_0.

Work W_e is unavailable, whereas work W_u represents the exergy B of the medium at the state 1, $W_u \equiv B$, whenever this work W_u is maximum. This means the change 1–2 in the photon gas occurs at constant entropy, and with $Q_{1-2} = 0$, according to the isentropic process equation (5.26), in which the pressure p of the photon gas is determined by formula (5.21).

Spanner (1964) introduced the concept of *maximum economic efficiency* η_s, in which, instead of using the useful work W_u, he applied absolute work W_{1-2} related to the initial internal energy U_1 of the radiation arriving to the considered leaf:

$$\eta_s = \frac{W_{1-2}}{U_1} = 1 - \frac{4}{3}\frac{T_2}{T_1} \tag{9.5}$$

where T_1 and T_2 are the absolute temperatures of the radiation matter at the beginning and at the end of the process, respectively.

However, if instead of the Spanner's efficiency η_s, one introduces a rather more justified efficiency η' defined with use of the useful work as:

$$\eta' = \frac{W_u}{U_1} \tag{9.6}$$

then the efficiency η' becomes just the exergy/energy radiation ratio ψ, expressed by equation (6.22), where $\psi = b_1/U_1$, and is in agreement with Petela (1961b, 1964):

$$\eta' = 1 + \frac{1}{3} \left(\frac{T_0}{T_1} \right)^4 - \frac{4}{3} \frac{T_0}{T_1} = \psi \tag{9.7}$$

The values of radiation exergy resulting from Petela's and Spanner's formulae can be compared. For example, the exergy of 1 m^3 of initial radiation at temperature T, according to Petela (b_P), is determined by formula (5.29):

$$b_P = \frac{a}{3} \left(3T^4 + T_0^4 - 4\,T_0\,T^3 \right) \tag{9.8}$$

whereas Spanner's result (b_S) is based on equation (9.5), using formula (5.13) for U_r and assuming the volume 1 m^3. Constant $a = 7.564 \times 10^{-16}$ J/(m^3 K^4). After using formula (5.13) and substituting $T_1 = T$ and $T_2 = T_0$, Spanner's result becomes as follows:

$$b_S = \frac{a}{3} \left(3T^4 - 4\,T_0\,T^3 \right) \tag{9.9}$$

Equations (9.8) and (9.9) are presented in Figure 9.1 for $T_0 = 300$ K. For the high values of radiation temperature T, both exergy values approach each other, so that $b_S \approx b_P$. For example, for solar radiation ($T \approx 6000$ K), $b_S = b_P = 0.9149$ J/m^3. However, for the lower values of temperature T, as shown in Figure 9.1, both exergy values b_S and b_P differ significantly not only by numbers but also by the algebraic sign—although according to the definition, the exergy should be positive. Both exergy values b_S and b_P reach a minimum at $T = T_o$ however, the minimal values are $b_P = 0$ and $b_S = -2.041 \times 10^{-6}$ J/m^3. For $T \to 0$, they differ: $b_S = 0$ and $b_P > 0$; ($b_P = 2.0423 \times 10^{-6}$ J/m^3).

Petela's equation (9.8) is more justified, because the exergy is measured by the useful work, with $b_P \equiv W_u$. However, Spanner's exergy is expressed by the absolute work, with $b_S \equiv W_{1-2}$, part of which is the unavailable work used for compression of the environment. In practice, using Spanner's equation (9.9), one does not incur any numerical error when evaluating the exergy of radiation at high temperatures. However, using this formula for the low-temperature radiation, the error can be significant, as shown in Figure 9.1.

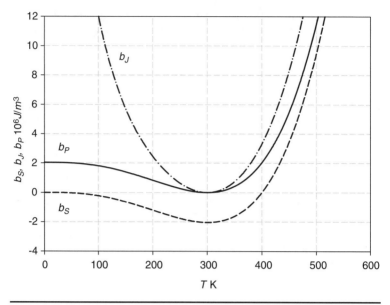

FIGURE 9.1 Comparison of black radiation exergy determined by Spanner (b_S), Petela (b_P), and Jeter (b_J).

9.3 Is Radiation Matter Heat?

In a system with the sun at surface temperature T_S, and the earth's surface at environment temperature T_0, work can be performed. The three theories of solar energy utilization are schematically shown in Figure 9.2. First, the sun and the environment are in direct contact with an ideal heat engine that consumes heat Q_S, rejects Q_{01}, and performs work W_1 at the Carnot efficiency of $\eta_{C1} = 1 - T_0/T_S$, with no exergy loss.

Second, the solar radiation, by its radiation pressure, generates work W_2 with the use of any ideal mechanical engine. The energy degradation, measured by exergy loss determined by equation (6.49), appears during emission of the solar radiation. The energy or exergy efficiency of this radiation-to-work conversion can be estimated, respectively, with the use of equation (6.20) or (6.23).

Third, solar radiation is absorbed at the surface of temperature T_a. An ideal heat engine (i) by direct contact with the surface, consumes heat Q_a, (ii) by direct contact with the environment, rejects heat Q_{03}, and (iii) performs work W_3 at the Carnot efficiency of $\eta_{C3} = 1 - T_0/T_a$. The exergy losses appear during emission and absorption of the solar radiation. These losses can be determined, respectively, by equations (6.53) and (6.70), together with equation (2.60). The energy and exergy efficiency of this radiation-to-heat conversion can be estimated with the use of equations (6.20) and (6.23).

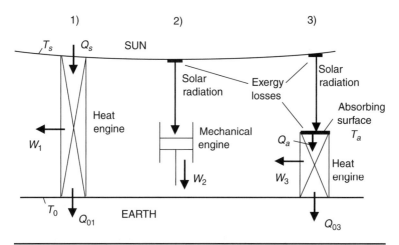

Figure 9.2   Utilization of solar energy (from Petela, 2003).

Jeter (1981), considering utilization of solar energy, arrived at a result corresponding to the first idea discussed above. The efficiency η_{C1} proposed by Jeter for estimation of the solar radiation exergy is unfair because this efficiency expresses the exergy not of the solar radiation but of the heat arriving from the interior of the sun to its surface. In addition, the idea of direct contact of the heat engine with the sun and earth is not realistic.

As shown in Figure 9.2, utilization of heat originating from the sun can be considered according to Jeter (first idea) or according to Petela (third idea). The two ideas are compared based on the more detailed scheme shown in Figure 9.3. The two situations correspond to the utilization of the thermal solar radiation, coming directly to the earth, with use of an ideal heat engine performing at the Carnot efficiency. In Jeter's situation (Figure 9.3a), the engine is supplied with heat q_J at temperature T_S, and this heat is equal to the energy $e_{S\omega}$ of solar emission arriving within the solid angle ω:

$$q_J = e_{S\omega} \tag{9.10}$$

The heat is converted to the work W_J at the Carnot efficiency η_{CS}, which is equal to the Jeter's conversion efficiency η_J:

$$\eta_J = \eta_{CS} = \frac{T_S - T_0}{T_S} = \frac{W_J}{q_J} \tag{9.11}$$

and the heat q_{0J}, at environment temperature T_0, is rejected. Thus, based on equation (9.11), e.g., the exergy of a 1 m^3 enclosed black radiation at temperature T can be expressed according to Jeter as $b_J = W_J = \eta_C \times q_J$. The exergy b_J calculated in such way as function of temperature T is shown in Figure 9.1 for comparison with Spanner's exergy b_S and Petela's exergy b_P. Jeter's values b_J are always positive, the largest ($b_J > b_P > b_S$), and also have the minimum at $T = T_0$.

a) Jeter's

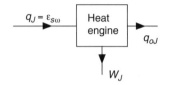

b) Petela's

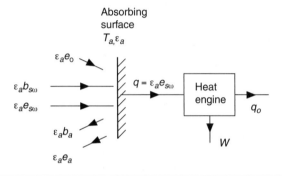

FIGURE 9.3 Comparison of Jeter's and Petela's interpretations of a conversion system (from Petela, 2003).

However, to obtain heat from solar radiation one has to first absorb the radiation and this absorption is not considered in Jeter's case. As indicated in Section 6.6, such absorption alone is impossible, by equation (6.56), without simultaneous emission. However, according to equation (6.70), the conversion of radiation into heat at simultaneous absorption and emission, when $T_S \neq T_a$, is possible although irreversible. The appropriate interpretation by Petela is presented in Figure 9.3b. The solar radiation, at temperature T_S, energy $e_{S\omega}$ and exergy $b_{S\omega}$, is first absorbed at the surface of emissivity ε_a and temperature T_a. Similarly to Figure 10.6 shown later, the absorbing surface receives also energy $\varepsilon_a e_0$ of zero exergy, as well as emits energy $\varepsilon_a e_a$ of exergy $\varepsilon_a b_a$. Then, heat q at temperature T_a, from the absorbing surface, is used in the engine to perform work W, whereas heat q_0 at temperature T_0, is rejected. Again, an ideal engine is assumed for which the Carnot efficiency η_{Ca} is:

$$\eta_{Ca} = \frac{T_a - T_0}{T_a} \qquad (9.12)$$

The exergy and energy conversion efficiencies for the considered system, in Figure 9.3b, according to (6.23) and (6.20), respectively, are:

$$\eta_B = \frac{W}{b_{S\omega}} \qquad (9.13)$$

and

$$\eta_E = \frac{W}{e_{s\omega}} \qquad (9.14)$$

Using data, e.g., from Table 10.1 discussed later, for calculation of solar radiation ($T_S = 6000$ K, $T_a = 350$ K, $T_0 = 300$ K, and $\varepsilon_a = 0.8$), the bands diagrams of exergy and energy balances are shown in Figure 9.4.

a) Jeter's

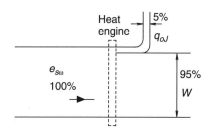

b) Petela's

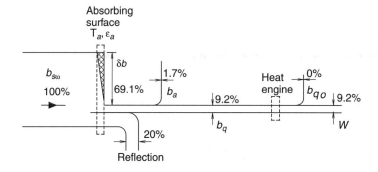

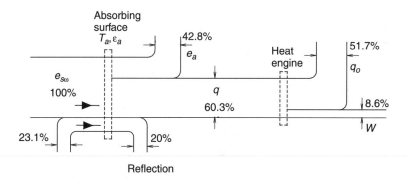

FIGURE 9.4 Jeter's and Petela's energy and exergy flow sheets for conversion systems, $e_{s\omega} = 1.59$ kW/(m² sr), $b_{s\omega} = 1.48$ kW/(m² sr) (from Petela, 2003).

The difference between the efficiencies η_J (Figure 9.4a) and η_B or η_E (Figure 9.4b) are significant (95% versus 9.2% or 8.6%, respectively). The pessimistic result of relatively low efficiencies η_B and η_E can change to the more optimistic result of increased efficiencies if solar radiation is utilized with the use of a lens. However, the difference between the efficiency concepts still remains essential.

9.4 Bejan's Discussion

Bejan (1987, 1997) provides a creative review and discussion of the radiation exergy problems. Regarding the approach by Petela (1964), the two main issues of Bejan are (i) the source of the initial radiation matter in the cylinder–piston system considered by Petela (1964), and (ii) what happens to the radiation matter rejected during expansion in the system. Directed with these issues, Bejan presented extensive illuminating considerations, mostly about the radiation exergy phenomena related to the cylinder–piston cycles.

The attempt by Petela (1961b) was to determine the formula for calculation of the exergy of radiation, based on the assumed definition of exergy. The exergy is a property of matter, like other properties (e.g., enthalpy, internal energy or entropy, etc.), and depends only on its instantaneous state determined by instantaneous parameters (temperature, pressure, etc.).

Wright et al. (2002) correctly stated that the radiation exergy does not depend on its source or fate. Therefore, the above two issues by Bejan should not be understood as a negation of Petela's radiation exergy formulae but rather can be recognized as the initiation of creative discussion of the radiation exergy concept in some interesting circumstances.

Bejan's conclusion is that all three theories (Spanner's, Jeter's, and Petela's) concerning the ideal conversion of thermal radiation into work, although obtaining different results, are correct. However, according to Petela, these three different results on the limiting energy efficiency of utilization of the radiation are correct but for the different and incomparable situations and only the Petela's situation is adequately representing the problem. All of the discussed efficiencies assume work as an output. However, Petela's work is equal to the radiation exergy; Jeter's work is the heat engine cycle work; and Spanner's work is an unavailable absolute work. As input, both Petela and Spanner assume the internal energy of radiation, whereas Jeter assumes heat (Table 9.1). It should be emphasized that heat from the sun is acquired not at the sun temperature but at the temperature of absorbing surface. In fact, from the sun arrives at the earth a photon gas of the exergy which should be accepted as the practical value of the solar radiation resource.

The numerical illustration of the three limiting energy efficiencies, as a function of radiation temperature T, is shown in Figure 9.5 for

Researcher	Input	Output	Unified efficiency expression $(T_2 = T_0)$
Spanner	Internal energy of radiation	Absolute work	$1 - \dfrac{4}{3}\dfrac{T_0}{T}$
Jeter	Heat	Net work of heat engine	$1 - \dfrac{T_0}{T}$
Petela[a]	Internal energy of radiation	Useful work = radiation exergy	$1 - \dfrac{4}{3}\dfrac{T_0}{T} + \dfrac{1}{3}\left(\dfrac{T_0}{T}\right)^4$

[a] In equation (9.69) of Bejan (1997) there is a misprint: "1" should be added to the right-hand side of this equation.

TABLE 9.1 Numerators (Output) and Denominators (Input) of the Limiting Energy Efficiency of Radiation Utilization by Three Different Researchers (from Petela, 2003)

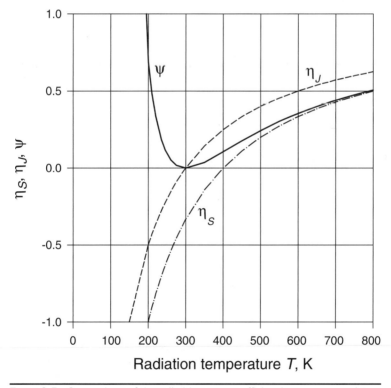

FIGURE 9.5 Comparison of three limiting energy efficiencies: η_J, η_s, and ψ ($T_0 = 300$ K) (from Petela, 2003).

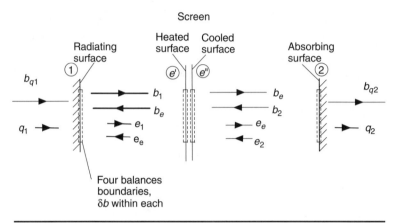

FIGURE 9.6 Scheme of radiation between two surfaces with a screen between them (from Petela, 2003).

comparison. The efficiencies are calculated with η_J from equation (9.11), η_s from equation (9.5) and ψ from equation (9.7).

It should be emphasized that in the analysis of radiation exergy problems, besides the cylinder–piston model, there also can be other models taken into account. To illustrate some possible operations with the radiation exergy concept, the example of radiative heat transfer with the screen between radiating surfaces (not necessarily involving the sun) can be considered according to Figure 9.6. This is a geometrical simple system of parallel planes, 1 and 2, at respective constant temperatures T_1 and T_2, separated with the parallel screen of the two side surfaces, heated (e') and cooled (e''). The screen is very thin so the conduction of heat across the screen can be neglected. Assuming simply that all the considered surfaces are perfectly black, the screen temperature T_e can be calculated from the energy conservation equation ($q_1 = q_2$), where q_1 and q_2 are heat fluxes supplied to surface 1 and rejected from surface 2, respectively. The following equation can be derived:

$$T_e = \sqrt[4]{\frac{T_1^4 + T_2^4}{2}} \qquad (9.15)$$

Heat q_1 can be calculated based on the radiation energy e_1 and e_e, or e_2, determined appropriately from equation (3.22), where $q_1 = e_1 - e_e$. The radiation exergy values b_1, b_e, and b_2 result, respectively, from equation (6.8). The exergy of heat is determined by equation (2.61) and the exergy losses δb, separately, for all four surfaces 1, e', e'', and 2, are from appropriately interpreted equation (2.60). The exemplary calculation results of the energy and exergy balances for the four surfaces are shown in Table 9.2 (for relatively low $T_1 = 1000$ K) and Table 9.3 (for about the sun temperature, $T_1 = 6000$ K). Input and output are

Variable	1. Radiating surface $T_1 = 1000$ K		2. Heated screen surface $T_e = 846$ K		3. Cooled screen surface $T_e = 846$ K		4. Absorbing surface $T_2 = 400$ K	
	+	−	+	−	+	−	+	−
Energy								
Heat (q_1)	100			100	100			100
Emitted radiation		205.25		105.25		105.25		5.25
Absorbed radiation	105.25		205.25		5.25		105.25	
Energy loss								
Total	205.25	205.25	205.25	205.25	105.25	105.25	105.25	105.25
Exergy								
Heat (b_{q1})	100			92.21	92.21			35.71
Emitted radiation		176.72		80.08		80.08		0.79
Absorbed radiation	80.08		176.72		0.79		80.08	
Exergy loss		3.36		4.43		12.92		43.58
Total	180.08	180.08	176.72	176.72	93.00	93.00	80.08	80.08

TABLE 9.2 Items (%) of the Four Energy and Exergy Balance Equations for the Low Temperature $T_1 = 1000$ K ($q_1 = 27.61$ kW/m^2, $b_{q1} = 19.33$ kW/m^2, $T_0 = 300$ K) (from Petela, 2003)

Variable	1. Radiating surface $T_1 = 6000$ K		2. Heated screen surface $T_e = 5045$ K		3. Cooled screen surface $T_e = 5045$ K		4. Absorbing surface $T_2 = 350$ K	
	+	−	+	−	+	−	+	−
Energy								
Heat (q_1)	100			100	100		100	
Emitted radiation		200.00		100.00		100.00		~0
Absorbed radiation	100.00		200.00		~0		100.00	
Energy loss								
Total	200.00	200.00	200.00	200.00	100.00	100.00	100.00	100.00
Exergy								
Heat (b_{q1})	100			99.00	99.00		15.04	
Emitted radiation		196.49		96.92		96.92		~0
Absorbed radiation	96.91		196.49		~0		96.92	
Exergy loss		0.42		0.57		2.08		81.88
Total	196.91	196.91	196.49	196.49	99.00	99.00	96.92	96.92

TABLE 9.3 Items (%) of the Four Energy and Exergy Balance Equations for About the Sun Temperature $T_1 = 6000$ K ($q_1 = 36.72$ MW/m², $b_{q1} = 68.55$ MW/m², $T_0 = 300$ K) (from Petela, 2003)

denoted, respectively, by the signs "+" and "−". Tables 9.2 and 9.3 each illustrate the two different, and both correct, viewpoints on the conversion processes of radiation—the energetic or exergetic views. Again, only the exergy approach demonstrates the degradation of energy.

9.5 Discussion by Wright et al.

The original interpretations and understanding of numerous aspects of heat radiation exergy, as well as derivation of many constructive conclusions, especially for engineers, were presented by Wright et al. (2002). Their work analyzed many doubts and misinterpretations of researchers, whose viewpoint was that application of Petela's exergetic formulae for the conversion of heat radiation fluxes is discussible. They outlined three questions that are the foundation for these misinterpretations. First: How can we define the environment for heat radiation? Second: How is motivated the Petela's assumption that reversible conversion of black radiation is possible? Third: How does the re-radiated emission affect the maximum work obtainable?

Referring to these three questions, Wright et al. (2002) presented the correct definition (agreeable with Petela's) of the environment and its role in considerations of radiation exergy. They perfectly read the intentions of Petela (1961, 1964) regarding determination of the environment for heat radiation exergy. Then they confirmed Petela's intention of formulating the principles of the reversibility of radiation flux conversion, using an ingenious argument based on conversion devices combined with a Carnot heat engine. Accordingly, with Petela, they understood that the re-radiated emission reduced the device's efficiency; however, it did not change the exergy of incident radiation and thus, did not change the limiting efficiency of theoretical utilization of the incident radiation. Their viewpoint was based on resolving fundamental questions such as the following problems: "inherent irreversibility, definition of environment, the effect of inherent emission, and the effect of concentrating source radiation."

Additionally, they rightly raised the necessity to restate any exergy balance equations involving radiative heat transfer, by introducing the available formulae for radiation exergy. In their own way, they explained the meaning of Petela's result for the nonzero value of the radiation flux and for the enclosed radiation when the radiation temperature approaches absolute zero. They evaluated the maximum exergetic efficiency for some different conversion processes.

9.6 Other Authors

The exergy approach to thermal radiation was developed by many other authors. Press (1976) analyzed diluted solar radiation, and for direct (undiluted) sunlight he derived his formula (4) which contains the expression ψ in a form in accordance with Petela's equation (6.22). He rejected Jeter's application of the Carnot efficiency (for solar radiation of temperature about 6000 K and environment temperature about 300 K) in a thermodynamic evaluation of direct sunlight.

Parrott (1978) obtained the obvious disagreement when comparing his formula for solar radiation to the formula by Petela (1964) derived for any direct radiation propagating within a solid angle 2π. Parrott (1979) belongs to the authors who erroneously accept the state function of exergy—to be exact, the radiation exergy/energy ratio—as also being a function of geometrical parameters, e.g., the angle of the cone subtended by the sun's disc. However, using the *availability* which was once one of the names of exergy, he obtained a result confirming the results of Petela (1964).

Landsberg and Tonge (1979) developed the results of some other researchers. They considered the photon density in diluted blackbody radiation by introducing a new function X depending on the dilution factor. Their final results were in agreement with the results by Petela (1964). The emissivity ε used by Petela (1964), they misinterpreted (their page 561) as the dilution factor. For any arbitrary (diluted, indirect) heat radiation, Petela (1962) derived formulae that do not need any emissivity but do require measured data on the radiation spectrum and angle of propagation.

Gribik and Osterle (1984), regarding the first discussion point, correctly Petela's derivation of formula (6.8), valid for any radiation emission from a perfectly gray surface (not only for solar radiation) by (a) calculating the useful work in the cylinder–piston system and (b) applying availability (exergy). However, their argument runs against the assumption that useful work represents the exergy of heat radiation, and as a result, they agreed with a concept of "literally destroyed radiation," which means absorption. Such absorption, without accompanying emission, as proved in Section 6.6, is impossible. They disapproved of Jeter's application of the Carnot efficiency for determination of radiation exergy. Thus, from the arguments on the maximum efficiency of the solar radiation utilization by Petela (1964), Spanner (1964), Parrott (1978, 1979), and Jeter (1981), they approved only of Spanner's result.

Gribik and Osterle (1986), regarding the first discussion point, correctly disagreed that radiation is heat. Radiation is the transport of energy, which can occur even in a vacuum. This is not a feature of other transport systems such as conduction or convection. To exchange heat between a radiation field and a substance the phenomena of emission and absorption have to occur. Both phenomena involve irreversibility, which is manifested only by exergy analysis, whereas energy analysis reveals no energy degradation. Therefore, from an energy viewpoint, radiation is recognized as heat.

Regarding the second discussion point, they do not agree that the practical value (exergy) of radiation should be applied to the human environment, which is filled with radiation at the environment temperature. By referring the radiation value to the temperature of absolute zero, they represented the energy viewpoint, which differs from the exergy viewpoint. They did not take into account that, in contrast to the environment temperature, the absolute temperature is not achievable and thus should not be taken into the consideration of practical value of radiation.

Boehm (1986) pointed out that neglecting the environment compression by Spanner is the cause of the difference between Petela's (1964) and Spanner's (1964) formulae. Boehm recognizes as an apparent flaw Spanner's negative efficiency values for radiation temperatures below the environment temperatures.

Badescu (2008) introduced exergy into the area of nuclear radiation. He considered only blackbody radiation. Based on statistical thermodynamics he discussed Petela's exergy/energy radiation ratio ψ (mistakenly called the *Petela–Landsberg–Press ratio*) as the one resulting from the four possible interpretative solutions of exergy estimation of radiation flux. This one solution confirms the Petela ψ formula; however, without knowing a priori the solution, e.g., based on derivation of Petela's exergy analysis of radiating surfaces, one would not know which of the four equations is the correct solution. Badescu's considerations can be recognized as a creative approach applying the exergy concept to nuclear technology. The rough estimation of exergy efficiency of a nuclear power station was outlined by Szargut and Petela (1965, 1968).

Recently, Fraser and Kay (2001) attempted to introduce the exergy of solar thermal radiation into considerations of ecosystems. Inspired by an environmental protection strategy, Wall (1993) suggested a "consumption tax" for use of nonrenewable energy resources. This pollution tax would be determined based on the exergy value of utilized resources. Obviously, a tax exemption for the use of solar radiation would stimulate its wider utilization.

9.7 Summary

The present chapter has enhanced understanding and accuracy of the exergy analysis of processes involving heat radiation. In the existing literature in this field, the most discussed formulae for heat radiation exergy are those derived by Petela (1961a), Spanner (1964), and Jeter (1981). The discrepancy between formulae by Petela and by Spanner arises because Spanner applied the absolute work instead of the useful work to express the maximum practically available work (exergy). The discrepancy between formulae by Petela and Jeter is because

Petela applied the exergy analysis, whereas Jeter developed the energy analysis in which the degradation of heat to radiation at the same temperature, is not revealed.

Based on analysis of all the available formulae for the exergy of radiation, it was concluded by Petela and some other researchers that for exergy of radiation matter existing at a certain instant, regardless of from where the radiation originated and regardless of what will happen to the radiation in the next instant, **the only justified formulae for estimation of radiation exergy are those derived by Petela for any enclosed photon gas and for any arbitrary radiation flux.**

The commonly known insensitivity of energy analysis to the quality of energy, in contrast to exergy, is one of the disadvantages of energy analysis. Only exergy clearly discloses degradation of energy in the processes of absorption and emission of radiation.

The traditional cylinder–piston model is often used in analyses of various classical thermodynamic problems. However, although the model can be used also in concepts of diversified radiation exergy, more attention needs to be paid to the possibility of using other models that involve the system of radiating surfaces on which the emission and absorption occur. The results for both kinds of models have to agree. Usually, the surface models do not raise doubts and thus can often be used for verification of the results for the cylinder–piston system.

Nomenclature for Chapter 9

B	exergy of radiation, J
b	radiation exergy, J/m^3
e	emission energy, W/m^2
p	radiation pressure or absolute static pressure of gas, Pa
Q	heat, J
q	heat rate, W/m^2
s	entropy of emission, W/(m^2 s)
T	absolute temperature, K
U	internal energy, J
V	volume, m^3
W	work, J or W/m^2

Greek

δb	exergy loss, W/m^2
ε	emissivity
η	efficiency
η'	efficiency in a certain case
ψ	ratio of emission exergy to emission energy
ω	solid angle within sun is seen from the earth, sr

Subscripts

a	absorbing surface
C	Carnot
e	compression of environment, or screen
J	Jeter
P	Petela
q	heat
S	Spanner or sun
u	useful
ω	related to solid angle within sun as seen from the earth
0	environment
$1, 2, 3$	denotations

CHAPTER 10

Thermodynamic Analysis of Heat from the Sun

10.1 Introduction

Solar radiation is the principal energy source for life on earth. This radiation establishes the temperatures of both the atmosphere and the earth's surface. However, human activity changes the conditions of the energy exchange between the sun and earth, and the observed tendency is a gradual increase in these temperatures. In the present chapter, the simplified explanation of the mechanism of such global warming is discussed with some rough quantitative estimation of the accounted parameters. The so-called *greenhouse effect* is described based on the simplified model of a canopy applied to increase the effectiveness of harvesting solar radiation, which, although rich, is much diluted.

The most common devices for utilization of solar radiation are cookers of different types. A simple solar parabolic cooker with the cylindrical shape of a trough is analyzed from the viewpoint of exergy. Supported by calculations, we discuss the methodology of detailed exergy analysis of the cooker, the distribution of exergy losses, and, for the example of the cooker surfaces, explain the general problem of how the exergy loss on any radiating surface should be determined if the surface absorbs radiation fluxes of different temperatures. An imagined surface is used to close the system of the cooker surfaces. Optimization is needed to increase both the energy and exergy efficiencies of the cooker.

Equations are derived for heat transfer between the three surfaces—cooking pot, reflector, and imagined surface making up the system. The mathematical model allows for theoretical estimation of the energy and exergy losses due to unabsorbed insolation, convective and radiative heat transferred to the environment, and, additionally, the estimation of exergy losses due to radiative irreversibilities on the surfaces, as well as the irreversibility of the useful heat transferred to the water.

The exergy efficiency of the cooker is determined to be relatively very low ($\sim$1%), and about ten times smaller than the respective energy efficiency, which is in agreement with experimental data from the literature. The influence of input parameters (e.g., geometrical configuration, emissivities of surfaces, heat transfer coefficients, and temperatures of the water and environment) on output parameters is determined, and the distribution of the energy and exergy losses is described.

Generally, heat from absorbed solar radiation is available at relatively low temperatures, which is comparable to the temperatures of waste heat from many other sources, e.g., exhaust gases from industrial boilers or other combustion installations. In the current world situation with a growing shortage of energy resources, heating by solar radiation, categorized as the recovery of such low-temperature waste heat, is often considered. However, to obtain the economically significant power of solar radiation, the accessibility of relatively large geographical areas exposed to solar radiation is required.

10.2 Global Warming Effect

The atmosphere is a medium with very irregular properties. It absorbs from solar radiation mostly visible wavelength energy (about 29%), whereas from the earth's radiation the atmosphere absorbs mostly infrared radiation (about 92%).

The earth receives energy from the sun by radiation. The earth reflects about one-third of the extraterrestrial solar radiation, and the remaining radiation, assumed to be 100%, arrives at the atmosphere, land, and oceans as shown in the approximate scheme of Figure 10.1a. This radiation absorbed by the atmosphere and the earth, mostly in the visible wavelengths, is re-radiated to space in an equal amount, mostly in the infrared wavelengths, so that the global thermal equilibrium is maintained. To this equilibrium also is contributed the heat exchanged by convection between the atmosphere and earth; however, such convection, for the simplicity of the discussion here, can be neglected.

The received energy can represent different absolute energy (in W) because equilibrium can be maintained at different temperatures on the earth's surface. It is usually estimated that the annual average temperature T_{earth} of the earth's surface is about 14°C (287 K), and at this temperature the earth emits $\sim$192%. The atmosphere, at a certain assumed effective temperature T_{atm}, emits energy to both the earth ($\sim$138%) and space ($\sim$83%).

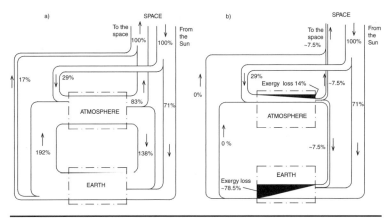

Figure 10.1 Simplified scheme of radiation energy (a) and exergy (b).

The greenhouse effect is applied in gardening. A greenhouse is built of glass walls; the solar radiation penetrating through the glass heats the ground inside, and then the confined air is heated from the ground. Based on this rough analogy, the term *greenhouse effect* is used also for the process in which infrared radiation is exchanged between the atmosphere the earth's surface. These two processes differ because air in a real greenhouse is trapped, whereas air in the environment, when warmed from the ground, rises and mixes with cooler air aloft. Such a difference can be practically demonstrated by opening holes in the greenhouse roof or walls, which will cause air exchange with the environment and thus cause a considerable drop in the air temperature. The analogy can be followed in that the glass roof of the greenhouse traps the infrared radiation to warm the greenhouse air, whereas the thick layer of atmosphere plays the same role for the earth.

It is estimated that in the absence of the greenhouse effect the earth's surface temperature would be decreased from about 14°C to about –19°C. It is believed that the recent warming of the lower atmosphere is the result of enhancing the greenhouse effect by an increase of the amount of gaseous, liquid, and solid ingredients of radiative properties different from such properties of air.

The global warming effect is not effectively described by the concept of exergy because exergy relates to environment temperature with no regard to how high is this temperature. For example, for any value of T_{earth} the exergy radiated from the earth's surface shown in Figure 10.1b will always be zero.

Example 10.1 For a very rough consideration of radiation processes related to 1 m² of the earth's surface, it can be assumed, e.g., that the yearly average solar irradiance $S = 100$ W/m² arrives in the atmosphere, which has transmissivity for solar radiation $\tau_S = 0.71$ (Figure 10.1). The radiation energy $S \times \tau_S$ is absorbed by the black surface of the earth at the yearly average temperature $T_0 = 287.16$ K ($14°$C). The remaining part of the irradiance $S \times (1 - \tau_S)$ is entirely absorbed by the atmosphere. The energy emission $e_0 = \sigma \times T_0^4 = S \times \tau_S$ from the earth's surface arrives at the same atmosphere for which transmissivity for low-temperature radiation $\tau_0 = 0.17$ (Figure 10.1). Thus, the atmosphere absorbs the energy amount S reduced by the amount of the earth's emission transmitted through the atmosphere. The atmosphere can be assumed as a body of emissivity $(1 - \tau_0)$ and emitting radiation to the black sky at temperature T_{sky}. The energy balance equation for the atmosphere is:

$$S - S\tau_S\tau_0 = (1 - \tau_0)\,\sigma \left(T_A^4 - T_{sky}^4\right) \tag{10.1}$$

Assuming that the sky temperature is equal to the environment temperature ($T_0 = T_{sky}$) the resultant temperature T_A of the atmosphere can be determined from formula (10.1) as $T_{atm} = 305.13$ K. The global warming effect can be considered in terms of two characteristic factors describing pollution of the atmosphere, such as the influence of τ_S and τ_0 on the change of environment temperature T_0. Based on equation (10.1) the two partial derivatives can be considered:

$$\frac{\partial T_0}{\partial \tau_S} = \frac{\tau_0}{(1 - \tau_0)} \frac{S}{4\sigma T_0^3} = 3.813 \cdot 10^{-2} \quad \text{K}/\% \tag{10.2}$$

$$\frac{\partial T_0}{\partial \tau_0} = \frac{S\tau_S - \sigma \left(T_A^4 - T_0^4\right)}{4\sigma T_0^3 (1 - \tau_0)} = -8.03 \cdot 10^{-2} \quad \text{K}/\% \tag{10.3}$$

For example, the increase from 0.71 to 0.72 of the transmissivity τ_S of the atmosphere for solar radiation causes, based on formula (10.2), the increase in environment temperature T_0 from 287.16 to 287.198 K.

In another example, based on equation (10.3), if transmissivity τ_0 of the atmosphere for the low-temperature radiation increases from 0.17 to 0.18, then from formula (10.3) the environment temperature T_0 decreases from 287.16 to 287.08 K.

10.3 Effect of a Canopy

The global warming effect often is compared to the effect of a greenhouse in which solar radiation is trapped by using a transparent canopy over the earth's surface absorbing solar radiation.

Approximate analysis of the effect of a canopy stretched above certain surface and screening this surface from direct solar radiation can be carried out for three typical situations presented schematically in Figure 10.2.

A black horizontal plate of surface area A located on the earth's surface can be exposed to direct solar radiation as shown schematically in Figure 10.2a. In the thermodynamic equilibrium state the irradiance S is spent on heat Q extracted at constant plate temperature T_p and on the convective (E_{p-0}) and radiative (E_{p-sky}) heat fluxes from the plate to the surroundings. The plate temperature T_p is controlled by

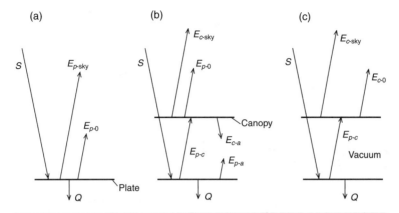

Figure 10.2 The three typical situations considered in a study of the canopy effect.

the appropriately arranged heat Q. The energy balance equation for the plate is:

$$S = Q + E_{p-\text{sky}} + E_{p-0} \tag{10.4}$$

where

$$E_{p-sky} = A\sigma \left(T_p^4 - T_{sky}^4 \right) \tag{10.5}$$

$$E_{p-0} = A h_{p-0} \left(T_p - T_0 \right) \tag{10.6}$$

and where h_{p-0} is the convective heat transfer coefficient. The harvesting of solar energy can be determined by the energetic efficiency η_E:

$$\eta_E = \frac{Q}{S} \tag{10.7}$$

or by the exergetic efficiency η_B:

$$\eta_B = \frac{Q}{\psi_c S} \left(1 - \frac{T_0}{T_p} \right) \tag{10.8}$$

where ψ_c is the exergy/energy ratio discussed in Section 6.5. It was shown there that, for direct radiation of the sun at its surface temperature 6000 K, the theoretical value of the ratio is $\psi = 0.933$. According to Gueymard (2004), the irradiance of the direct solar radiation arriving at the earth is 1366 W/m². Because the irradiance values applied in the present canopy consideration are smaller than the exergy/energy ratio, it can be assumed for a smaller irradiance temperature, e.g., $\psi_c = 0.9$.

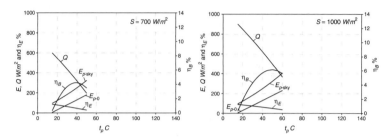

Figure 10.3   Plate exposed to solar radiation in the cases of $S = 700$ W/m^2 (left) and $S = 1000$ W/m^2 (right).

For example, assuming $A = 1$ m^2, $h_{p-0} = 5$ W/(m^2 K), and T_{sky} determined by formula (6.66) in which $T_0 = 287.16$ K (14°C), Figure 10.3 shows the calculation results for the two different values of irradiance, $S = 700$ W/m^2 (left) and $S = 1000$ W/m^2 (right). With increasing plate temperature T_p, the energetic efficiency η_E decreases, whereas the exergetic efficiency η_B has a maximum.

In the second situation, shown in Figure 10.2b, the plate of the surface area 1 m^2 is a fragment of a very large and flat surface of the same uniformly distributed temperature T_p and radiative properties. The plate is screened from solar radiation by a large, flat, and horizontal canopy. Material of the canopy can transmit complete solar radiation to the plate (i.e., canopy transmissivity $\tau_{\text{sol}} = 1$), although low-temperature emission from the plate is entirely absorbed by the canopy (canopy transmissivity $\tau_{\text{pla}} = 0$). The extreme values of these two transmissivities are assumed to show better the effect of the canopy on exchanged radiative heat. Due to the very small thickness of the canopy, both its surfaces—that the one exposed to the sun and the one exposed to the plate—have the same canopy temperature T_c.

In the thermodynamic equilibrium state shown in Figure 10.2b, the irradiance S is spent on heat Q extracted at a constant plate temperature T_p and on the convective (E_{p-a}) and radiative (E_{p-c}) heat fluxes from the plate to the canopy. The plate temperature T_p is controlled by the appropriately arranged heat Q. The canopy temperature T_c is constant for the given plate temperature T_p and is distributed uniformly over the surfaces of the canopy. The energy balance equation for the plate is:

$$S = Q + E_{p-c} + E_{p-a} \tag{10.9}$$

where

$$E_{p-c} = \sigma \left(T_p^4 - T_c^4 \right) \tag{10.10}$$

$$E_{p-a} = A h_{p-a} \left(T_p - T_a \right) \tag{10.11}$$

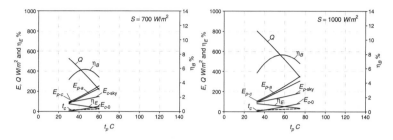

Figure 10.4 Plate under a canopy in the environment air in the cases of $S = 700$ W/m^2 (left) and $S = 1000$ W/m^2 (right).

and where h_{p-a} is the respective convective heat transfer coefficient and T_a is the temperature of air between the plate and the canopy. Simplifying, it is assumed that $T_a = T_0$. The energy balance of the canopy is:

$$E_{p-c} = E_{c-\text{sky}} + E_{c-0} + E_{c-a} \qquad (10.12)$$

where

$$E_{p-a} = E_{c-0} = h_{p-a}\left(T_p - T_0\right) \qquad (10.13)$$

and where $h_{p-a} = h_{c-a}$ are the respective convective heat transfer coefficients. The harvest of solar energy in the considered situation can be determined by the energetic efficiency η_E and exergetic efficiency η_B determined, respectively, from formulae (10.7) and (10.8).

For example, assuming $h_{p-a} = h_{c-a} = 5$ W/(m^2 K), Figure 10.4 shows the calculation results for the two different values of irradiance, $S = 700$ W/m^2 (left) and $S = 1000$ W/m^2 (right). As in situation (a), also in situation (b): with increasing plate temperature T_p the energetic efficiency η_E decreases, whereas the exergetic efficiency η_B has the maximum.

The third possible situation, shown in Figure 10.2c, is the same as the previous situation (b), except that between the plate and canopy is a vacuum; thus, in this space, heat convection does not occur. The energy balance equations for the plate and the canopy are thus:

$$S = Q + E_{p-c} \qquad (10.14)$$

$$E_{p-c} = E_{c-\text{sky}} + E_{c-0} \qquad (10.15)$$

The energetic efficiency η_E and exergetic efficiency η_B are determined, respectively, also from formulae (10.7) and (10.8). Figure 10.5 shows the calculation results for the two different values of irradiance, $S = 700$ W/m^2 (left) and $S = 1000$ W/m^2 (right). As in situation (b),

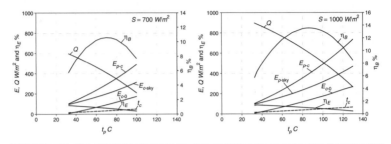

Figure 10.5 Plate under a canopy with a vacuum between the canopy and the plate in the cases of S = 700 W/m² (left) and S = 1000 W/m² (right).

also in situation (c): with increasing plate temperature T_p, the energetic efficiency η_E decreases, whereas the exergetic efficiency η_B has the maximum.

To summarize the comparative discussion of the three models (Figure 10.2): The irradiated black plate (a), the plate under the canopy (b), and the plate under the vacuum and canopy (c) were all considered under simplifying assumptions of extreme values of surface properties to better emphasize the canopy idea. The comparison of Figures 10.3–10.5 illustrates the benefits of applying a canopy to increase the effect of trapping solar radiation. The amount of exergy (practical value) of absorbed heat grows gradually through the three considered situations from (a) to (c).

10.4 Evaluation of Solar Radiation Conversion into Heat

Solar radiation can be converted to heat for various applications including, e.g., cooking, driving a sterling engine, melting, etc. A simple introduction to the potential of solar radiation for heating as well as determination of heat temperature in possible applications can be outlined as follows.

Formula (6.8) can be applied for calculation of solar radiation exergy. However, the model of the two infinite surfaces (Figure 6.1) applied for derivation of formula (6.8) is inadequate for the earth–sun configuration. Any absorbing surface on the earth, in relation to the sun in zenith, can be considered according to the modified scheme shown in Figure 10.6. From the sun, the black ($\varepsilon = 1$) radiation of exergy b_ω, energy e_ω, and entropy s_ω, within the solid angle ω, arrives at the absorbing surface. These three fluxes are absorbed on the earth by the absorbing surface at temperature T_a and emissivity ε_a.

However, as was proven by equation (6.56), absorption alone is impossible. Therefore, one has to take into consideration that the absorbing surface, in the solid angle 2π, emits its own radiation fluxes of exergy b_a, energy e_a, and entropy s_a, and obtains, in the solid angle

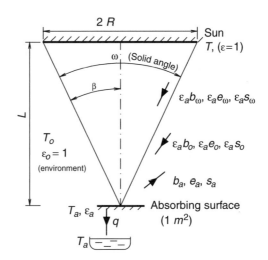

FIGURE 10.6
Scheme of
radiation exchange
between sun and
absorbing surface
on earth (from
Petela, 2003).

$2\pi-\omega$, the radiation fluxes of exergy b_0, energy e_0, and entropy s_0 from the environment at temperature T_0 (assumed to be equal to the sky temperature, $T_{sky} = T_0$) and at assumed emissivity $\varepsilon_0 = 1$. (It was proven previously that the value of the environment emissivity does not affect the results.)

In the present discussion, solar radiation is considered to be non-polarized, uniform, and black at temperature $T = 6000$ K, arriving at the earth within the solid angle ω. Exergy b_ω of such radiation can be calculated from formula (7.50) as follows:

$$b_\omega = \frac{b}{\pi} \int\limits_{\varphi=0}^{\varphi=2\pi} d\varphi \int\limits_{\vartheta=0}^{\beta} \cos\beta \, \sin\beta \, d\beta \qquad \text{(a)}$$

where b is determined from formula (6.8), and β and φ are the angle coordinates (i.e., azimuth and declination) of directions included within the range of a solid angle ω inside which from any point of the adsorbing surface the sun's surface is visible. After calculation of the both the integrals in equation (a) one obtains:

$$b_\omega = b\frac{R^2}{L^2} \qquad \text{(b)}$$

where the exergy b of the black radiation emitted by the sun within the solid angle of 2π is calculated according to formula (6.8):

$$b = \frac{\sigma}{3}\left(3T^4 + T_0^4 - 4T_0 T^3\right) \qquad \text{(c)}$$

The absorbing surface obtains also the black ($\varepsilon_0 = 1$) radiation energy e_o from the environment at temperature T_0 within a solid angle $2\pi-\omega$, but only a portion determined by emissivity ε_a of the adsorbing surface is absorbed by the surface:

$$e_0 = \varepsilon_a \sigma T_0^4 \left(1 - \frac{R^2}{L^2}\right) \tag{d}$$

According to the definition, the exergy of environment radiation is zero:

$$b_0 = 0 \tag{e}$$

The absorbing surface of emissivity ε_a and temperature T_a radiates its own emission e_a in the whole solid angle 2π:

$$e_a = \varepsilon_a \sigma T_a^4 \tag{f}$$

The respective exergy b_a of the absorbing surface, again according to formula (6.8), is:

$$b_a = \varepsilon_a \frac{\sigma}{3} \left(3T_a^4 + T_0^4 - 4T_0 T_a^3\right) \tag{g}$$

The energy emission e_ω arriving from the sun at the absorbing surface within the solid angle ω is:

$$e_\omega = e \frac{R^2}{L^2} \tag{h}$$

where the sun emission e in the whole solid angle 2π is:

$$e = \sigma T^4 \tag{i}$$

The temperature T_a of the absorbing surface remains constant because the heat q, at the heat source temperature T_a, from the absorbing surface is extracted in the amount resulting from the emitted and absorbed flux from this surface. In other words, heat q is determined from the following energy conservation equation for the absorbing surface remaining at the steady state:

$$q = \varepsilon_a e_\omega - e_a + e_0 \tag{j}$$

The exergy b_q of this heat q is determined by the Carnot efficiency for the heat sources temperatures T_a (hot) and T_0 (cold):

$$b_q = q \frac{T_a - T_0}{T_a} \tag{k}$$

Based on definition (6.34), the conversion efficiency η_B of the exergy b_ω of the sun's radiation into the exergy b_q of the heat source can be introduced:

$$\eta_B = \frac{b_q}{b_\omega} \tag{l}$$

For further considerations, the entropy fluxes are also introduced. Entropy s_ω of the solar radiation arriving at the absorbing surface:

$$s_\omega = s\frac{R^2}{L^2} \tag{m}$$

where the entropy s of solar radiation propagating within solid angle 2π, is:

$$s = \frac{4}{3}\sigma T^3 \tag{n}$$

Other entropy fluxes, s_a for the emission of the absorbing surface and s_0 for the absorbed environment radiation, are, respectively:

$$s_a = \varepsilon_a\frac{4}{3}\sigma T_a^3 \tag{o}$$

$$s_0 = \varepsilon_a\frac{4}{3}\sigma T_0^3\left(1 - \frac{R^2}{L^2}\right) \tag{p}$$

The overall process occurring at the absorbing surface is irreversible and the respective exergy loss δb is determined according to formula (2.60). The overall entropy growth Π for all processes involved consists of the entropy of generated heat (+), disappearing entropy of solar radiation (–), emitted entropy of absorbing surface (+), and disappearing entropy of environment radiation (–):

$$\Pi = \frac{q}{T_a} - \varepsilon_a s_\omega + s_a - s_0 \tag{r}$$

The correctness of the presented consideration can be verified by checking if the exergy balance equation for the absorbing surface is fulfilled. For the steady state, exergy input consists of the net exergy of solar radiation and environment absorbed by the considered surface. Exergy output is equal to both the exergy of emitted radiation from the considered surface and to the exergy of heat. The conservation equation is completed by the exergy loss:

$$b_\omega - (1 - \varepsilon_a)b_\omega + b_0 = b_a + b_q + \delta b \tag{s}$$

To illustrate the problem of the exergy balance equation and to compare it to the respective energy conservation equation:

$$e_\omega - (1 - \varepsilon_a)e_\omega + e_0 = e_a + q \tag{t}$$

Item	Comments	% exergy	% energy
Input	Sun	100	100
	Environment	0	23.1
	Total	100	123.1
Output	Reflection	20	20
	Emission	1.70	42.8
	Heat (efficiency)	9.23	60.3
	Loss	69.07	0
	Total	100	123.1

TABLE 10.1 Comparison of the Exergy and Energy Balances of a Surface Absorbing Solar Radiation ($\varepsilon_a = 0.8$) (from Petela, 2003)

some data is shown in Tables 10.1 and Table 10.2, both for $T_0 = 300$ K, $T_a = 350$ K, and $T = 6000$ K, and for $b_\omega = 1.484$ kW/(m^2 sr) and $e_\omega = 1.59$ kW/(m^2 sr). Table 10.1 is for $\varepsilon_a = 0.8$, whereas Table 10.2 is for $\varepsilon_a = 1.0$.

Analogously to the exergy efficiency η_B, the energy efficiency η_E can also be used, according to formula (6.32):

$$\eta_E = \frac{q}{e_\omega} \tag{10.16}$$

Using equations (h)–(j), (d), and (f) in (10.16):

$$\eta_E = \varepsilon_a \left[1 - \frac{T_a^4 - T_0^4 \left(1 - \dfrac{R^2}{L^2} \right)}{T^4 \dfrac{R^2}{L^2}} \right] \tag{10.17}$$

This is to note that for $R/L \to 1$ formula (10.17) comes to the formula (6.32) which expresses the energy conversion efficiency η_E for a case of the two infinite parallel planes.

Item	Comments	% exergy	% energy
Input	Sun	100	100
	Environment	0	28.88
	Total	100	128.88
Output	Reflection	20	0
	Emission	2.12	53.50
	Heat (efficiency)	11.54	75.38
	Loss	86.34	0
	Total	100	128.88

TABLE 10.2 Comparison of the Exergy and Energy Balances of a Surface Absorbing Solar Radiation ($\varepsilon_a = 1$) (from Petela, 2003)

For the case considered in Table 10.1, the values of efficiencies can be interpreted as $\eta_B = 9.23\%$ and $\eta_E = 60.3\%$, or, respectively, for Table 10.2, as $\eta_B = 11.54\%$ and $\eta_E = 75.38\%$. The values of the energetic and exergetic efficiencies differ significantly and, except for reflected radiation by the absorbing surface, other items of both balances are also very different.

Using formulae (b)–(d) and (f)–(k) in (l), the exergy conversion efficiency of solar radiation into heat can be determined as follows:

$$\eta_B = 3\,\varepsilon_a\,\frac{T_a - T_0}{T_a}\;\frac{T^4 - T_0^4 - \left(T_a^4 - T_0^4\right)\frac{L^2}{R^2}}{3T^4 + T_0^4 - 4T_0T^3} \tag{10.18}$$

It is worth noting again that for $L/R \to 1$, formula (10.18) becomes formula (6.36). The larger is the ratio L/R, the smaller is the efficiency. The increased emissivity ε_a of the absorbing surface increases the exergy conversion efficiency η_B. To determine the optimal temperature $T_{a,\mathrm{opt}}$, the following condition is used:

$$\frac{\partial \eta_B}{\partial T_a} = 0 \tag{10.19}$$

which leads to the equation:

$$4T_{a,\mathrm{opt}}^5 - 3T_o\,T_{a,\mathrm{opt}}^4 - T_0\,T^4\frac{R^2}{L^2} - T_0^5 + T_0^5\frac{R^2}{L^2} = 0 \tag{10.20}$$

For example, if the solar radiation is considered at $T_0 = 300$ K, $T = 6000$ K, $R = 6.955 \times 10^8$ m, and $L = 1.495 \times 10^{11}$ m, then $T_{a,\mathrm{opt}} \approx 369.9$ K (96.9°C). If the environment temperature drops to $T_0 = 273$ K, then $T_{a,\mathrm{opt}} \approx 352.8$ K (79.8°C).

The emissivity value ε_a has no effect on the optimal temperature $T_{a,\mathrm{opt}}$ of the surface. The T_a optimum, at the unchanged exergy b_ω of solar radiation, results from the fact that with increasing T_a the heat q decreases, whereas the Carnot efficiency, $\eta_{C,a} = 1 - T_0/T_a$, of this heat, increases.

From equation (10.20) the optimal (exergetic) temperature of the absorbing surface $T_{a,\mathrm{opt}}$ can be calculated for a given configuration (R and L) and the sun temperature T and environment temperature T_0. Analysis shows that the temperature T_a of the absorbing surface can be considered practically only in the range $T_0 \le T_a \le T_{a,\mathrm{max}}$. Temperature T_a smaller than T_0 requires additional energy to generate surroundings colder than the environment, whereas for $T_a > T_{a,\mathrm{max}}$ the heat q becomes negative because the radiation of the absorbing surface to the environment is larger than the heat received from solar radiation.

For the sun–earth configuration shown in Figure 10.6, the calculated $T_{a,\mathrm{opt}}$ is relatively low and so is the respective exergy conversion

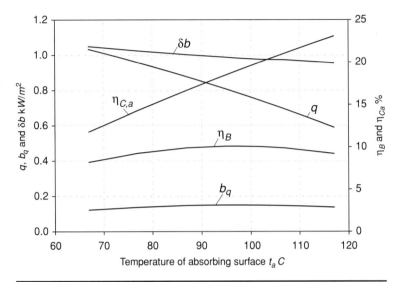

FIGURE 10.7   Effect of varying temperature T_a on the absorbing surface at constant $T = 6000$ K, $T_o = 300$ K, and $\varepsilon_a = 0.8$.

efficiency, as shown in Figure 10.7. However, temperature $T_{a,\text{opt}}$ can be significantly increased and the efficiency improved, e.g., by focusing the solar radiation with a thin lens. For example, if we have a perfect lens of diameter $2 \times R = 4$ cm and the sun is seen from a focal distance of $L = 0.5$ m, then from equation (10.20) the calculated optimal

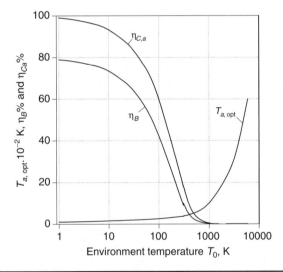

FIGURE 10.8   Effect of varying environment temperature T_0 at constant $T = 6000$ K and $\varepsilon_a = 0.8$.

(exergetic) temperature is $T_{a,\text{opt}} = 741.5$ K, for which, from equation (10.17), $\eta_E = 43.8\%$ at $T_o = 300$ K and $T = 6000$ K.

The desire of man to travel cosmically motivates consideration of the environment temperature in a wide range, theoretically $0 < T_0 < T$. This is shown in Figure 10.8. With decreasing environment temperature T_0, the optimal temperature $T_{a,\text{opt}}$ of the absorbing surface continuously diminishes and the exergy conversion efficiency η_B grows, approaching 80% for $T_0 \rightarrow 0$. At the same time the Carnot efficiency $\eta_{C,a} = 1 - T_0/T_a$, also grows and reaches 100% for $T_0 \rightarrow 0$.

10.5 Thermodynamic Analysis of the Solar Cylindrical–Parabolic Cooker

10.5.1 Introductory Remarks

In this section, we consider the cylindrical–parabolic cooker as an example of how thermodynamic analysis can be developed for a process in which radiation plays a role.

One of the most popular areas of the application of solar radiation, especially in countries with an abundance of solar energy, is cooking. The literature on solar cookers is extensive. For example, Kundapur (1988) presents a detailed review of different types of solar cookers. The performance of solar energy collectors based on exergy was analyzed by Fujiwara (1983). Maximization of solar energy collection from an available geographical area and then cooling the collector, including the problem of energy storage by melting phase-changing materials, are discussed by Bejan (1997). Aspects of solar cookers such as standard testing and performance evaluations were reported by Funk (2000). Ozturk (2004) for the first time determined experimentally the exergetic efficiency of a solar parabolic cooker of the cylindrical trough shape. This type of cooker is discussed here as exemplary for theoretical analysis.

A solar cylindrical–parabolic cooker (SCPC) is a device driven by solar radiation, which generally, especially when compared to energy efficiency, has very low exergy efficiency. There is practically little one can do in order to improve its performance. The performance of the SCPC can be enhanced only a little by appropriate design of the geometrical configuration and optical properties of the surfaces used to exchange heat by radiation.

The principles of radiative heat transfer can be found in many textbooks on heat transfer, e.g., Holman (2009), so the present consideration, according to Petela (2005), will focus on exergy analysis, the methodology for which is outlined in Chapter 4. Analysis of the conversion process of energy, which conserves itself totally regardless of its quality, serves well for design calculations, whereas exergy analysis, which takes into consideration the quality of energy, serves mostly for practical estimation and analysis of the process.

The main reason for the low efficiency of devices driven by solar radiation lies in the impossibility of full absorption of the real value of insolation. To obtain high-quality energy, at a high temperature, the absorbing surface has to be at a high temperature, at which a large amount of energy would be emitted from the surface to the environment. This factor influences both energy and exergy efficiencies. The low-exergy performance efficiency of SCPC, and of other devices driven by solar radiation, is caused by a significant degradation of energy. The relatively high temperature ($\sim$6000 K) of solar radiation is degraded to a relatively low temperature, e.g., to the temperature T_w of heated water, which is not much higher than the environment temperature T_0 ($T_w \approx T_0$).

The effect of such degradation, which causes a significant difference between energy and exergy efficiencies, can be illustrated, e.g., by simple consideration of the ratio r of the exergy growth to the energy growth of water when it is preheated from initial temperature T_w to the higher temperature by ΔT. The definition of exergy from formula (2.45) and the entropy of substance from (2.38) at constant pressure can be used in the consideration. Referring to 1 kg of water with specific heat c, the growth of the water exergy b_w divided by the growth of the water enthalpy h_w is:

$$r = \frac{(b_w)_{T_w + \Delta T} - (b_w)_{T_w}}{(h_w)_{T_w + \Delta T} - (h_w)_{T_w}}$$

$$= \frac{c(T_w + \Delta T - T_0) - T_0 c \ln \frac{T_w + \Delta T}{T_0} - c(T_w - T_0) + T_0 c \ln \frac{T_w}{T_0}}{c(T_w + \Delta T - T_0) - c(T_w - T_0)}$$

and after rearranging:

$$r = 1 - \frac{T_0}{\Delta T} \ln \left(1 + \frac{\Delta T}{T_w} \right) \tag{10.21}$$

For example for $\Delta T = 20$ K, based on equation (10.21) and for T_w smaller than the temperature for boiling water (100°C), the values of r are shown in Figure 10.9 for two different environment temperatures T_0 (280 and 320 K, respectively). The exergy/energy growth ratio r is very small; however, it increases with the growing water temperature T_w and with decreasing environment temperature T_0.

For rough estimation, the ratio r multiplied by the exergy/energy radiation ratio ψ can be recognized as the ratio of exergetic η_B and energetic η_E efficiencies of utilization of solar radiation for heating, $\eta_B / \eta_E = \psi \times r$. For example, assuming $\psi_S = 0.933$ for solar radiation and taking into account the values of r from Figure 10.9, one can

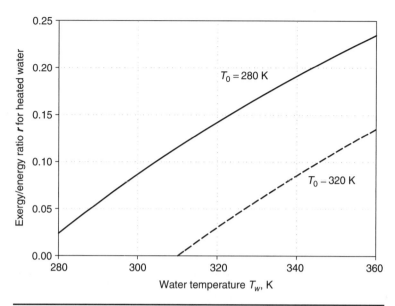

FIGURE 10.9 The exergy/energy growth ratio r for heated water as a function of water temperature T_w and environment temperature T_0 ($\Delta T = 20$ K).

estimate the efficiencies ratio $\eta_B / \eta_E \approx 0.1$. In the next section, exergy analysis of the SCPC, including radiative, convective, and conductive heat transfer, is developed to reveal the exergy losses distribution causing the low exergy efficiency of the SCPC.

10.5.2 Description of the SCPC

An SCPC is schematically shown in Figure 10.10. The cylindrical cooking pot filled with water is surrounded by the cylindrical–parabolic reflector. The frame supporting the reflector and cooking pot is not shown. The considered system of exchanging energy consists of three long surfaces of length L. The outer surface 3 of the cooking pot has an area A_3. The inner surface 2 of the reflector has an area A_2. The system is made up of the imagined plane surface 1 of area A_1.

The imagined surface 1, which represents the ambience and the insolation supplied to the considered system, is defined by transmissivity $\tau_1 = 1$ (and thus reflectivity $\rho_1 = 0$), absorptivity $\alpha_1 = 0$, and emissivity $\varepsilon_1 = 0$.

However, the effective emission of the imagined surface 1 can be determined as the insolation I calculated as follows:

$$I = 2.16 \cdot 10^{-5} A_1 \varepsilon_S \sigma T_s^4 \tag{10.22}$$

where $2.16 \cdot 10^{-5}$ is the solid angle within which the sun is seen from the earth, ε_S is the emissivity of the sun's surface (assumed as $\varepsilon_S = 1$),

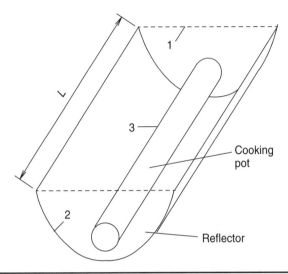

Figure 10.10 The scheme of the SCPC (from Petela, 2005).

σ is the Boltzmann constant for black radiation, and T_S is the absolute temperature of the sun's surface (assumed as $T_S = 6000$ K). Formally, it can be assumed that the emission of surface 1 is $E_1 = I$.

It is assumed that surfaces 2 and 3 have uniform temperatures T_2 and T_3, respectively; uniform reflectivities, ρ_2 and ρ_3, respectively, different from zero; and the emissivities of the surfaces, ε_2 and ε_3, respectively, are:

$$\varepsilon_2 = 1 - \rho_2 \tag{10.23}$$

$$\varepsilon_3 = 1 - \rho_3 \tag{10.24}$$

Thus, the emissions of surfaces 2 and 3 are:

$$E_2 = A_2 \varepsilon_2 \sigma T_2^4 \tag{10.25}$$

$$E_3 = A_3 \varepsilon_3 \sigma T_3^4 \tag{10.26}$$

The geometric configuration of the SCPC can be described by the value ϕ_{i-j} of the nine view factors for the three surfaces 1, 2, and 3.

10.5.3 Mathematical Model for Energy Analysis of the SCPC

The calculations are carried out only for the 1 m section of the SCPC length, which is significantly long ($L >> 1$). The following known quantities are assumed as the input data:

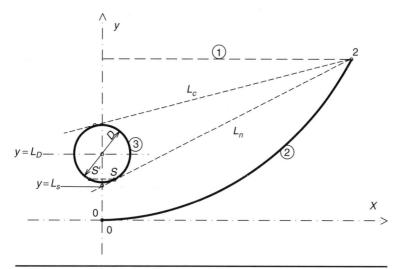

Figure 10.11 The scheme for calculation of the geometrical configuration of both the SCPC surfaces and the view factors (from Petela, 2005).

- outer diameter D of the cooking pot and its location L_D (shown in Figure 10.11);
- dimensions x_2 and y_2 of the parabolic reflector (shown in Figure 10.11);
- heat transfer coefficients k_2 and k_3;
- emissivities ε_2 and ε_3 of surfaces 2 and 3;
- absolute temperature of the sun's surface $T_S = 6000$ K;
- absolute water temperature T_w (average of the inlet and outlet temperatures);
- absolute environment temperature $T_0 = 293$ K.

The equations below are introduced to determine the following unknown quantities (output data):

- surfaces areas A_1, A_2, A_3;
- all view factors, φ_{i-j};
- reflectivities ρ_2 and ρ_3 (from the assumed emissivities ε_2 and ε_3, respectively);
- emissions E_2, E_3, and insolation I ($I = E_1$);
- convective heat $Q_{2,c}$ from the reflector to the environment;
- radiative heat $Q_{2,r}$ from the outer side of the reflector to the environment;

- convective heat $Q_{3,c}$ from surface 3 to the environment;
- radiosity of the three surfaces J_1, J_2, and J_3;
- absolute temperatures T_2 and T_3 of surfaces 2 and 3, respectively;
- energy efficiency of the SCPC, expressed by the water enthalpy change $Q_{3,u}$.

To derive mathematical equations describing the energy exchange in the three-surface system of the SCPC at a certain instant in which the SCPC remains in thermal equilibrium, the following energy conservation equations for each successive surface can be written:

$$J_1 = \varphi_{2-1}J_2 + \varphi_{3-1}J_3 + Q_{2,c} + Q_{2,r} + Q_{3,c} + Q_{3,u} \tag{10.27}$$

$$\varepsilon_2 \left(\varphi_{1-2}J_1 + \varphi_{2-2}J_2 + \varphi_{3-2}J_3 \right) = E_2 + Q_{2,c} + Q_{2,r} \tag{10.28}$$

$$\varepsilon_3 \left(\varphi_{1-3}J_1 + \varphi_{2-3}J_2 + \varphi_{3-3}J_3 \right) = E_3 + Q_{3,c} + Q_{3,u} \tag{10.29}$$

The magnitudes J_1, J_2, and J_3 are the radiosity values for surfaces 1, 2, and 3, respectively, and the values of ϕ_{i-j} are the respective view factors. The radiosity expresses the total radiation that leaves a surface and includes emission of the considered surface as well as all reflected radiation arriving from other surfaces of the system. The concept of radiosity is convenient for energy calculation; however, it cannot be used for exergy considerations because it does not distinguish between the temperatures of the components of the radiosity. The two independent equations for radiosity are included in the calculation:

$$J_1 = I \tag{10.30}$$

$$J_2 = E_2 + \rho_2 \left(\varphi_{1-2} \, J_1 + \varphi_{2-2} \, J_2 + \varphi_{3-2} \, J_3 \right) \tag{10.31}$$

The radiosity J_1 of the imagined surface 1 equals the insolation I.

It is assumed that the reflector is very thin so the uniform temperature T_2 prevails throughout the whole reflector thickness as well as on the inner and outer sides of the reflector. Thus, the heat $Q_{2,c}$ transferred from both sides of the reflector is:

$$Q_{2,c} = 2 \, A_2 h_2 (T_2 - T_o) \tag{10.32}$$

and the heat radiating from the outer side of reflector to the environment is:

$$Q_{2,r} = A_2 \varepsilon_2 \sigma (T_2^4 - T_o^4) \tag{10.33}$$

where h_2 is the convective heat transfer coefficient and T_0 is the environment temperature.

Heat $Q_{3,c}$ transferred by convection from surface 3 to the environment is:

$$Q_{3,c} = A_3 h_3 (T_3 - T_o) \tag{10.34}$$

and the useful heat $Q_{3,u}$ transferred through the wall of the cooking pot is:

$$Q_{3,u} = A_3 k_3 (T_3 - T_w) \tag{10.35}$$

where h_3 is the convective heat transfer coefficient, T_w is the absolute temperature of water in the cooking pot, and k_3 is the heat transfer coefficient, which takes into account the conductive heat transfer through the cooking pot wall and convective heat transfer from the inner cooking pot surface to the water.

The equations system (10.22)–(10.35) can be solved by successive iterations.

Energy analysis of the SCPC can be carried out based on evaluation of the terms in the following energy conservation equation for the whole SCPC:

$$\varphi_{2-1} J_2 + \varphi_{3-1} J_3 + Q_{2,c} + Q_{2,r} + Q_{3,c} + Q_{3,u} = I \tag{10.36}$$

The first two terms in equation (10.36) represent radiation energy escaping from the SCPC due to the radiosities of surfaces 2 ($\varphi_{2,1} J_2$) and 3 ($\varphi_{3,1} J_3$). Dividing both sides of equation (10.36) by I, the percentage values β of the equation terms can be obtained, e.g., for heat $Q_{2,c}$, the corresponding $\beta_{2,c}$ is:

$$\beta_{2,c} = \frac{Q_{2,c}}{I} \tag{a}$$

however, the term with $Q_{3,u}$ determines the energetic efficiency

$$\eta_E = \frac{Q_{3,u}}{I} \tag{10.37}$$

Therefore, equation (10.36) can be also written as:

$$\sum \beta + \eta_E = 1 \tag{10.38}$$

10.5.4 Mathematical Consideration of the Exergy Analysis of an SCPC

Exergy analysis usually gives an additional basis for the quality interpretation of the process being examined. For the considered SCPC, we calculate the exergy of radiating fluxes, the overall exergy efficiency of the SCPC process, and the exergy losses during irreversible component phenomena occurring in the SCPC.

It is convenient to determine an exergy B of radiation emission at temperature T by multiplying its emission energy E by the characteristic exergy/energy ratio ψ, defined by formula (6.22):

$$B = E\psi \tag{10.39}$$

For example, the exergy efficiency η_B of the SCPC is the ratio of the exergy of the useful heat $Q_{3,u}$, at temperature T_3, and the exergy of solar emission at temperature T_S:

$$\eta_B = 100 \frac{Q_{3,u}\left(1 - \frac{T_0}{T_3}\right)}{I\,\psi_S} \tag{10.40}$$

where ψ_S is the exergy/energy ratio for the solar emission of temperature T_S, which for $T_S = 6000$ K and $T_0 = 293$ K is $\psi_S = 0.9348$.

Reflection and transmition of radiation are reversible, so the exergy losses in the SCPC are considered only for the following component phenomena:

- Simultaneous emission and absorption of radiation at surfaces 2 (δB_2) and 3 (δB_3). There is no exergy loss at the imagined surface 1 ($\delta B_1 = 0$) because neither absorption nor emission occurs but only transmission of radiation, which is reversible. Other surfaces, 2 and 3, are solid and thus produce the irreversible effects of radiation.

- Irreversible transfer of convection heat Q_{2c} from both sides of surface 2 to the environment (δB_{Q2c}).

- Irreversible transfer of radiation heat Q_{2r} from the outer side of surface 2 to the environment (δB_{Q2r}).

- Irreversible transfer of heat Q_{3u} from surface 3 to water (δB_{Q3u}), due to the temperature difference $T_3 - T_w$,

- Irreversible transfer of convection heat Q_{3c} from surface 3 to the environment (δB_{Q3c}).

- The exergy δB_1 escaping through surface 1, resulting from reflections from the SCPC surfaces to the environment. This loss is sensed only by the SCPC and is not irreversible because theoretically it can be used elsewhere. This loss consists of the radiation exergies B_{1-1}, B_{2-1}, and B_{3-1} at the three different temperatures (T_S, T_2, and T_3):

$$\delta B_1 = B_{1-1} + B_{2-1} + B_{3-1} \tag{10.41}$$

Analogously to the energy conservation equation, the exergy balance equation can be applied. When relating all the equation terms to the exergy input, which is the exergy $I \times \psi_S$ of solar radiation entering the SCPC system, the following conservation equation can be written:

$$\xi_{B11} + \xi_{B21} + \xi_{B31} + \xi_{BQ2c} + \xi_{BQ2r} + \xi_{Q3u} + \xi_{Q3c} + \xi_{B2} + \xi_{B3} + \eta_B = 100$$
(10.42)

where any percentage exergy loss ξ is calculated as the ratio of the loss to the exergy input, e.g., for the convection heat $Q_{2,c}$ one obtains:

$$\xi_{Q2c} = \frac{\delta B_{Q2c}}{I \psi_S}$$
(b)

In the considered system of nonblack surfaces the energy striking a surface is not totally absorbed, and part of it is reflected back to other surfaces. The radiant energy can thus be reflected back and forth between surfaces many times. To simplify further considerations of such a multi-reflections effect, it is assumed that surface 3 is black, ($\varepsilon_3 = 1$). Thus, as the imagined surface 1 was previously assumed to be black ($\varepsilon_1 = 1$), the only nonblack surface in the exergetic analysis of the SCPC system is surface 2 ($\varepsilon_2 < 1$).

The nine exergy losses appearing in equation (10.42) can be categorized into three groups. The first group contains the exergy losses (external) related to heat transfer, (ξ_{BQ2c}, ξ_{BQ2r}, ξ_{Q3u}, ξ_{Q3c}). The second group (ξ_{B11}, ξ_{B21}, ξ_{B31}) determines the exergy fluxes (external losses) escaping from the SCPC, and the third group contains the exergy losses (internal) due to irreversible emission and absorption on surface (ξ_{B2}, ξ_{B3})

Consider the losses of the **first group**. The exergy loss δB_{Q2c}, due to the convection transfer of heat $Q_{2,c}$ from the two sides of the reflector to the environment is equal to the exergy of heat $Q_{2,c}$:

$$\delta B_{Q2c} = Q_{2,c} \left(1 - \frac{T_0}{T_2} \right)$$
(10.43)

The exergy loss δB_{Q2r} due to the radiation transfer of heat $Q_{2,r}$ from the outer side of the reflector to the environment is equal to the exergy of heat $Q_{2,r}$:

$$\delta B_{Q2r} = Q_{2,r} \left(1 - \frac{T_0}{T_2} \right)$$
(10.44)

The external exergy loss δB_{Q3u} due to the transfer of the useful heat $Q_{3,u}$ from surface 3 through the cooking pot wall to water is equal to the difference of the exergy of this heat at temperature T_3 and at temperature T_w:

$$\delta B_{Q3u} = Q_{3,u} \left(\frac{T_3 - T_0}{T_3} - \frac{T_w - T_0}{T_w} \right) \tag{10.45}$$

The exergy loss δB_{Q3c}, due to convective heat transfer from surface 3 to the environment, is determined similarly:

$$\delta B_{Q3c} = Q_{3,c} \left(\frac{T_3 - T_0}{T_3} - \frac{T_0 - T_0}{T_0} \right) \tag{10.46}$$

where, obviously, the second fraction in the brackets of equation (10.46) is zero.

Consider the losses of the **second group**. As results from formula (10.41) the external exergy loss δB_1 is equal to the escaping exergy of three unabsorbed emissions at temperatures $T_1 = T_S$, T_2, and T_3 reflected to the environment. By multiplying these emissions respectively by the exergy/energy ratio the three losses can be expressed as follows:

$$B_{1-1} = Q_{1-1}\psi_S \tag{10.47}$$

$$B_{2-1} = Q_{2-1}\psi_2 \tag{10.48}$$

$$B_{3-1} = Q_{3-1}\psi_3 \tag{10.49}$$

where ψ_S, ψ_2, and ψ_3 are calculated from formula (6.22) for T_S, T_2, and T_3, respectively, and Q_{1-1}, Q_{2-1}, and Q_{3-1} are the sums of the unabsorbed portions of the respective emissions of surfaces 1, 2, and 3. Thus, heat Q_{1-1} represents energy portions from many reflections of emission E_1 of temperature T_S, at the concave surface 2 and arriving at surface 1:

$$Q_{1-1} = E_1\varphi_{1-2}\rho_2\varphi_{2-1} + E_1\varphi_{1-2}\rho_2\varphi_{2-2}\rho_2\varphi_{2-1}$$
$$+ E_1\varphi_{1-2}\rho_2\varphi_{2-2}\rho_2\varphi_{2-2}\rho_2\varphi_{2-1} + \cdots \tag{10.50}$$

The portions in equation (A3.2) can be expressed as the sum of the terms of the infinite geometric progression with the common ratio $\phi_{2-2} \times \rho_2$, thus:

$$Q_{1-1} = E_1\varphi_{1-2}\,\rho_2\varphi_{2-1}\frac{1}{1 - \varphi_{2-2}\rho_2} \tag{10.51}$$

Heat $Q_{2-1}(T_2)$ represents the portion $E_2 \times \varphi_{2-1}$ of emission E_2 of surface 2, which directly arrives at surface 1, as well as the portions in results of many reflections of emission E_2 at surface 2, arriving at surface 1:

$$Q_{2-1} = E_2\varphi_{2-1} + E_2\varphi_{2-2}\rho_2\varphi_{2-1} + E_2\varphi_{2-2}\rho_2\varphi_{2-2}\rho_2\varphi_{2-1} + \cdots \tag{10.52}$$

and

$$Q_{2-1} = E_2\varphi_{2-1}\frac{1}{1 - \varphi_{2-2}\rho_2} \tag{10.53}$$

Heat $Q_{3-1}(T_3)$ represents the portion $E_3 \times \varphi_{3-1}$ of emission E_3 of surface 3 that arrives at surface 1 as direct radiation, as well as the portions in results of many reflections of emission E_3 at surface 2:

$$Q_{3-1} = E_3\varphi_{3-1} + E_3\varphi_{3-1}\rho_2\varphi_{2-1} + E_3\varphi_{3-1}\rho_2\varphi_{2-2}\,\rho_2\varphi_{2-1}$$
$$+ E_3\varphi_{3-1}\rho_2\varphi_{2-2}\rho_2\varphi_{2-2}\rho_2\varphi_{2-1} + \cdots \tag{10.54}$$

and

$$Q_{3-1} = E_3\left(\varphi_{3-1} + \varphi_{3-2}\rho_2\varphi_{2-1}\frac{1}{1 - \varphi_{2-2}\rho_2}\right) \tag{10.55}$$

Calculation of the internal exergy losses from the **third group** can be based either on the determination of the overall entropy growth used in the Guoy–Stodola equation (2.60) or determined from the exergy balance equation for the considered surface in the steady state. The latter method will be used here.

The considered surface and the imagined heat source connected to the surface both have to be assumed. For each surface, 2 or 3, the arriving emissions at three different temperatures need to be taken into account. It is assumed that these emissions are absorbed by the surface and transferred as heat to the imagined heat source. Then, immediately, this heat is taken from the source to generate the emission of the surface at its temperature. Thus, e.g., the exergy balance equation for surface 2 can be interpreted as including the following:

The input exergy entering surface 2, represented by terms due to:

- emission $\psi_S \times Q_{1-2}$ arriving from surface 1;
- emission $\psi_2 \times Q_{2-2}$ arriving from surface 2;
- emission $\psi_3 \times Q_{3-2}$ arriving from surface 3;
- heat $E_2(1 - T_0/T_2)$ needed for emission of surface 2 and delivered from the heat source.

The output exergy leaving surface 2, represented by terms due to:

- emission $Q_{1-2} \times (1 - T_0/T_2)$ of surface 1 converted as the heat absorbed by the heat source;
- emission $Q_{2-2} \times (1 - T_0/T_2)$ of surface 2 converted as the heat absorbed by the heat source;

- emission $Q_{3-2} \times (1 - T_0/T_2)$ of surface 3 converted as the heat absorbed by the heat source;
- emission $E_2 \times \psi_2$ of surface 2.

The exergy balance equation can be written in the form of exergy loss δB_2 equal to the difference of the exergy input and output:

$$\delta B_2 = \psi_3 Q_{3-2} + \psi_S Q_{1-2} + \psi_2 (Q_{2-2} - E_2)$$
$$- (Q_{3-2} + Q_{1-2} + Q_{2-2} - E_2) \left(1 - \frac{T_0}{T_2}\right) \qquad (10.56)$$

Analogically, the exergy loss δB_2 is:

$$\delta B_3 = \psi_2 Q_{2-3} + \psi_S Q_{1-3} + \psi_3 (Q_{3-3} - E_3)$$
$$- (Q_{1-3} + Q_{2-3} + Q_{3-3} - E_3) \left(1 - \frac{T_0}{T_3}\right) \qquad (10.57)$$

Applying again the formula for the sum of the terms of the infinite geometric progression, the values or required heat can be determined as follows.

Heat Q_{1-2}, at temperature T_S, is the sum of the portions of emission of surface 1 reaching surface 2, thus:

$$Q_{1-2} = E_1 \varphi_{1-2} \varepsilon_2 \frac{1}{1 - \varphi_{2-2} \rho_2} \qquad (10.58)$$

Heat Q_{2-2} at temperature T_2 is the sum of the portions of emission of surface 2 reaching surface 2, thus:

$$Q_{2-2} = E_2 \varphi_{2-2} \varepsilon_2 \frac{1}{1 - \varphi_{2-2} \rho_2} \qquad (10.59)$$

Heat Q_{3-2} at temperature T_3 is the sum of the portions of emission of surface 3 reaching surface 2, thus:

$$Q_{3-2} = E_3 \varphi_{3-2} \varepsilon_2 \frac{1}{1 - \varphi_{2-2} \rho_2} \qquad (10.60)$$

Heat Q_{13}, at temperature T_S, which is the sum of the totally absorbed insolation that reaches surface 3 at view factor φ_{1-3}, and the totally absorbed insolation parts reflected from surface 2, can be determined as follows:

$$Q_{1-3} = I \varphi_{1-3} + I \varphi_{1-2} \rho_2 \varphi_{2-3} \frac{1}{1 - \varphi_{2-2} \rho_2} \qquad (10.61)$$

Heat Q_{2-3} at temperature T_2 is the sum of the portions of emission of surface 2 reaching surface 3, thus:

$$Q_{2-3} = E_2 \varphi_{2-3} \frac{1}{1 - \varphi_{2-2}\rho_2} \tag{10.62}$$

Heat Q_{3-3} at temperature T_3 is the sum of the portions of emission of surface 3 reflected from surface 2 to surface 3:

$$Q_{3-3} = E_3 \varphi_{3-2}\rho_2\varphi_{2-3} \frac{1}{1 - \varphi_{2-2}\rho_2} \tag{10.63}$$

As already shown in this section, the assumption of black surfaces 1 and 3 for exergetic consideration requires many more equations compared to the energetic consideration developed in Section 10.5.3 for the system in which only one black surface 1 was assumed. Obviously, exergetic consideration of a system with only one black surface 1 would require significantly more equations compared to the consideration presented in this section.

Example 10.2   Calculative analyses of the exemplary SCPC are performed in the following steps.

i. Calculation of the surface areas. We consider the three surfaces (shown in Figure 10.11) that take part in the heat exchange. The considerations are carried out for a 1-m long section of the SCPC.

Surface 1. The imagined surface 1 has the area A_1 determined by the coordinate x_2 of point 2:

$$A_1 = 2x_2 \tag{a}$$

Surface 2. The surface A_2 is equal to the length of the parabolic arc of the reflector and using the coordinates x_2 and y_2 of point 2:

$$A_2 = \sqrt{4 x_2^2 + y_2^2} + \frac{y_2^2}{2 x_2} \ln \frac{2 x_2 + \sqrt{4 x_2^2 + y_2^2}}{y_2} \tag{b}$$

If the distance L_s shown in Figure 10.11 is negative, $L_s < 0$, then, when determining radiation heat exchange between surfaces 1 and 2, the length of the parabolic arc, not seen from surface 1, has to be subtracted from A_2. Thus, the three unknowns x_s, y_s, and L_s, defined by Figure 10.11, have to be determined from the following three relations. The proportionality:

$$\frac{x_2}{y_2 - L_s} = \frac{x_s}{y_s - L_s} \tag{c}$$

the equation of the circle written for point S:

$$x_s^2 + (y_s - L_D)^2 = \left(\frac{D}{2} \right)^2 \tag{d}$$

and the equation of the straight line, tangent to the circle at point S, in which one of the two derivatives of the circle equation for point S is used:

$$\frac{x_s}{\sqrt{\left(\frac{D}{2}\right)^2 - x_s^2}} = \frac{y_2 - y_s}{x_2 - x_s} \tag{e}$$

where D is the outer diameter of cooking pot, L_D is the distance of the circle center from the origin of the coordinate system, and x_s and y_s are the coordinates of the tangency point S.

The considerations in the present example are carried out for $L_s \geq 0$, and this condition has to be watched when considering variation of the coordinates x_2 and y_2.

Surface 3. The surface area A_3 of the cooking pot is equal to the length of the circle:

$$A_3 = \pi D \tag{f}$$

However, there is a circle arc on surface 3, between points S and S' (location of point S' is symmetric to point S), which is not seen from surface 1. Thus, the length S_{cir} of this arc:

$$S_{cir} = D \, \text{arc} \sin \frac{2 \, x_s}{D} \tag{g}$$

has to be subtracted from A_3 to obtain the surface area A_{31} really involved in the radiation heat exchange between surface 1 and 3:

$$A_{31} = \pi D - S_{cir} \tag{h}$$

ii. View factors. For the considered system of surfaces shown in Figure 10.11, it is assumed that:

- The temperature of any surface is uniform over the entire surface area. This is equivalent to the assumption that the conductivity of the surface material is very large.

- The view factor for any surface is uniform over the entire surface area. This means that the surfaces are not too close to each other.

The view factor, for radiation transferred from any surface i to any surface j, is defined as the fraction φ_{i-j} of the total radiation leaving the surface i, which reaches surface j. In the considered SCPC there are nine view factors φ_{i-j}, where $i = 1, 2, 3$ and $j = 1, 2, 3$. From the energy conservation law the following three equations can be derived:

$$\sum_i \varphi_{j-i} = 1 \tag{i}$$

Making use of the reciprocity relations:

$$A_i \varphi_{i-j} = A_j \varphi_{j-i} \tag{j}$$

Surfaces 1 and 3 are not concave, thus:

$$\varphi_{1-1} = \varphi_{3-3} = 0 \qquad \text{(k)}$$

The ninth equation, to determine φ_{1-3}, is derived based on Polak's crossed-string method described in Section 7.5.1. The method can be applied when the considered surfaces are the sides of parallel and infinitely long cylinders, not necessarily circular but not concave. In practice the method can be applied not only to infinitely long cylinders but also to finite cylinders that are sufficiently long. The method introduces the two lengths of imagined strings, shown in Figure 10.11, which cross (L_c) or do not cross (L_n) when they gird the surfaces:

$$\varphi_{1-3} = \frac{L_c - L_n}{L_1} \qquad \text{(l)}$$

The length L_n is the straight distance between points S and 2, and L_c is the length of the girding string from point S' to point 2. The length L_1, representing surface 1, is:

$$L_1 = A_1 \qquad \text{(m)}$$

As a result from Figure 10.11, the other lengths can be determined as follows:

$$L_n = \sqrt{(x_2 - x_s)^2 + (y_2 - y_s)^2} \qquad \text{(n)}$$

$$L_c = L_n + \frac{\pi D}{2} - S_{\text{cir}} \qquad \text{(o)}$$

iii. Calculation for energy analysis. Based on the results of the mathematical model discussed in Section 10.5.3, the energy analysis can be carried out. Table 10.3 illustrates the responsive trends of the output data to the change in some input parameters. For comparison, the exergy efficiency η_B of the SCPC, determined by equation (10.40) in Section 10.5.4, is also shown in Table 10.3.

The values in column 3 of Table 10.3 are considered as the reference values for studying the influence of varying input parameters on the output data for the SCPC. Therefore, each of the next columns (4–14) corresponds to the case in which the input is changed only by the value shown in a particular column, whereas the other input parameters remain at the reference level.

The most interesting performance criteria of the SCPC are its capacity ($Q_{3,u}$) and the energy and exergy efficiencies (η_E and η_B). For the assumed variation of each of the input parameters shown in columns 4–14, the capacity $Q_{3,u}$ fluctuates in the range of 89.46–149.77 W, whereas the energy efficiency ranges from 5.64% to 9.23% and is always about ten times larger than the exergy efficiency ranging from 0.54% to 0.83%.

Table 10.3 can serve as the basis for studying the significance of the SCPC dimensions, surface properties, heat transfer coefficients, and operation temperatures. For example, reducing the heat transfer coefficient h_2 (column 8), which means, e.g., improving insulation of the outer surface of the reflector, has a relatively small effect on the growth of the efficiencies, and the capacity of the SCPC becomes only a little larger. Variation of the input value k_3 simulates the heat transfer condition for the useful heat $Q_{3,u}$, which can be transferred to the heated water flowing through the cooking pot at various speeds.

Quantity	Units	Reference value					Changed input								
1	2	3	4	5	6	7	8	9	10	11	12	13	14	15
Input:														
x_2	m	0.5	0.6											At the reference data the perfect insulation of the outer surface of the reflector is applied
y_2	m	0.4		0.3										
L_D	m	0.1			0.08	0.08								
D	m	0.1												
h_2	$W/m^2\,K$	6					10							
k_3	$W/m^2\,K$	3500						4000						
h_3	$W/m^2\,K$	6							10					
T_0	K	293								303				
T_w	K	320									330			
ε_2		0.2											0.1	
ε_3		0.9												0.97
Output:														
φ_{31}		0.376	0.397	0.41	0.40	0.380	0.376	0.376	0.376	0.376	0.376	0.376	0.376	0.439
φ_{32}		0.624	0.603	0.59	0.60	0.620	0.624	0.624	0.624	0.624	0.624	0.624	0.624	0.561
φ_{13}		0.094	0.086	0.11	0.10	0.077	0.094	0.094	0.094	0.094	0.094	0.094	0.094	0.123
φ_{12}		0.906	0.914	0.89	0.89	0.923	0.906	0.906	0.906	0.906	0.906	0.906	0.906	0.877
φ_{22}		0.178	0.147	0.12	0.19	0.195	0.178	0.178	0.178	0.178	0.178	0.178	0.178	0.134
φ_{23}		0.146	0.126	0.15	0.14	0.116	0.146	0.146	0.146	0.146	0.146	0.146	0.146	0.145
φ_{21}		0.676	0.727	0.73	0.66	0.689	0.676	0.676	0.676	0.676	0.676	0.676	0.676	0.721

294

Table 10.3 Illustration of the Energy Analysis and the Trends Responsive to the Change of Some Input Parameters ($T_S = 6000$ K, $I = 1586.39$ W/m^2) (from Petela, 2005)

		1	2	3	4	5	6	7	8	9	10	11	12	13
A_1	m^2	1	1.2	1	1	1	1	1	1	1	1	1	1	1
A_2	m^2	1.341	1.507	1.22	1.34	1.341	1.341	1.341	1.341	1.341	1.341	1.341	1.341	1.217
A_3	m^2	0.314	0.314	0.31	0.31	0.251	0.314	0.314	0.314	0.314	0.314	0.314	0.314	0.314
Q_1	W	416.8	466.1	412	415	390.6	423.0	416.5	416.5	403.3	399.2	323.5	427.4	396.8
$-Q_{2c}$	W	215.0	250.7	199	215	220.7	230.9	215.0	215.0	199.0	217.7	120.1	213.4	213.5
$-Q_{2r}$	W	21.88	25.56	20.3	21.9	22.49	13.76	21.88	21.88	22.33	22.17	5.928	21.69	0
$-Q_{3u}$	W	128.5	138.8	141	127	106.5	127.2	128.5	94.53	149.8	89.46	146.4	141.2	132.2
$-Q_{3c}$	W	51.12	51.13	51.1	51.1	40.90	51.11	51.09	85.09	32.30	69.90	51.14	51.14	51.12
T_3	K	320.12	320.13	320	320	320.12	320.11	320.10	320.09	320.14	330.08	320.13	320.13	320.12
T_2	K	306.4	306.9	307	306	306.7	301.6	306.4	306.4	315.4	306.5	300.5	306.3	319.5
E_2	W	133.9	151.5	122	168	134.5	125.7	133.9	133.9	150.3	134.2	61.92	133.7	158.4
E_3	W	168.3	168.3	168	134	134.6	168.3	168.2	168.2	168.3	190.2	168.3	181.4	168.3
J_1	W	1586	1904	1586	1585	1586	1586	1586	1586	1586	1586	1586	1586	1586
J_2	W	1617	1862	1488	1617	1645	1607	1617	1617	1636	1630	1753	1609	1646
J_3	W	206.9	208.1	208	207	166.0	206.9	206.9	206.8	207.2	229.1	208.9	192.9	207.3
η	%	8.097	7.290	8.88	8.00	6.714	8.018	8.101	5.959	9.441	5.639	9.226	8.900	8.336
η_B	%	0.731	0.658	0.80	0.72	0.606	0.724	0.731	0.538	0.538	0.676	0.833	0.803	0.752

Growth of the environment temperature (column 10), which depends, e.g., on the time of day, causes the growth in energy efficiency and the capacity $Q_{3,u}$, whereas the exergy efficiency decreases significantly.

Temperature T_w (column 11) represents the stage of cooking; the later the stage, the higher T_w can be expected. Growth in T_w causes a decrease in the capacity and in both energy and exergy efficiencies. It appears that with increased specific exergy of water the emission E_3 increases simultaneously, which reduces the capacity $Q_{3,u}$.

Column 15 represents the theoretical case of perfect insulation of the outer surface of the reflector, ($Q_{2,r} = 0$ and $Q_{2,c}$ is reduced by half). In this case both the energy and exergy efficiencies increase reaching values of 8.34% and 0.752%, respectively, whereas the capacity ($Q_{3,u}$) reaches 132.2 W.

iv. Calculations for exergy analysis. Similar to the results of energy analysis shown in Table 10.3, the exergy analysis for assumed $\varepsilon_3 = 1$ shows all the terms appearing in equation (10.42), as presented in Table 10.4.

Similar to the analysis of data in Table 10.3, the output data of Table 10.4 allows for estimating the responsive trends of the exergy losses and the energy and exergy efficiencies, due to the change of the input quantities. Table 10.4 also presents how the exergy losses are distributed depending on the changes in the input parameters. From the exergy viewpoint the largest loss is that of the escaping exergy B_{1-1}, which in the considered cases ranges from 57.069% to 65.565% (ξ_{B11}). The other reversible losses B_{2-1} and B_{3-1} are significantly smaller—below 0.0282% (ξ_{B21}) and below 0.2567% (ξ_{B31}), respectively. As mentioned before, the quantities B_{1-1}, B_{2-1}, and B_{3-1} represent the exergy losses of the considered SCPC; however, they are reversible and, potentially, might be utilized somewhere else.

In the second place are the losses occurring due to the irreversible phenomena at surfaces 2 and 3 caused by the simultaneous absorption and emission of radiation of various temperatures. These losses δB_2 and δB_3, ranging from 10.587 to 21.144% (ξ_{B2}) and from 16.817 to 22.254% (ξ_{B3}), respectively, express the degradation of radiation.

The irreversible loss δB_{Q2c} occurs due to heat lost to the environment and because the temperature T_2 is relatively low, below 0.654% (ξ_{BQ2c}).

The loss δB_{Q3c}, below 0.53% (ξ_{BQ3c}), is relatively small because heat from surface 3 to the environment occurs at a small temperature difference.

The loss δB_{Q3u}, below 0.0051%, (ξ_{BQ3u}), is the smallest because the temperature difference $T_3 - T_w$ is very small, and thus degradation of heat $Q_{3,u}$, during transfer from the cooking pot wall to water, can be neglected.

The exergy of the useful heat $Q_{3,u}$ is represented by the exergy efficiency η_B, which ranges from 0.6038% to 0.9448%. For comparison, the values of the energy efficiency of the SCPC, for the cases presented in Table 10.4, are also shown.

v. Discussion and comparison of the results of energy and exergy analyses. As mentioned before, both energy and exergy analyses can be used together for a full estimation of the considered process. The energy analysis serves mainly for design purposes, whereas the exergy analysis presents an image of the process quality.

Comparison of the energy and exergy balances for the considered SCPC, at the assumed $\varepsilon_3 = 1$, is presented in Table 10.5. In both analyses the radiation escaping from the SCPC is estimated at a relatively high level (energy: 68.34 + 4.43 = 72.77% and exergy 57.069 + 0.026 + 0.132 = 57.226%). The energy analysis allows for splitting the escaping radiation loss according to the radiosity of surface 2 and 3, whereas the exergy analysis makes this split according to the temperature (T_S, T_2, and T_3) of the escaping emissions.

Quantity	Units	Reference value	Changed input										
1	2	3	4	5	6	7	8	9	10	11	12	13	14
Input:													At the reference data the perfect insulation of the outer surface of the reflector is applied
x_2	m	0.5	0.6										
y_2	m	0.4		0.3									
L_D	m	0.1			0.08								
D	m	0.1				0.08							
h_2	W/m² K	6					10						
k_3	W/m² K	3500						4000					
h_3	W/m² K	6							10				
T_o	K	293								303			
T_w	K	320									330		
ε_2		0.2										0.1	
Output:													
ξ_{B11}	%	57.069	60.270	57.458	57.220	60.254	57.069	57.069	57.069	57.069	57.069	65.565	57.069
ξ_{B21}	%	0.0257	0.0272	0.0249	0.0257	0.0282	0.0144	0.0257	0.0257	0.0234	0.0264	0.0040	0.0257
ξ_{B31}	%	0.1316	0.1134	0.1355	0.1311	0.1075	0.1316	0.1314	0.1313	0.0538	0.2567	0.1418	0.1316
ξ_{B2}	%	20.433	20.008	18.979	20.426	21.144	20.673	20.433	20.433	20.531	20.438	10.587	20.433
ξ_{B3}	%	20.528	17.895	21.548	20.397	16.817	20.533	20.529	20.530	21.024	20.180	22.254	20.528
ξ_{BO2c}	%	0.6189	0.6244	0.5852	0.6186	0.6540	0.4349	0.6189	0.6189	0.5151	0.6357	0.1963	0.3095
ξ_{BO2r}	%	0.0629	0.0636	0.0596	0.0629	0.0666	0.0259	0.0629	0.0629	0.0575	0.0647	0.0097	0
ξ_{BO3u}	%	0.0038	0.0037	0.0045	0.0037	0.0032	0.0037	0.0033	0.0022	0.0051	0.0018	0.0048	0.0038
ξ_{BO3c}	%	0.2922	0.2438	0.2925	0.2923	0.1339	0.2923	0.2919	0.4861	0.1171	0.5298	0.2926	0.2923
η_B	%	0.8340	0.7506	0.9126	0.8238	0.6922	0.8262	0.8343	0.6409	0.6038	0.7980	0.9448	0.8340
η	%	9.241	8.317	10.112	9.128	7.670	9.155	9.245	7.101	10.600	6.654	10.469	9.241

TABLE 10.4 Illustration of the Exergy Analysis and of the Trends Responsive to the Change of Some Input Parameters (T_S = 6000 K, I = 1586.39 W/m², ε_3 = 1) (from Petela, 2005)

Description	Energy		Exergy	
	Expression	%	Expression	%
Input:				
Insolation (radiosity $J_1 = I$)	($I = 1586.38$ W)	100	($I \cdot \psi_S =$ 1483.1W)	100
Total		100		100
Output:				
Escaping radiosity from surface 2	$\beta\varphi_{21J2}$	68.34		
Escaping radiosity from surface 3	$\beta\varphi_{31J3}$	4.43		
Escaping fraction of emission E_1 ($E_1 = I$)			ξ_{B11}	57.069
Escaping fraction of emission E_2			ξ_{B21}	0.026
Escaping fraction of emission E_3			ξ_{B31}	0.132
Radiation irreversibility on surface 2			ξ_{B2}	20.433
Radiation irreversibility on surface 3			ξ_{B3}	20.528
Transfer of convective heat from surface 2	β_{Q2c}	13.41	ξ_{BQ2c}	0.619
Transfer of radiative heat from surface 2	β_{Q2r}	1.36	ξ_{BQ2r}	0.063
Transfer of convective heat from surface 3	β_{Q3c}	3.22	ξ_{BQ3c}	0.292
Irreversibility of transferred useful heat Q_{3u}			ξ_{BQ3u}	0.004
Useful heat Q_{3u} delivered to water	η	9.24	η_B	0.834
Total		100		100

TABLE 10.5 Comparison of the Energy and Exergy Balance Terms for the Considered SCPC ($\varepsilon_3 = 1$) (from Petela, 2005)

According to the energy analysis the heat losses to the environment are relatively high; by convection 13.41% ($Q_{2,c}$), 3.22% ($Q_{3,c}$), and by radiation 1.36% ($Q_{3,r}$), whereas the exergy estimation of these losses is relatively very low— 0.619%, 0.292%, and 0.063 %, respectively.

As shown in Table 10.5, the energy analysis does not reveal any degradation losses at surfaces 2 and 3 or during transfer of the useful heat, in contrast to the exergy analysis which, respectively, estimates the first two losses as being relatively high, 20.433% and 20.528%, and the third loss as very small, 0.004%.

vi. Some comments on the possible optimization of the SCPC surfaces configuration. The characteristic dimensions of the SCPC can be discussed separately. Table 10.3 shows that separate variation of x_2 and y_2 is quite effective. As

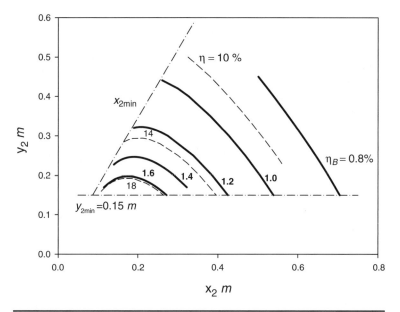

FIGURE 10.12 The influence of the reflector dimensions x_2 and y_2 on the energy efficiency η and exergy efficiency η_B of the SCPC ($\varepsilon_3 = 1$) (from Petela, 2005).

shown, reducing the *openness* x_2 and *depth* y_2 of the reflector causes the growth in energy efficiency η_E of the SCPC, although the capacity $Q_{3,u}$ grows only as y_2 is increased. The increase in the efficiencies also comes with the increase of D or L_D.

The problem of maximization of the energy efficiency as a function of four dimensions, $\eta(x_2, y_2, L_D, D)$, is beyond the scope of this discussion. However, for example, based on simpler analyses, $\eta_B(x_2, y_2)$ and $\eta_E = (x_2, y_2)$, the fragments of the curves of the constant exergy and energy efficiencies in the coordinates system (x_2, y_2), at constant values of dimensions $D = 0.1$ m and $L_D = 0.1$ m, are presented in Figure 10.12. The area of the greatest efficiency values corresponds to the small values x_2 and y_2. However, x_2 cannot be chosen to be too small ($x_2 \geq x_{2min}$), because the characteristic distance L_S becomes negative, which means that the reflector surface is not fully utilized, as its lower part is not exposed to the radiation of surface 1. Parameter y_2 also cannot be too small ($y_2 \geq y_{2min} = L_D + D$), because the assumed configuration of the three surfaces in the presented example (the cooker does not stick out above the reflector rim) would not be preserved. Additionally, for the x_2 and y_2 values close to their minimum values, there is an increase in the significance of the distribution of local values of the view factor, which weakens assumptions about the uniform distribution of this factor, applied in the present consideration.

Optimal values of x_2 and y_2, at constant L_D and D, can define the shape of the cooker surface profile for which the energy efficiency is maximal. However, it is worth emphasizing that based on some additional calculations, it has been found that the optimal profile, when scaled proportionally, still ensures the same value of the maximum efficiency, and the capacity of the SCPC increases by the scale-up factor. For example, it can be concluded that the space (volume) occupied by one big SCPC of surface area A_1 can be reduced by the application of the cluster of number i of small SCPCs of which each has the surface area $A_{1,i}$, where $i \approx A_1 / A_{1,i}$. However, due to the i-times decreased diameter D, to maintain the same flow rate of the heated fluid, the flow velocity of this fluid also has to be increased i times, which would result in a greater pressure drop of the fluid. Thus, for example, the minimum summary cost of a reduced cooker size and the increased cost of a pumping fluid can be optimized.

10.5.5   Conclusion Regarding the Solar Cylindrical–Parabolic Cooker

The theoretical energy and exergy analyses and the distribution of exergy losses for a solar cylindrical–parabolic cooker were presented. Also shown was the exergy analysis of the radiating surface absorbing many radiation fluxes of different temperatures. The imagined surface was introduced into the analysis to close the considered system of cooker surfaces. The optimization possibilities for the surface configuration, to increase both energy and exergy efficiencies of the cooker, were discussed.

It was shown that from the energy viewpoint the low efficiency is mainly due to the escape of a large amount of insolation that is not absorbed, and additionally due to heat loss to the environment. Exergy efficiency is even lower than energy efficiency, mainly due to the large exergy of the escaping insolation and additionally due to degradation of the insolation absorbed on the surfaces of the reflector and the cooking pot. The energy efficiency of the SCPC is relatively low, i.e., it ranges from 6% to 19%, and the exergy efficiency is even lower by about ten times. The presented theoretically calculated values of the efficiencies are relatively close to the values experimentally measured by Ozturk (2004).

The influence of the geometric configuration of the cooker on its performance was outlined. By applying variation only of the *openness* (x_2) and *depth* (y_2) of the SCPC, it was shown that energy efficiency above 18%, and exergy efficiency above 1.6% could both be reached. It can be confirmed by calculation that the determined optimal surface profile of the considered SCPC can be scaled up—at the unchanged optimal efficiencies—to the SCPC with all the dimensions changed proportionally. However, to maintain the same capacity, an appropriate adjustment of the flow velocity of the heated fluids is required. Optimization of the geometry of the cooker, including additional dimensional parameters, can be considered in the future.

Nomenclature for Chapter 10

A surface area, m^2
B radiation exergy, W
b exergy of emission density, W/m^2
c specific heat of water, W/(kg K)

D	outer diameter of cooking pot, m
E	emission energy, W
e	emission density, W/m^2
h	convective heat transfer coefficient, $W/(m^2\ K)$
I	direct insolation, W
J	radiosity, W
k	heat transfer coefficient, $W/(m^2\ K)$
L	SCPC length, m
L	mean distance from the sun to the earth, $L = 1.495 \times 10^{11}$ m
L_c	crossed string length, m
L_D	distance of cooking pot from the reflector bottom, m
L_n	not crossed string length, m
L_s	intersection ordinate, m
m	water mass flow rate, kg/s
Q	heat delivered or extracted from surface, W
q	heat, W/m^2
R	radius of the sun, $R = 6.955 \times 10^8$ m
r	exergy/energy growth ratio
S	solar irradiance, W/m^2,
S, S'	tangency points
S_{cir}	length of circle arc, m
SCPC	solar cylindrical–parabolic cooker
s	entropy of the emission density, $W/(m^2\ K)$
x_s, y_s	coordinates of tangency point S (Fig. 10.11) defining shape and size of the reflector, m
x_2, y_2	coordinates of point 2 (Fig. 10.11), m
T	absolute temperature, K
t	temperature,°C

Greek

α	absorptivity
β	percentage energy loss
β	flat angle coordinates (azimuth), deg
δB	exergy loss, W
η_E	energy efficiency of the SCPC, harvesting of solar radiation
η_B	exergy efficiency of the SCPC, harvesting of solar radiation
φ	view factor
φ	flat angle coordinate (declination), deg
ε	emissivity
ψ	maximum exergy/energy radiation ratio
Π	overall entropy growth, $W/(m^2\ K)$
ρ	reflectivity
σ	Boltzmann constant for black radiation, $\sigma = 5.667 \cdot 10^{-8}\ W/(m^2\ K^4)$
ξ	percentage exergy loss
τ	transmissivity
ω	solid angle, sr

Subscripts

atm	atmosphere
a	air
a	absorbing surface
c	canopy
C	Carnot
earth	earth
i	initial or successive number
j	successive number
max	maximum
opt	optimum
pla	exposed to the plate
p	plate
Q	heat
q	heat
S	solar
sol	exposed to the sun
sky	sky
w	water
ω	solid angle
0	environment
1	imagined surface making up the SCPC surfaces system
2	inner surface of reflector
3	outer surface of cooking pot

## CHAPTER 11

Thermodynamic Analysis of a Solar Chimney Power Plant

11.1 Introduction

Analysis of a solar chimney power plant (SCPP) is presented in this chapter according to Petela (2009). The SCPP is typical of many possible examples of power plants driven by solar radiation. The overall process of power generation in the SCPP is very complex. Up to the present date, only selected aspects have been studied. The present study attempts to develop an analysis of the total SCPP process. The complexity of such a thermodynamic object forces many simplifying assumptions. This necessity for simplification should not seem discouraging because, e.g., the efficiency of the Carnot cycle was also derived with far-reaching simplifications and despite this has a fundamental significance in thermodynamics.

Although not easy to prove, it is supposed that the proposed mathematical thermodynamic model has enough information to determine the effects of varying input parameters on the SCPP output parameters, especially determining the trends for these effects.

The proposed model involves some magnitudes that, although they do not precisely determine a real situation (e.g., the effective temperature of a surface or the average convective coefficients of heat transfer), they must, however, not be assumed constant, i.e. they have a certain freedom to vary and respond to show their approximate values and trends of variation.

We introduce the following characteristic elements:

- formulation of the energy balance of the total SCPP;
- application of the exergy balance for interpretation of component processes;

- application of the eZergy balance for estimation of the effect of gravity;
- exchange of radiation energy and exergy between the chimney and deck;
- distinguishing the energy, exergy, and eZergy losses to the environment and sky;
- the concept of a convective–radiative effective temperature for the surfaces.

The purpose of the present study is not to perform optimization and only the optimum ratio $r_T = 2/3$ for the relative pressure drop in a turbine is assumed. Here we outline the methodology of exergy analysis applied to the SCPP and develop possible different thermodynamic interpretations of processes occurring within it. Based on the energy and exergy balances, the distribution of solar input between the SCPP components is determined. The applied concept of mechanical exergy of air (air eZergy) permits additional interpretation. Based on the eZergy of air, the positive gravity input is determined for all the components in which air plays a role. Discussion of the meaning, significance, and interpretation of the concept of gravity input, which can be positive, negative, or neutral, remains still open.

Obviously, there are many more problems to study in the future, preferably within the more complex thermodynamic model of the SCPP.

11.2 Description of the Plant as the Thermodynamic Problem

A typical SCPP consists of a circular greenhouse-type collector with a tall chimney at its center. Air flowing radially inward under the collector deck is heated from the collector floor and deck, and enters the chimney through a turbine.

Figure 11.1 depicts an example of an SCPP selected for the present study. Draft-driven environmental air (point 0) enters the collector through the gap of height H_e. The collector floor of diameter D_f is under the transparent deck, which declines appropriately to ensure the constant radial cross-sectional area for the radially directed flow of the air. The assumption of a constant cross-sectional area in the collector means that $\pi \times D_f \times H_e = \pi \times D_1 \times H_1 = \pi \times D_1^2/4$, and so the assumed value H_e allows for calculation of the inlet turbine diameter $D_1 = (4 \times H_e \times D_f)^{1/2}$ and height $H_1 = D_1/4$. The collector floor preheats air from state 0 to state 1 (state 1 prevails in the zone denoted by a dashed line). The preheated air (state 1) then expands in the turbine to state 2. The turbine inlet and outlet diameters are D_1 and D_2,

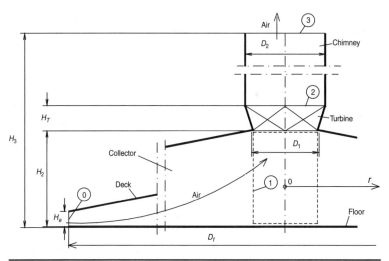

FIGURE 11.1 Scheme of the considered SCPP (from Petela, 2009).

respectively. The height of the turbine is H_T ($H_1 + H_T = H_2$). Expanded air leaves the SCPP (at point 3) through the chimney at height H_3.

For the established geometrical parameters of the collector–turbine–chimney system, and for the constant thermodynamic input data such as solar radiation intensity and environment parameters, the system spontaneously self-models itself in response to the actual situation. This means that the buoyancy effect determines the flow rate of air through the system as well as all the air parameters, temperature, and pressure along the path of the air flow.

Exact thermodynamic consideration of the SCPP requires the involvement of many different theoretical areas such as, e.g., thermodynamics, heat transfer, and fluid mechanics. Exergy analysis especially requires taking into account the temperature distribution over surfaces (e.g., floor, deck, or chimney), and requires complex expression of the exergy of nonequilibrium radiation emitted from these surfaces—each element of the surface has a different temperature and radiates different exergy.

Energy analysis is a little simpler because the energy of radiation from a surface is calculated with a simpler formula, and the energy fluxes need not be followed separately for each particular temperature during successive reflections between gray surfaces. (The energy of radiation, in contrast to exergy, does not distinguish between radiation fluxes of different temperatures.)

Complex, irregular (i.e., selective) spectra of radiating bodies (e.g., air, floor, partly transparent deck, etc.) make consideration additionally difficult. There are problems with a locally considered view factor for radiation transfer: varying conditions for convective coefficients (e.g., the dependence of materials properties on thermodynamic parameters; specific heat of air, viscosity coefficient, etc.); distribution of air flow velocity, in at least a two-dimensional space, including possible turbulent behavior of air; conservation of momentum; and energy conversion from kinetic to potential forms.

The processes can occur in an unsteady mode, especially when starting, restarting, or shutting down periods. The boundary conditions and the environment conditions can vary significantly. All such factors exacerbate the difficulties of calculations.

Generally, there are three methods for obtaining the solution in a specific study of thermodynamic objects.

First, there is the *analytical method*, which is expected to produce a solution of differential equations in an analytical form (usually successful in a very simplified case). Then, from the solution of the differential problem, ordinary solvable equations can be obtained. Usually, the introduction of many simplifying assumptions allows us to pass over the stage of formulating differential equations and directly develop regular algebraic equations that can be solved if the number of unknowns is not larger than the number of derived equations. The present study belongs in this category.

Second, there is the *numerical method*, which solves numerically by developing differential equations into finite difference equations. This allows for significantly fewer simplifying assumptions. However, the numerical method as a replacement for the analytical approach brings some inadequacy via the discrete use of variables.

Third, there is the *similarity theory method*. According to this method the characteristic dimensionless simplexes (i.e., similarity criteria) are extracted from differential equations. The criteria are used for derivation of mutual relations based on experimental data from the appropriately programmed measurements. The relations are fragmentary solutions, and have a specific meaning for a particular integrals of a differential problem. In order to formulate an interpretative model for a process the similarity theory may apply experimental data obtained on a laboratory, pilot, or commercial scale. This method can be applied to a SCPP.

SCPPs belong to the category of thermodynamic objects for which the fully developed application of any of these three methods is practically very difficult, especially when exergy is involved. Therefore, for engineering purposes certain simplified models of processes are usually formulated. These models ignore some less important component phenomena, providing a better chance for an achievable solution. The purpose of even a simplified solution is to ensure the possibility of measurable estimation of basic relations between process parameters, and also the possibility for the design and selection of optimal versions, all without uncertain guesses.

Obtaining a complete and exact mathematical description of the SCPP from a thermodynamic viewpoint is very complex. Formulation of the description based on mathematical analysis leads to differential equations for mass, momentum, energy conservation, and exergy balance for which the more details of the processes that are included into the analysis, the more difficult they are to solve. However, the more details that are considered, the higher is the significance of the obtained solution. Therefore, the optimization problem depends on selection of the optimal precision degree of the thermodynamic analysis.

Simplified models often need to be applied based on an assumption eliminating many of the real features of the thermodynamics problem. Some models simplify more than others. For each model, the expectation of exact results diminishes with the growing number of assumptions. However, simplified models usually describe at least the trends and the intensity of the response from dependent object parameters to the varying input parameters of the object. Such results are often quite satisfactory for practical purposes.

The idea of an SCPP was initiated by Schlaich in the late 1970s. The first 36-kW pilot SCPP plant began operation in Manzanares, near Madrid, Spain. The chimney was 195 m high and the radius of the collector was 120 m. Gradually, the SCPP concept was developed further and many publications discussed various aspects of a solar plant including complex processes of heat transfer and fluid mechanics. For example, Pasumarthi and Sherif (1998a,b) developed the mathematical model of a collector for studying the temperature and flow velocity of heated air and determining the expected SCPP efficiency and power. Padki and Sheriff (1992) considered the effects of geometrical parameters on the chimney performance. The temperature and pressure fields for flowing air were studied both analytically, e.g., by Ming et al. (2006), and analytically and numerically, e.g., by Pastohr et al. (2004), estimating also the varying temperatures of the ground and heat transfer coefficients.

Many studies also focus on the pressure drop across the turbine as a part of the total available pressure difference in the system. The pressure drop was evaluated, e.g., by Haaf et al. (1983), Mullet (1987), Schlaich (1995), Gannon and Backström (2000), Bernardes et al. (2003), and recently by Von Backström and Fluri (2006). Thermodynamic variables regarding the chimney dependent on wall friction, additional losses, internal drag, and cross-sectional area for chimneys as tall as 1500 m were studied by Backtröm and Gannon (2000). Exergy interpretation of the chimney process was analyzed by Petela (2008d). However, it is worth noting that, due to the complexity of the object, most existing publications consider certain selected aspects of SCPP, but none consider the entire SCPP with all its components.

In the following sections, the methodology for full thermodynamic analysis including eZergy is outlined by a simplified interpretative mathematical model of the SCPP.

11.3 The Main Assumptions for the Simplified Mathematical Model of the SCPP

In the existing literature there is no detailed, readily applicable, or consistent data on all the SCPP parameters. Therefore, the required data are generated by the proposed simplified model, which is considered under the following main assumptions:

(i) The floor has no heat loss to the soil. It is perfectly insulated and is perfectly black (emissivity $\varepsilon_f = 1$). Thus, there is no solar energy reflected from the floor. It is worth noting that a further simplification, not applied in the present consideration, could be the assumption that the floor material be of almost infinitely large conductivity, which then could motivate the assumption of a constant temperature of the floor in the entire collector.

(ii) The deck material is prepared in such a way that it is almost perfectly transparent for solar radiation (transmissivity $\tau_d = 0.95$) and the remaining part (5%) of solar radiation arriving at the deck is reflected. However, the deck material absorbs perfectly (absorptivity $\alpha = 1$) any low-temperature radiation, e.g., from the floor. Thus, consideration of multi-reflected radiation fluxes is simplified. In addition, the deck is thin enough that heat conducted through the deck occurs at a zero temperature gradient. The properties of the deck are assumed so as to better expose the effect of trapping solar radiation energy within the collector.

(iii) Variation of floor temperatures with the radial coordinate r was analyzed by Pastohr et al. (2004). Depending on the model used, they received relatively significantly different distributions for the temperature, though it was not clear why. The literature data on temperature distribution at the deck surface is very limited. Therefore, for the present study, a certain *effective temperature* T_{eff} for the floor or deck is applied, according to consideration in Section 6.10.2, which includes potentials for the heat transfer by conduction and radiation:

$$A \left(h_{\text{average}} \, T_{\text{eff}} + \sigma T_{\text{eff}}^4 \right) = \int_A h_{\text{local}} \, T \, dA + \sigma \int_A T^4 \, dA \qquad (11.1)$$

where h_{average} and h_{local} are the average and local convective heat transfer coefficients, respectively, σ is the Boltzmann constant for black radiation, A is the surface area, and T is the local surface temperature. As some computation estimates indicate, the concept of effective temperature can be applied only when the surface temperature varies within a not too large range, as in the case of the floor and deck considered.

(iv) Chimney material is perfectly black. The chimney wall is thin, thus there is no temperature gradient along the wall thickness and both sides of the chimney (inner and outer) have the same temperature constant along the chimney height.

(v) Distribution of air temperature is represented by certain effective temperatures defined according to equation (11.1), however, with excluded radiative heat transfer.

(vi) Air is considered to be an ideal gas, the parameters for which fulfill the state equation $p = \rho \times R \times T$, equation (2.1a), and the specific heat is assumed to be constant (i.e., average, not varying with temperature).

(vii) Air is almost perfectly transparent for radiation (transmissivity $\tau_a \approx 1$ and emissivity $\varepsilon_a \approx 0$). Air can exchange heat only by convection or conduction.

(viii) Air flow in the entire SCPP is frictionless. The relative air pressure drop r_T during expansion inside a turbine is estimated differently by many authors, as discussed by Backström and Fluri (2006). The drop is considered in the range from 0.66 (e.g., by Mullet, 1987) to 0.97 (by Bernardes et al., 2003). According to investigation by Backström and Fluri (2006), "for maximum fluid power, the optimum ratio" $r_T = 2/3$. Therefore in the present consideration it is assumed that:

$$\frac{p_1 - p_2}{p_1 - p_3} = r_T = \frac{2}{3} \tag{11.2}$$

As mentioned in Section 11.1, optimization of the SCPP is not the objective of the present work and only the suggested optimum value r_T is assumed for analysis of the SCPP. For illustration purposes, based on later calculation results (Table 11.1, column 4), the distribution of the environment pressure and the pressure of air along its flow within the SCPP are shown in Figure 11.2. The air pressure of the environment (solid thin line) drops from p_0 at the zero level, to p_3 at the level H_3 of the chimney exit. The air pressure inside the SCPP drops from p_0 to p_1 at the collector outlet (dashed line), and it is assumed that the same pressure p_1 prevails also at the inlet to the turbine. Then, within the turbine, air pressure (thick solid line) drops from p_1 to p_2 during adiabatic (isentropic) expansion generating power. Air from turbine flows upward and its pressure (dotted line) achieves value p_3 at the chimney exit.

(ix) Using average values of gravitational acceleration and air density along the height H_3:

$$p_3 = p_0 - \frac{g_0 + g_3}{2} \frac{\rho_0 + \rho_{03}}{2} H_3 \tag{11.3}$$

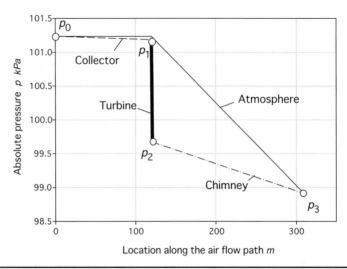

Figure 11.2 Distribution of the absolute pressure in the considered SCPP (from Petela, 2009).

where the following approximations, used by Petela (2008b), were applied: $g_{H3} = g_0 - 3.086 \times 10^{-6} \times H_3$ and $\rho_{H3} = \rho_0 - 9.973 \times 10^{-5} \times H_3$. At the earth's surface the atmospheric pressure $p_0 = 101.235$ kPa and gravitational acceleration $g_0 = 9.81$ m/s².

(x) The momentum conservation equation for the air flow within the collector is derived as:

$$p_0 - p_1 = \rho_{a1} w_1^2 \qquad (11.4)$$

where ρ_{a1} and w_1 are the density and flow velocity, respectively, of air at point 1.

(xi) The deck and chimney radiate to the space of sky temperature T_{sky}. Therefore, for a clear sky, formula (6.76) is applied.

(xii) In order to obtain the fair comparison basis, the reference state for calculation of energy is the same as for exergy: environment temperature $T_0 = 288.14$ K, (15°C), and environment pressure $p_0 = 101.235$ kPa.

Additional assumptions are discussed in the following sections.

11.4 Energy Analysis

Energy analysis is based on the energy conservation equations. The energies E are used in six equations written successively for: floor surface, air in collector, collector (including floor, air, and deck), turbine, chimney, and chimney surface:

$$E_{s-f} = E_{f-a} + E_{f-d} \tag{11.5}$$

$$E_{f-a} + E_{d-a} = E_{a1} + E_{w1} + E_{p1} \tag{11.6}$$

$$E_{s-f} = E_{a1} + E_{w1} + E_{p1} + E_{d-sky} + E_{d-0} + E_{d-ch} \tag{11.7}$$

$$E_{a1} + E_{w1} + E_{p1} = E_{a2} + E_{w2} + E_{p2} + E_P \tag{11.8}$$

$$E_{a2} + E_{w2} + E_{p2} + E_{d-ch}$$
$$= E_{a3} + E_{w3} + E_{p3} + E_{ch-0} + E_{ch-sky} + E_{ch-gr} \tag{11.9}$$

$$E_{a-ch} + E_{d-ch} = E_{ch-0} + E_{d-sky} + E_{ch-gr} \tag{11.10}$$

Energies E have the following subscripts:

s–f	- solar radiation arriving at the floor;
f–a	- convection heat from floor to air;
f–d	- energy exchanged by radiation between floor and deck;
d–a	-convection heat from deck to air;
d–sky	- energy exchanged by radiation between deck and sky;
d–0	- convection heat from deck to atmosphere;
d–ch	- energy exchanged by radiation between deck and chimney;
ch–0	-convection heat from chimney surface to atmosphere;
ch–sky	- energy exchanged by radiation between chimney surface and sky;
ch–gr	- energy exchanged by radiation between chimney surface and ground;
a–ch	- heat transferred from chimney air to the chimney surface;
$1a$, $2a$, $3a$	- enthalpy of air at point 1, 2, and 3;
$w1$, $w2$, $w3$	- kinetic energy due to the air flow velocity w_1, w_2, and w_3;
$p1$, $p2$, $p3$	- potential energy of air at point 1, 2 and 3;
P	- turbine power.

Kinetic energies are calculated as $E_w = m \times w^2/2$, where m is the air mass flow rate, $m = 0.25 \times \pi \times D_1^2 \times w_1 \times \rho_{a1}$. Enthalpy of the air is $E_a = m \times c_p \times (T_a - T_0)$, where c_p is the specific heat of air at constant pressure.

The potential energy of the considered air, at its constant density ρ, depends on the altitudinal variation of atmospheric air density and gravitational acceleration. The solution of differential formula on potential energy E_p, J/kg, equal to potential exergy, can be determined based on formula (f) in Section 2.6.2:

$$E_p = m \left\{ -\frac{1}{a_4 \rho} \left[\frac{a_2}{6 \, a_4} (\rho - a_3)^3 + \frac{a_1}{2} (\rho - a_3)^2 \right] \right\} \qquad (11.11)$$

where a_1, a_2, a_3, and a_4 are constant values (see Nomenclature).

Total solar energy received by the floor is:

$$E_{S-f} = \tau_d \varepsilon_f S A_d \qquad (11.12)$$

where S, W/m^2, is the solar radiosity at the earth surface, τ_d is the transmissivity of the deck, and ε_f is the floor emissivity, ($\varepsilon_f = 1$).

Energy exchanged by radiation between deck and chimney is:

$$E_{d-ch} = \varepsilon_d \varphi_{d-ch} \frac{\pi}{4} \left[D_f^2 - (c_D D_2)^2 \right] \sigma \left(T_{d,\text{eff}}^4 - T_{ch}^4 \right) \qquad (11.13)$$

where $\sigma = 5.6693 \times 10^{-8}$ W/(m^2 K^4) is the Boltzmann constant for black radiation, $T_{d,\text{eff}}$ is the effective temperature of the deck, and c_D is the factor to account on thickness of the chimney wall. The view factor φ_{d-ch} can be calculated from the reciprocity relation:

$$\varphi_{d-ch} \frac{\pi}{4} \left[D_f^2 - (c_D D_2)^2 \right] = \varphi_{ch-d} \pi c_D D_2 (H_3 - H_2) \qquad (11.14)$$

It can be derived that $\varphi_{ch-d} = 0.5 \times (90 - \beta)/90$, where the angle β is determined by $\tan \beta = 2 \times H_3 / D_f$.

Energy exchanged by radiation between the floor and deck:

$$E_{f-d} = A_d \sigma \left(T_{f,\text{eff}}^4 - T_{d,\text{eff}}^4 \right) \qquad (11.15)$$

where $T_{f,\text{eff}}$ is the effective temperature of the floor and surface area $A_d = \pi \times (D_f^2 - D_1^2)/4$.

The following formulae are applied for convection heat transfer from:

floor to air:

$$E_{f-a} = A_d h_{f-a} \left(T_{f,\text{eff}} - T_{a,\text{eff}} \right) \qquad (11.16)$$

deck to air:

$$E_{d-a} = A_d h_{d-a} \left(T_{d,\text{eff}} - T_{a,\text{eff}} \right) \qquad (11.17)$$

deck to environment:

$$E_{d-0} = A_d h_{d-0} \left(T_{d,\text{eff}} - T_0 \right) \qquad (11.18)$$

chimney to environment:

$$E_{ch-0} = A_{ch} h_{ch-0} \left(T_{ch} - T_0 \right) \qquad (11.19)$$

and chimney air to chimney wall:

$$E_{a-ch} = \pi D_2 \left(H_3 - H_2\right) h_{a-ch} \left(\frac{T_{a,2} + T_{a,3}}{2} - T_{ch}\right) \tag{11.20}$$

where h is the respective coefficient and the chimney surface $A_{ch} = \pi \times c_D \times D_2 \times (H_3 - H_2)$. The coefficient h_{a-ch} is determined as $h_{a-ch} = Nu \times k/D_2$, where $k = 0.0267$ W/(m K) is thermal conductivity of air and the Nusselt number $Nu = 0.023 \times Re^{0.8} \times Pr^{0.4}$; the Prandtl number for air is $Pr = 0.7$; and the Reynolds number $Re = w_2 D_2/\nu$ (kinematic viscosity coefficient for air $\nu = 1.6 \times 10^{-5}$ m^2/s).

In a similar way h_{f-a} is the determined coefficient. Although the air flow is driven by the buoyancy effect, the forced convection mechanism of the air flow is assumed. Thus, the calculations are based on the Reynolds number instead of the Grashof number. In calculations the average flow velocity of air is assumed. The effective diameter D_{eff} for the air flow was assumed as the average ratio of the respective flow cross-sectional areas (A_1) multiplied by four, to the respective perimeter lengths L_0 or L_1; $D_{eff} = 2 \times A_1/(1/L_0 + 1/L_1)$. It was assumed that $h_{a-d} = h_{f-a}$.

The following formulae are applied for energy exchange by radiation between:

floor and deck:

$$E_{f-d} = A_d \sigma \left(T_{f,\text{eff}}^4 - T_{d,\text{eff}}^4\right) \tag{11.21}$$

deck and chimney:

$$E_{d-ch} = \varphi_{d-ch} A_d \sigma \left(T_{d,\text{eff}}^4 - T_{ch}^4\right) \tag{11.22}$$

deck and sky:

$$E_{d-\text{sky}} = \varphi_{d-\text{sky}} A_d \sigma \left(T_{d,\text{eff}}^4 - T_{\text{sky}}^4\right) \tag{11.23}$$

chimney and sky:

$$E_{ch-\text{sky}} = \varphi_{ch-\text{sky}} A_{ch} \sigma \left(T_{ch}^4 - T_{\text{sky}}^4\right) \tag{11.24}$$

chimney and ground beyond the floor:

$$E_{ch-gr} = \varphi_{ch-gr} A_{ch} \sigma \left(T_{ch}^4 - T_{gr}^4\right) \tag{11.25}$$

where the view factors fulfill the following relations:

$$\varphi_{d-\text{sky}} + \varphi_{d-ch} = 1 \tag{11.26}$$

$$\varphi_{ch-\text{sky}} + \varphi_{ch-d} + \varphi_{ch-gr} = 1 \tag{11.27}$$

The view factors φ_{ch-d} and φ_{d-ch} are determined based on equation (11.14), whereas the configuration of the chimney relative to the sky determines the view factor $\varphi_{ch-sky} = 0.5$.

Calculation of temperature $T_{a,2}$ is based on the equation for the isentropic expansion in a turbine at an assumed isentropic exponent κ for air and the internal efficiency of the turbine η_T. Conversion of the energy of air into electric power occurs at an overall efficiency η_p, which additionally includes mechanical and electric efficiencies of the turbine–generator unit.

The energy calculations were carried out with additional assumptions. The air temperature distribution in the collector was assumed to be linear and thus $T_{a,\text{eff}} = (T_0 + T_{a1})/2$. The diameter ratio $D_1/D_2 = 0.95$. Based on additional calculations the air temperature drop in the chimney can be estimated as proportional to the chimney surface and inversely proportional to the air mass rate, $T_{a,2} - T_{a,3} = 0.154 \times D_2 \times H_3/m$.

We explore an example of computation using the presented model as follows. Applying the consideration to the pilot plant at Manzanares the floor diameter is $D_f = 240$ m and the chimney height is $H_3 = 195$ m. Other data are as follows:

$s = 800 \text{ W/m}^2$	$\eta_T = 0.7$	$H_e = 0.3$ m
$\kappa = 1.4$	$h_{d-0} = 5 \text{ W/(m}^2 \text{ K)}$	$c_p = 1000 \text{ J/(kg K)}$
$h_{ch-0} = 7 \text{ W/(m}^2 \text{ K)}$	$c_D = 1.015$	$H_T = 1$ m
$T_{gr} = T_0$	$R = 287.04 \text{ J/(kg K)}$	

The model responses to the input parameters are presented in Tables 11.1 and 11.2. Column 4 of the tables presents the basic input case (i.e., reference case), the results for which are compared to the results of other input cases represented by columns 5–10 and which are discussed in Section 11.7. Column 4, with relatively low power, presents the closest case to the power reported for the Manzanares pilot plant. The energy results of this column are used for the respective bands diagram (Figure 11.3). The diagram shows how the solar radiation energy arriving at the deck $E_S = 39.05$ MW, reduced by 5% reflection, is distributed between five SCPP components: collector air, floor, deck, turbine, and chimney.

In the diagram the energy streams E, W, are represented by their percentage values e related to the solar radiation energy E_S. The floor (blackbody) fully absorbs the solar radiation (95.00%) transmitted through the deck and converts this radiation energy into the energy at the level of temperature $T_{f,\text{eff}}$. Part of this energy ($e_{f-d} = 77.19\%$) radiates to the deck and the rest $e_{f-a} = 17.81\%$ is transferred by convection to heated air in the collector. The power performed by the turbine is relatively small ($E_P = 0.23$ MW) mostly due to the small mass flow rate of air ($m = 276$ kg/s) and due to the small pressure drop during

#	Quantity	Units	Ref. value	Mono-variant changes of input parameters and resulting output			
1	2	3	4	5	6	7	8
2	**Input**						
3	S	W/m^2	**800**	**850**	800	800	800
4	H_3	m	**195**	195	**200**	195	195
5	D_f	m	**240**	240	240	**250**	240
6	He	m	**0.3**	0.3	0.3	0.3	**0.35**
7	**Output**						
8	p1	Pa	101,233.66	101,233.93	101,233.67	101,233.83	10,1231.82
9	p2	Pa	99,686.13	99,686.22	99,646.76	99,686.19	99,685.52
10	p3	Pa	98,912	98,912	98,853	98,912	98,912
11	w1	m/s	1.10	0.99	1.09	1.03	1.67
12	m	kg/s	276	245	274	268	501
13	Tfeff	K	388.3	394.5	388.4	389.3	381.7
14	Tdeff	K	329.8	333.8	329.9	330.6	325.0
15	Taeff	K	303.18	304.62	303.20	304.02	299.20
16	Ta1	K	318.19	321.08	318.24	319.89	310.25
17	Tch	K	292.43	292.90	292.35	292.65	291.81
18	**Energy**						
19	ea1	%	22.99	21.09	22.92	21.75	30.78

TABLE 11.1 Responsive Trends of Some Output Parameters to Changes of Some Input Parameters (from Petela, 2009)

#	Quantity	Units	Ref. value	Mono-variant changes of input parameters and resulting output			
1	2	3	4	5	6	7	8
20	ea2	%	22.24	20.45	22.15	21.07	29.45
21	ea3	%	20.75	19.05	20.63	19.67	27.84
22	ew1	%	4.63e-4	3.11e-4	4.56e-4	3.64e-4	1.95e-3
23	ew2	%	1.93e-4	1.30e-4	1.91e-4	1.52e-4	8.12e-4
24	ew3	%	3.88e-4	2.61e-4	3.83e-4	3.05e-4	1.64e-3
25	ep1	%	0.3994	0. 4020	0.3988	0.3995	0.3871
26	ep2	%	0.5118	0.5046	0.5139	0.5055	0.5396
27	ep3	%	0.5278	0.5129	0.5300	0.5180	0.5951
28	efa	%	17.81	16.23	17.76	16.89	23.73
29	efd	%	77.19	78.77	77.24	78.11	71.27
30	eda	%	5.577	5.259	5.562	5.258	7.435
31	ed0	%	26.05	26.82	26.07	26.52	23.05
32	edsky	%	44.21	45.27	44.24	45.00	39.51
33	edch	%	1.35	1.42	1.36	1.34	1.26
34	echO	%	0.90	0.94	0.90	0.89	0.83
35	echsky	%	1.70	1.64	1.74	1.61	1.78
36	echgr	%|	0.230	0.241	0.235	0.224	0.212

	EP	kW	229	203	234	222	423
37							
38	$eP(\eta_E)$	%	0.64	0.53	0.65	0.57	1.18
39	**Exergy**						
40	ba1	%	1.2443	1.2439	1.2423	1.2393	1.2434
41	ba2	%	0.08	0.27	0.06	0.20	-0.83
42	ba3	%	-0.61	-0.34	-0.65	-0.44	-1.95
43	bw1	%	5.148e-4	3.46e-4	5.07e-4	4.05e-4	2.16e-3
44	bw2	%	2.15e-4	1.44e-4	2.12e-4	1.69e-4	9.03e-4
45	bw3	%	4.31e-6	2.89e-6	4.25e-6	3.39e-6	1.82e-5
46	bp1	%	0.4438	0.4467	0.4431	0.4439	0.4301
47	bp2	%	0.5687	0.5607	0.5710	0.5617	0.5995
48	bp3	%	0.00586	0.00570	0.005889	0.00576	0.00661
49	bfa	%	5.10	4.86	5.09	4.88	6.46
50	bfd	%	17.24	18.58	17.26	17.62	14.89
51	bda	%	0.783	0.798	0.782	0.750	0.937
52	bd0	%	3.657	4.072	3.664	3.781	2.907
53	bdsky	%	2.237	2.640	2.242	2.343	1.623
54	bdch	%	0.114	0.131	0.115	0.115	0.095
55	bch0	%	0.0145	0.0168	0.0144	0.0151	0.0115

Table 11.1 Responsive Trends of Some Output Parameters to Changes of Some Input Parameters (from Petela, 2009) (Continued)

| # | Quantity | Units | Ref. value | Mono-variant changes of input parameters and resulting output | | | |
1	2	3	4	5	6	7	8
56	$bchsky$	%	-0.044	-0.041	-0.046	-0.042	-0.049
57	$bchgr$	%	0.182	0.215	0.189	0.176	0.143
58	Δbf	%	72.66	76.55	77.65	77.51	78.65
59	Δba	%	4.20	3.97	4.19	3.94	5.73
60	Δbd	%	10.4496	10.9423	10.4600	10.6287	9.3263
61	ΔbT	%	0.33	0.28	0.34	0.29	0.60
62	Δbch	%	1.21	1.10	1.23	1.16	1.70
63	$bP\,(\eta_B)$	%	0.70	0.59	0.73	0.63	1.31
64	**eZergy**						
65	$za1$	%	8.96	8.37	8.94	8.57	11.42
66	$za2$	%	8.82	8.25	8.79	8.44	11.11
67	$za3$	%	8.24	7.68	8.19	7.88	10.56
68	zGa	%	7.28	6.68	7.26	6.89	9.75
69	zGT	%	0.89	0.75	0.91	0.80	1.60
70	$zGch$	%	0.673	0.593	0.675	0.632	1.160

TABLE 11.1 Responsive Trends of Some Output Parameters to Changes of Some Input Parameters (from Petela, 2009) (*Continued*)

#	Quantity	Units	Ref. value	Mono-variant changes of input parameters and resulting output	
1	2	3	4	9	10
2	*Input*				
3	S	W/m^2	**800**	**850**	800
4	H$_3$	m	**195**	195	195
5	D$_f$	m	**240**	240	**250**
6	He	m	**0.3**	**0.35**	**0.35**
7	*Output*				
8	p1	Pa	101,233.66	101,232.6	101,232.3
9	p2	Pa	99,686.13	99,685.76	99,685.68
10	p3	Pa	98,912	98912	98912
11	w1	m/s	1.10	1.47	1.54
12	m	kg/s	276	438	480
13	Tfeff	K	388.3	388.5	383.2
14	Tdeff	K	329.8	329.2	326.1
15	Taeff	K	303.18	300.45	299.97
16	Ta1	K	318.19	312.75	311.79
17	Tch	K	292.43	292.30	292.06
18	*Energy*				
19	ea1	%	22.99	28.17	29.02
20	ea2	%	22.24	27.07	27.84
21	ea3	%	20.75	25.55	26.33
22	ew1	%	4.63e-4	1.24e-3	1.46e-3
23	ew2	%	1.93e-4	5.18e-4	6.10e-4
24	ew3	%	3.88e-4	1.04e-3	1.23e-3
25	ep1	%	0.3994	0.3975	0.3924
26	ep2	%	0.5118	0.5364	0.5357
27	ep3	%	0.5278	0.5802	0.5846
28	efa	%	17.81	21.53	22.38
29	efd	%	77.19	73.47	72.62
30	eda	%	5.577	7.040	7.032
31	ed0	%	26.05	24.16	23.74
32	edsky	%	44.21	40.92	40.59
33	edch	%	1.35	1.35	1.26

TABLE 11.2 Responsive Trends of Some Output Parameters to Changes of Some Input Parameters (from Petela, 2009)

#	Quantity	Units	Ref. value	Mono-variant changes of input parameters and resulting output	
1	**2**	**3**	**4**	**9**	**10**
34	ech0	%	0.90	0.89	0.83
35	echsky	%	1.70	1.71	1.69
36	echgr	%|	0.230	0.227	0.209
37	EP	MW	229	368	404
38	eP (η_E)	%	0.64	0.96	1.03
39	**exergy**				
40	ba1	%	1.2443	1.2614	1.2506
41	ba2	%	0.08	−0.45	−0.58
42	ba3	%	−0.61	−1.40	−1.58
43	bw1	%	5.148e-4	1.38e-3	1.62e-3
44	bw2	%	2.15e-4	5.76e-4	6.77e-4
45	bw3	%	4.31e-6	1.16e-5	1.37e-5
46	bp1	%	0.4438	0.4416	0.4360
47	bp2	%	0.5687	0.5960	0.5952
48	bp3	%	0.00586	0.00645	0.00650
49	bfa	%	5.10	6.18	6.17
50	bfd	%	17.24	16.38	15.42
51	bda	%	0.783	0.98	0.91
52	bd0	%	3.657	3.349	3.071
53	bdsky	%	2.237	2.022	1.756
54	bdch	%	0.114	0.112	0.098
55	bch0	%	0.0145	0.0139	0.0123
56	bchsky	%	−0.044	−0.045	−0.046
57	bchgr	%	0.182	0.178	0.143
58	Δbf	%	72.66	77.44	78.41
59	Δba	%	4.20	5.45	5.39
60	Δbd	%	10.45	9.91	9.58
61	ΔbT	%	0.33	0.49	0.53
62	Δbch	%	1.21	1.50	1.58
63	bP (η_B)	%	0.70	1.07	1.15
64	**eZergy**				

TABLE 11.2 Responsive Trends of Some Output Parameters to Changes of Some Input Parameters (from Petela, 2009) (*Continued*)

#	Quantity	Units	Ref. value	Mono-variant changes of input parameters and resulting output	
1	2	3	4	9	10
65	za1	%	8.96	10.62	10.87
66	za2	%	8.82	10.38	10.61
67	za3	%	8.24	9.84	10.08
68	zGa	%	7.28	8.92	9.19
69	zGT	%	0.89	1.32	1.41
70	zGch	%	0.673	1.00	1.06

TABLE 11.2   Responsive Trends of Some Output Parameters to Changes of Some Input Parameters (from Petela, 2009) (*Continued*)

the air expansion. The percentage power of the turbine $e_P = 0.64\%$ represents the energy efficiency η_E of the SCPP. The exhausted energy (enthalpy) of air from the chimney is $e_{a3} = 20.75\%$, whereas the exhausted potential and kinetic energies are small; $e_{p3} = 0.52\%$ and $e_{w3} = 3.87 \times 10^{-4}\%$, respectively. The other SCPP energy losses are by radiation and convection heat transferred from the deck and chimney to the sky and environment. Solar energy reflected from the deck is assumed as $e_R = 5.00\%$.

11.5 Exergy Analysis

Data obtained from energy analysis can be used for the design, operation, and evaluation of the SCPP. However, the same processes, with the same data, can be also interpreted based on exergy analysis with the use of exergy balance equations. Exergy B in these equations has subscripts corresponding to E in energy analysis. The five separate exergy equations can be written for floor, deck, air in collector, turbine, and chimney.

These five exergy equations are analogous to five energy equations (11.5)–(11.9) and differ by the additional members, ΔB, representing the respective irreversible exergy losses:

$$B_{S-f} = B_{f-a} + B_{f-d} + \Delta B_f \tag{11.28}$$

$$B_{f-d} = B_{d-a} + B_{d-\text{sky}} + B_{d-0} + B_{d-ch} + \Delta B_d \tag{11.29}$$

$$B_{f-a} + B_{d-a} = B_{a1} + B_{w1} + B_{p1} + \Delta B_a \tag{11.30}$$

$$B_{a1} + B_{w1} + B_{p1} = B_{a2} + B_{w2} + B_{p2} + B_P + \Delta B_T \tag{11.31}$$

$$B_{a2} + B_{w2} + B_{p2} + B_{d-ch} = B_{a3} + B_{w3} + B_{p3} + B_{ch-0}$$
$$+ B_{ch-\text{sky}} + B_{ch-gr} + \Delta B_{ch} \tag{11.32}$$

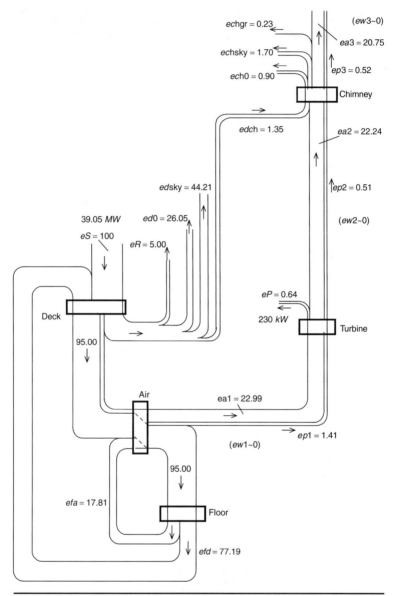

Figure 11.3 Energy balance of the SCPP according to Table 11.1, column 4 (the values are expressed in percent) (from Petela, 2009).

Exergy of solar radiation can be estimated for the radiation temperature slightly smaller than 6000 K. According to data, e.g., in Table 6.1, the exergy B_S can be assumed as being about 90% of radiation energy, $B_S \approx 0.9 \times E_S$.

Generally, radiation exergy B of a surface at temperature T, emissivity ε, and surface area A is determined based on formulas (6.10) applied for the whole area A:

$$B = \varphi A \varepsilon \frac{\sigma}{3} \left(3T^4 + T_0^4 - T_0 T^3\right) \tag{11.33}$$

where φ is the view factor accounting for the geometrical configuration of the considered surface in relation to a surface irradiated by the considered surface. Based on the conclusion of Section 7.5.3, formula (7.77), the exergy of radiation exchanged between any two different surfaces at different temperature T_x and T_y, can be determined by application of formula (11.33) for both considered surfaces, which leads to the following formula:

$$B_{x-y} = B_x - B_y = \varphi_{x-y} A_x \varepsilon_{x-y} \frac{\sigma}{3} \left[3\left(T_x^4 - T_y^4\right) - 4 T_0 \left(T_x^3 - T_y^3\right)\right] \tag{11.34}$$

where ε_{x-y} is the effective emissivity depending on emissivities ε_x and ε_y of the respective surfaces and is calculated identically to radiation energy exchange. The effective emissivity simplifies to $\varepsilon_{x-y} = 1$ when the emissivities $\varepsilon_x = \varepsilon_y = 1$. Formula (11.34) is used appropriately for calculations of the five radiation exergies: B_{f-d}, B_{d-sky}, B_{d-ch}, B_{ch-sky}, and B_{ch-gr}.

The physical exergy of air (B_{a1}, B_{a2}, and B_{a3}) is calculated based on the common formula (2.50):

$$B_a = m \left[c_p \left(T_a - T_0\right) - T_0 \left(c_p \ln \frac{T_a}{T_0} - R \ln \frac{P}{P_0}\right)\right] \tag{11.35}$$

where c_p and R are the specific heat and individual gas constant, respectively. Obviously, exergy of air entering the collector is zero, $B_{a0} = 0$, because air is taken from the environment.

Exergy B of convective heat transferred from a surface at temperature T to air (environmental or heated) is calculated based on the energy E of this heat from the common formula (2.61):

$$B = E \left(1 - \frac{T_0}{T}\right) \tag{11.36}$$

Formula (11.36) is used appropriately for calculations of the four exergies B_{f-a}, B_{d-a}, B_{d-0}, and B_{ch-0}. Potential exergies of air are equal potential energies ($B_{p1} = E_{p1}$, $B_{p2} = E_{p2}$, and $B_{p3} = E_{p3}$). Kinetic exergies of air are equal kinetic energies ($B_{w1} = E_{w1}$, $B_{w2} = E_{w2}$, and $B_{w3} = E_{w3}$).

The results in column 4 of Table 11.1 concerning exergy are used for the respective bands diagram (Figure 11.4). The diagram shows how solar radiation exergy arriving at deck $B_S = 32.41$ MW, reduced by the 5% reflection, is distributed between five SCPP components:

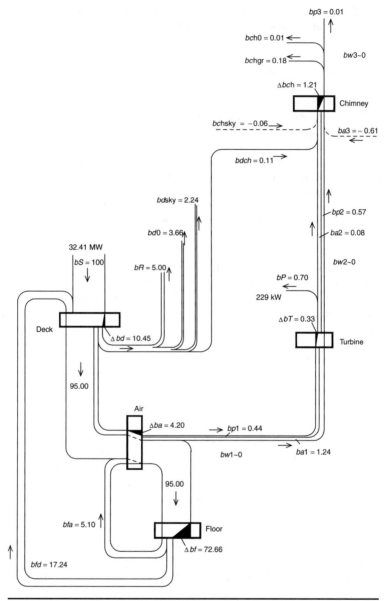

Figure 11.4 Exergy balance of the SCPP according to Table 11.1, column 8 (the values are expressed in percent) (from Petela, 2009).

collector air, floor, deck, turbine, and chimney. In the diagram the exergy streams B, W, are represented by their percentage values b related to the solar radiation exergy B_S. Exergy considerations disclose large degradation of solar radiation. The floor fully absorbs the received high-temperature radiation exergy and converts it to the exergy at the lower temperature $T_{f,\text{eff}}$. Part of this $T_{f,\text{eff}}$ exergy ($b_{f-d} = 17.24\%$) radiates to the deck and another part $b_{f-a} = 5.10\%$ is transferred by convection to heated air in the collector. The remaining large part ($\Delta b_f = 72.16\%$) is lost during irreversible processes of absorption and emission at the floor surface.

The power B_p performed by a turbine is the same as in the energy balance, $B_P = E_P = 0.23\,\text{MW}$ (i.e., exergy of work is equal to the work). Percentage power of turbine $b_P = 0.70\%$ represents the exergy efficiency η_B of the SCPP. Exergy efficiency is slightly higher than energy efficiency because the same power is related to radiation exergy, which is smaller than radiation energy. The exhausted exergy of air from the chimney is negative $b_{a3} = -0.61\%$, whereas the exhausted potential and kinetic exergies are small, $b_{p3} = 0.01\%$ and $b_{w3} = e_{w3}$, respectively. The SCPP loses exergy due to irreversibility and from radiation and convection heat transferred from the deck and chimney to the sky and environment. Solar exergy reflected from the deck is $b_R = e_R = 5.00\%$.

The negative value of the physical exergy (b_a) of air can be stated as follows. The possibility of negative exergy was mentioned in Section 2.6.1. Exergy is a measure of the parameters of matter departing from the equilibrium state with the human environment. Therefore, exergy should always be positive whenever the parameters are different from the parameters of the environment. The exergy of matter within the system (*kind of certain "exergy of internal energy"*) is really always positive, whereas exergy of matter exchanged with the system (*kind of "exergy of enthalpy"*) can be negative for certain combinations of matter parameters. This can happen especially for air, e.g., when the air temperature is not much higher relative to the environment, and the air pressure is lower than atmospheric pressure.

11.6 Exergy Analysis Using the Mechanical Exergy Component for a Substance

Another interpretation of the processes is possible by applying the concept of mechanical exergy, which is one of the exergy components of a substance. As explained in Section 2.6.2, mechanical exergy takes into consideration the theoretical possibility that exergy (i.e., maximum possible work) of a substance can be executed not only at the earth's surface level but also at different altitudes to which the substance can be brought by way of the buoyancy effect. This effect occurs when the density ρ of a considered substance differs from density ρ_0 of the environment, $\rho \neq \rho_0$.

As discussed in Section 4.5.4, if mechanical exergy is applied, then on the input side of the exergy balance equation the gravity input term is introduced. Without such input the equation would not be completed. Thus, the gravity input is calculated from the exergy balance equation in which mechanical exergy of the substance is used. In such an equation the exergy loss is the same as in the traditional exergy balance, i.e., it is calculated from the Guoy–Stodola law.

It is presumed that the value of the gravity input (negative, positive, or zero) expresses the effect of the gravitational field on the processes in which any substance takes part. In the SCPP, five components are considered: floor, deck, air in collector, turbine, and chimney. Exergy balance equations for floors and decks do not contain terms for the substance. These equations contain only terms for radiation and convection heat for which the gravitational effect has not been considered so far. Therefore, the exergy balances for floors and decks remain unchanged. However, the three modified exergy balance equations are considered for heating air in a collector, turbine, and chimney.

We call the mechanical exergy of a substance *eZergy*, denoted by Z, W, or z, %, to easily distinguish it from traditional exergy (B or b), e.g., used in Section 11.5. (The only substance considered in the SCPP is air.) Denotations of other exergy magnitudes remain in the present section unchanged because their values are unchanged. However, eZergy of air generally differs from exergy of air, $Z_a \geq B_a$, which results from the definition of mechanical exergy (eZergy):

$$Z_a = \max\left(B_p + B_H, B_a\right) \tag{11.37}$$

where B_p is the potential exergy ($B_p = E_p$) and B_a is the traditional physical exergy of air calculated from formula (11.35). Magnitude B_H is the physical exergy also calculated based on equation (11.35); however, for the environment parameters (temperature T_H and pressure p_H) prevailing at the altitude H:

$$B_H = m\left[c_p\left(T_a - T_H\right) - T_H\left(c_p \ln \frac{T_a}{T_H} - R \ln \frac{P}{P_H}\right)\right] \tag{11.38}$$

In relation to the formula applied by Petela (2008) the approximations for wider ranges of the considered atmospheric parameters and altitude are applied according to Petela (2009b) as follows:

$$H = 1.215485 \cdot 10^6 - 1.214 \cdot 10^6 \, \rho_a^{6.02353 \cdot 10^{-3}} \tag{11.39}$$

and of the approximated atmospheric parameters at altitude H:

$$T_H = 288.16 - 0.0093\, H + 3.2739 \cdot 10^{-7} H^2 - 2.9861 \cdot 10^{-12} H^3 \tag{11.40}$$

$$p_H = 101{,}235\, e^{1.322 \cdot 10^{-4} H} \tag{11.41}$$

Except for three equations—(11.30)–(11.32)—the equations (11.28)–(11.36) from the discussion of exergy in Section 11.5 also remain valid in this section. Equations (11.30)–(11.31) change, respectively, as follows:

$$G_a + B_{f-a} = Z_{1a} + B_{w1} + B_{a-d} + \Delta B_a \qquad (11.42)$$

$$G_T + Z_{1a} + B_{w1} = Z_{2a} + B_{w2} + B_P + \Delta B_T \qquad (11.43)$$

$$G_{ch} + Z_{2a} + B_{w2} + B_{d-ch} = Z_{3a} + B_{w3} + B_{ch-gr} + B_{ch-sky} + B_{ch-0} + \Delta B_{ch}$$

$$(11.44)$$

where G_a, G_T, and G_{ch} are the gravity inputs in the eZergy balance equations for the collector air, turbine, and chimney, respectively. Note that in the eZergy balance equation the potential exergy does not appear as a separate term because this exergy is interpreted by the eZergy of substance.

The results in column 4 of Table 11.1 concerning exergy and eZergy are used for the respective bands diagram (Figure 11.5). Again, the diagram shows how the solar radiation exergy $B_S = Z_S = 32.41$ MW, reduced by 5% reflection, arriving at the deck is distributed between five SCPP components: collector air, floor, deck, turbine, and chimney in the case of using substance eZergy. In the diagram (Figure 11.5) the eZergy streams Z, W, are represented by their percentage values z, related to the solar radiation exergy Z_S. The part of the diagram (Figure 11.5) related to the floor and deck is the same as in Figure 11.4, because substance does not appear in the balances of the floor or deck. Also, degradations of solar radiation and convective heat are the same as shown in Figure 11.4, and the power performed by the turbine is unchanged (0.23 MW). The percentage power of the turbine $z_P = b_P = 0.70\%$ represents the eZergy efficiency of the SCPP. Specificity of the diagram (Figure 11.5) shows the relatively large eZergies of air $z_{a1} = 8.96\%$, $z_{a2} = 8.82\%$, and $z_{a3} = 8.24\%$. As a result, the gravity inputs are $z_{Ga} = 7.28\%$ for air in the collector, smaller for the turbine $z_{GT} = 0.89\%$, and the smallest for the chimney $z_{Gch} = 0.67\%$.

11.7 Trends of Response for the Varying Input Parameters

The derived mathematical model of the SCPP responds to the input data as shown by the computation results in Tables 11.1 and 11.2. The results can be used to illustrate the trends of the output data in response to changes in some input parameters. The values in column 4 of Table 11.1 are considered to be the reference values for studying the influence of the varying input parameters on the output data. Therefore, each of the next columns (5–8) corresponds to the case in which the input is changed only by the value shown in a particular column (bold type), whereas the other input parameters remain at the reference level.

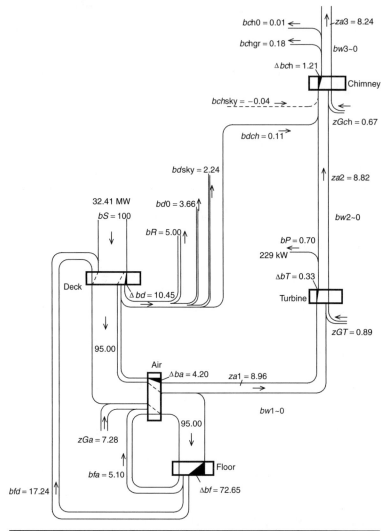

Figure 11.5   eZergy balance of the SCPP according to Table 11.1, column 8 (the values are expressed in percent) (from Petela, 2009).

For example, column 7 of Table 11.1 corresponds to a change in the floor diameter D_f, which increases from 240 to 250 m. The 10-m D_f increase causes the 1-K increase of the effective temperature $T_{f,\text{eff}}$ of the floor from 388.3 to 389.3 K. At the same time:

w_1 decreases from 1.10 to 1.03 m/s
m decreases from 276 to 268 kg/s
E_P decreases from 229 to 222 kW

Quantity	$H_3 = 195$ m						
	$D_f = 240$ m			Quantity, $S = 800$ W/(m^2 K)			
INPUT							
S, W/(m^2 K)	800	850	850	D_f, m	240	250	250
H_e, m	0.3	0.3	0.35	H_e, m	0.3	0.3	0.35
OUTPUT							
e_P, %	0.64	0.53	0.96	e_P, %	0.64	0.57	1.03

TABLE 11.3 Analysis of the Influence of S and D_f on the SCPP Energy Efficiency (from Petela, 2009)

b_{a3} increases from −0.61 to −0.44%
b_{p3} decreases from 0.00586 to 0.00576%
z_{Ga} decreases from 7.28 to 6.89%, etc.

The model responses can also be the basis for deducing many other observations. For example, it results from Table 11.1 that increased solar radiation S from 800 (column 4) to 850 W/(m^2 K) (column 2) causes a seemingly unexpected decrease in energy efficiency from 0.64% to 0.53%. However, it can be deduced that increased S requires an adjustment in the SCPP dimensions to better utilize the increased S. Thus, it results from Table 11.2 that if with growing S (from 800 to 850), at the same time the air entrance height H_e is adjusted, e.g., from 0.3 (column 4) to 0.35 (column 9), then the energy efficiency grows much higher (e.g., reaches 0.96%) compared to the case of H_e growing to the same value as in column 8 but at the unchanged $S = 800$ W/(m^2 K). Similar reasoning about adjusting the SCPP dimensions to the input situation can also be carried out for the increasing floor diameter D_f. The reasoning for both input (S and D_f) is illustrated by Table 11.3.

It can also be noted that there is seemingly an unexpected negative exergy of radiation exchanged between the chimney and sky (b_{ch-sky}). Because exergy of radiation at a temperature below T_0 is positive, this effect, possibly disclosed only by exergy, occurs because T_{sky} and is more below T_0 than T_{ch} is above T_0.

Generally, as a consequence of many assumptions, the calculated quantitative responses to varying SCPP input are of limited certainty. However, perhaps better, certainty can be expected to determine the direction of trends found in response to varying input parameters. These trends result from Table 11.1 because the values and algebraic signs of the partial derivative $\partial M/\partial N$ are interpreted by the finite differences:

$$\frac{\partial M}{\partial N} \approx \frac{\Delta M}{\Delta N} = \frac{M(i_M, j_M = 4) - M(i_M, j_N)}{N(i_N, j_N = 4) - N(i_N, j_N)} \quad (11.45)$$

where M and N are any output and input variables, respectively, $i_M = 8, \ldots, 70$ is a row number from column 1; $i_N = 3, \ldots, 6$ is a row number from column 1; $j_M = 4, \ldots, 8$ is a column number from row 1; and $j_N = 4, \ldots, 8$ is the column number from row 1. For example, consider the direction trend of a change in power E_P (M) in response to a change of diameter D_f (N). Substituting into formula (11.45), respectively, for $i_M = 37$, $i_N = 4$, and $j_N = 7$, $\partial M / \partial N = \partial E_P / \partial D_f = (229 - 222)/(0.3 - 0.35) = -140$ kW/m, where the minus sign means that power E_P decreases with growing diameter D_f.

Nomenclature for Chapter 11

A	surface area, m^2
a_1	$= 9.7807$ m/s^2 , constant of equation (11.11)
a_2	$= -3.086 \times 10^{-6}$ 1/s^2, constant of equation (11.11)
a_3	$= 1.217$ kg/m^3, constant of equation (11.11)
a_4	$= -9.973 \times 10^{-5}$ kg/m^4, constant of equation (11.11)
B	exergy, W
b	exergy, %
c_D	chimney wall thickness coefficient
c_p	specific heat at constant pressure, J/(kg K)
D	diameter, m
E	energy, W
e	energy, %
G	gravity input, W
g	gravitational acceleration, m/s^2
H	height or altitude, m
H_e	height of air inlet, m
H_T	height of turbine, m
h	convective heat transfer coefficient, W/(m^2 K)
i	successive number of row in Table 11.1
j	successive number of column in Table 11.1
k	thermal conductivity W/(m K)
M	output
m	air mass flow rate, kg/s
N	input
Nu	Nusselt number
P	power, W
Pr	Prandtl number
p	absolute static pressure, Pa
R	287.04 J/(kg K), individual gas constant (for air)
Re	Reynolds number
r	radial coordinate
r_T	relative pressure drop in turbine
S	solar radiosity, W/m^2
SCPP	solar chimney power plant

T absolute temperature, K
w flow velocity, m/s
Z mechanical exergy (eZergy), W
z mechanical exergy (eZergy), %

Greek

α absorptivity
β angle, deg
ΔB exergy loss, W
Δb exergy loss, %
ε emissivity
φ view factor
η efficiency
η_T internal efficiency of turbine
κ isentropic exponent
ν kinematic viscosity coefficient, m^2/s
ρ density, kg/m^3
σ $= 5.6693 \times 10^{-8}$ W/$(m^2 K^4)$: Boltzmann constant for
         black radiation
τ transmissivity

Subscripts

a air
B exergetic
ch chimney
d deck
E energetic
eff   effective
f floor
G gravity input
H altitude
M output
N input
P turbine power
p potential
R reflected
Q heat
S solar
sky   sky
T turbine
w velocity (kinetic)
x, y different surface
0 environment
$1, 2, 3$ localities shown in Figure 11.1

CHAPTER 12

Thermodynamic Analysis of Photosynthesis

12.1 Objectives of the Chapter

The present chapter focuses on a successive example of how thermodynamic analysis can be developed for a process in which radiation plays a role. The considered photosynthesis is also an example of a chemical process involving two kinds of matter—substance and radiation—that abides by the same thermodynamics equations of conservation. In a process composed of complex endothermic physical and complex chemical subprocesses, nonorganic substances are converted into the oxygen and solid organic substance of green plants. In the presence of water, this process of photosynthesis, which is so important for the life on earth, absorbs carbon dioxide from the environment air and returns oxygen.

The green substance is the initial link in the carbonization process, which at high temperature and pressure in a deposit layer occurs during millions years and gradually converts the green substance, through the stages of wood, pit, coal, anthracite which all are the natural fuels existing even to day. In such a long-term process, although the calorific value and chemical exergy of these fuels grow, the ratio of the chemical exergy to the calorific value diminishes, which means that, from an exergy viewpoint, the chemical value is larger than the energetic usability, e.g., for heating by combustion. For example, to illustrate the problem, the Grout-Apfelbeck's diagram of energetic value, for the series of solid natural fuels, was supplemented by Petela (1966) with the lines of constant exergy.

Simplified analysis of photosynthesis is developed according to Petela (2008a). Classic thermodynamic analysis of photosynthesis consists of a threefold study that consists of (1) energy analysis (i.e., the energy conservation equation is developed to estimate the energy effects of the process); (2) entropy analysis (i.e., the changes of entropy are used to estimate the irreversibility of the component processes); and (3) exergy analysis (which is developed for thermodynamic evaluation of both kinds of matter).

We can demonstrate how the available radiation of the energy spectrum can be utilized selectively by nature partly for driving chemical reactions and partly for heat. Radiation makes possible the chemical reactions due to the physical processes, from which the main process can be the diffusion of gaseous substrates and products. The growth of the green substance on a leaf surface is considered here as the result of the photosynthesis mechanism utilizing solar radiation. The rate of photosynthesis depends on the external thermodynamic conditions that can be controlled based on the disclosed mechanism of the photosynthesis.

Because photosynthesis is extremely complex, the problem of determining the influence of the controlled thermodynamic input parameters, including CO_2 fertilization, is approached with many simplifications. The purpose of the thermodynamic analysis developed here is to obtain or confirm preliminary conclusions and to inspire the further development of thermodynamic analysis of photosynthesis, in particular with the application of exergy. Application of eZergy is meanwhile postponed.

12.2 Simplified Description of Photosynthesis

Photosynthesis is the process by which the energy of photosynthetically active radiation (PAR), i.e., within the wavelength range 400–700 nm, is used to split gaseous carbon dioxide and liquid water and recombine them into gaseous oxygen and a sugar called *glucose*. The PAR range differs slightly from the radiation range (380–780 nm) related to the perception of light by the human eye; see CIE (1987) and Gates (1980).

The photochemical reaction of photosynthesis cannot occur without the presence of chlorophyll and is a complex two-stage process. First, the radiation-dependent process (i.e., the photochemical reaction) occurs. This requires direct radiant energy, which excites the photoactive (i.e., energy-carrying) molecules used in the second process—the light-independent process (i.e., dark reactions).

The overall process of photosynthesis is a series of complex reactions. In simplified terms, first water and CO_2 are consumed, and then oxygen is released. During the second stage, intermediate carbon–hydrogen molecules are consumed and sugar is generated. However, for the present analysis only the following endothermic overall reaction of photosynthesis is considered:

$$6\ H_2O + 6\ CO_2 \rightarrow C_6H_{12}O_6 + 6\ O_2 \tag{12.1}$$

Photosynthesis occurs commonly around us and is essential for life on earth. Detailed aspects of photosynthesis are discussed in many publications, for example, Borror (1960), Marchuk (1992), Vermaas (1998), Campbell et al. (1999a, 1999b), and Purves et al. (2003). Of course, photosynthesis occurs according to the Second Law of Thermodynamics, as is discussed, for example, by Brittin and Gamow (1961), Yourgrau and van der Merve (1968), and, with more detail, by Jurevic and Zupanovic (2003), who introduce their models for the mechanisms of photosynthesis. Based on statistical thermodynamics, Jørgensen and Svirezhev (2004) developed a thermodynamic theory of ecological systems and also evaluated characteristic variables and data for mathematical biology.

The present chapter approaches the analysis of photosynthesis based on the application of classic engineering thermodynamics and outlines the methodology of exergetic consideration of photosynthesis. Analysis of the subject is complex due to the involvement of many different scientific disciplines (e.g., chemistry, thermodynamics with exergy analysis, radiation, heat and mass transfer, etc.). Based on simplifying assumptions, a certain instantaneous model situation is considered to obtain only preliminary energy, entropy, and exergy viewpoints.

12.3 Some Earlier Work About Photosynthesis

From the viewpoint of classical thermodynamics, the literature on photosynthesis is relatively poor. Classical thermodynamic analysis seems to be too rough a tool to assess the complex chemical mechanisms of photosynthesis. Significantly more literature exists on the problems underlying the chemical mechanisms of photosynthesis, but this is beyond the scope of this book.

Energy analysis of green plants has been discussed previously. For example, Spanner (1964) considered the leaf as a heat engine (Section 9.2). His considerations of the "economic" efficiency of photosynthesis led to the (inadequate) formula for the exergy/energy ratio of radiation discussed later by Petela (2003).

However, the more precise approach of the exergy analysis of photosynthesis became possible only after the theory of the exergy of thermal radiation was developed by Petela (1962). Exergy analysis of plants was applied for the first time by Szargut and Petela (1965a). In their calculations, an exergy balance was carried out for 1 hectare area of forest over a 1-year period. It was assumed that (1) the annual average exergy of solar radiation arriving at the earth's surface is 10% of the exergy of extraterrestrial solar radiation (i.e., arriving at the highest layer of the earth's atmosphere), (2) the annual average water precipitation is 0.7 m/year, (3) the calorific value of wood is 7635 kJ/kg, and (4) the wood mass increases by 1350 kg. The exergy of wood was calculated from the ratio of chemical exergy to calorific value determined by the authors as 1.31. The exergy degree of perfection (the concept discussed in Section 4.6.2) of the forest vegetation was relatively very small (0.033%), and the respective energy degree of perfection was 1.4 times smaller. The growth of wood mass was related to the total radiation, not only to the PAR part of the whole spectrum.

Photosynthesis was later analyzed using the standard energy approach. However, exergy analyses have seldom been applied. For example, Bisio and Bisio (1998) discussed the approximate evaluation of the exergy efficiency and exergy losses, whereas Reis and Miguel (2006) analyzed the separate day and night processes. An approach to photosynthesis, based on statistic thermodynamics and exergy, was developed by Jørgensen and Svirezhev (2004).

In comparison to the approach of other researchers, the approach presented here applies the exergy analysis of photosynthesis discussed in terms of classic engineering thermodynamics. The analysis considers more details, involves more process parameters, and assumes that the diffusion of gases controls the process.

12.4 Assumptions Defining the Simplified Mathematical Model of Photosynthesis

Figure 12.1 presents a simplified scheme of photosynthesis. The system is defined by the system boundary and contains a leaf surface layer in which biomass is created at temperature T. Diffusion of gaseous substances and convective heat transfer occurs through the gaseous boundary layer at the leaf surface. The boundary layer is not considered for radiation fluxes because it is assumed that air in this layer is transparent to radiation. The leaf surface absorbs part of the incident

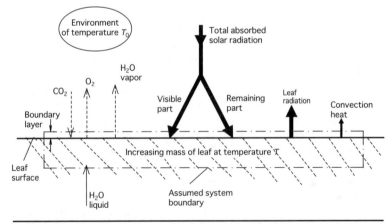

Figure 12.1 Simplified scheme of substances and radiation fluxes in photosynthesis (from Petela, 2008a).

solar radiation and emits its own "leaf" radiation of temperature T. The absorbed radiation is expended on the metabolism processes of the leaf and on maintaining the leaf temperature T above the environment temperature T_0.

Liquid water from the leaf body, at temperature T, enters the considered system. Only a relatively small amount of this water is active in the assimilation of the CO_2, which diffuses into the leaf from the external environment. The excess of water is transpired in the form of vapor diffusing from the leaf to the environment. Oxygen produced during photosynthesis also diffuses into the external environment. The water vapor and oxygen exiting the boundary layer, as well as the CO_2 entering the boundary layer, have environment temperature T_0 at the respective environment mole concentrations $z_{H_2O,0}$, $z_{O_2,0}$, and $z_{CO_2,0}$.

Both the potential and kinetic components of exergy are neglected because the flow velocities of these substances are very small, and the leaf is insignificantly above the earth's surface. Only the chemical and physical components of the energy and exergy of the substances are considered. The physical component reflects the physical parameters of the substance (e.g., temperature and pressure). The chemical component reflects the chemical composition of the substance and is determined based on the *devaluation reaction* as discussed in Section 2.7.

The energy, entropy, and exergy of radiation are calculated based on considerations in Sections 7.1–7.3, respectively.

The complex photosynthesis mechanisms occurring in the leaf are not considered. Only the overall effects described by equation (12.1) and the matter fluxes observed around the leaf (Figure 12.1) are analyzed. The following conditions are also assumed:

i. Considered is a conventional horizontal unitary (1 m^2) surface of the leaf during a certain instant at the determined constant conditions.

ii. To determine the actual energy arriving at the leaf's surface, the solar radiation energy of the spectrum measured at the highest layer of the atmosphere is multiplied by a certain weakening factor γ. The larger is the γ, the smaller is the weakening ($\gamma \leq 1$). In the proposed simplified model the weakening factor is not studied, nor is it concretely defined. Only certain possible values of γ are used in calculations. Some data, discussed, e.g., by Iqbal (1983), Muneer et al. (2004), and Gueymard (2008) can contribute to determination of the γ factor.

iii. Cloudy situations are not analyzed.

iv. Solar radiation arrives directly from the sun at its zenith (solar radiation is perpendicular to the horizontal surface of the considered leaf), within the solid angle determined by the diameter of the sun and its distance from the earth. The reduced effect due to the nonperpendicular radiation can be expressed by the appropriate value of factor γ.

v. Sufficient chlorophyll necessary for photosynthesis is available. Any change in the chlorophyll concentration during photosynthesis, as well as the thermodynamic effect of such change, are neglected.

vi. The surroundings of the considered leaf consists only of the surfaces at temperature T_0 and of absorptivity $\alpha_0 = 1$ (the sun's surface area is neglected as being seen within a relatively small solid angle).

vii. Mixtures of substances in the system are ideal; the components do not mutually interact. Therefore, mixture properties are the respective sums of the component properties. For example, the biomass contained within the leaf structure is an ideal solution of solid $C_6H_{12}O_6$ and water.

viii. The environment air contains only N_2, O_2, CO_2, and H_2O. The dry environment air contains 79.07% N_2, 20.9% O_2, and 0.03% CO_2. The sum of all mole fractions of the air components is $z_{N_2,0} + z_{O_2,0} + z_{CO_2,0} + z_{H_2O,0} = 1$, where $z_{H_2O,0}$ is determined by the relative humidity φ_0 of the air and the saturation pressure p_{s0} for the environment temperature T_0: $z_{H_2O,0} = \varphi_0 \cdot p_{s0}$. In terms of radiation it is assumed that the diatomic gases have transmissivity 100% and the concentration of the triatomic components (CO_2 and H_2O) is relatively small and they also have transmissivity 100%.

ix. The considered leaf has uniform temperature T; there is no heat transfer within the considered surface layer. According to the evaluation by Jørgensen and Svirezhev (2004) there is a several degree difference Δ between the leaf temperature and environment temperature.

x. The liquid water required for photosynthesis is available in a sufficient amount.

xi. In the considered conditions (as explained later, e.g., in Figure 12.8), the rate of sugar production is limited by the effectiveness of diffusion of gases, not by the reaction kinetics depending on temperature.

xii. The generated sugar has only chemical exergy b_{ch} resulting from chemical reaction (12.1). The component of the sugar exergy gained as a result of ordering is neglected. The *ordering* means the structure of the biomass according to genetic plan, considered, e.g., by Jørgensen and Svirezhev (2004). Also neglected are any eventual component exergies, e.g., exergy due to the presence and necessity of chlorophyll as well as anything that would reflect a mystery factor of life, etc. The inclusion of any of these additional exergy components, some mentioned by Jørgensen and Svirezhev (2004), would also require the inclusion of the respective component enthalpies in the energy conservation equation from which the rate of the sugar production is calculated. Without such adjustment the calculated rate would be so large that the exergy efficiency, taking into account the discussed exergy components, could be unfairly larger, possibly even larger than 100%.

Absorption means receiving radiant energy that is not necessarily converted to heat. The terms *absorb* and *absorptivity* are interpreted respectively to term of absorption. The term *radiation* means the "emitting process" or "emitted product"—a photon gas.

Yet more assumptions, in particular on some properties of substances and radiation are discussed in the following Sections. Evaluation of the significance of each simplification is not studied here in detail. However, it can be expected that some simplifications will not qualitatively affect the final conclusions, although the quantitative results will be affected remarkably. For example, the weakening factor γ can be determined through deep theoretical analysis, using precise data about the diffusive loss of radiation energy with an accounting of the complex time-dependent geometrical configuration of the leaf relative to the sun. Thus, γ can be finally estimated at the specific value included in the range of possible values in the presented model. In other words, the problem lies mostly not in the model effectiveness but in determination of the actual value of γ to be applied in the model. Obviously, radiation arriving at the leaf can be accurately determined by measuring the radiation spectrum directly at the leaf surface.

12.5 Properties of Substance

12.5.1 Energy of Substance

The substances in the system (Figure 12.1) are gaseous CO_2, O_2, and H_2O (assumed to be ideal), liquid water, and the leaf substance (biomass). The enthalpies of the gases are zero because at the system boundary they have environment temperature T_0. However, for the liquid H_2O the water vapor is the reference phase. Therefore, the enthalpy of liquid water is equal to the sum of the negative value of the latent heat of vaporization at temperature T_0 and the temperature difference $\Delta = T - T_0$ multiplied by the specific heat of water.

The generated biomass is assumed to be a mixture of liquid water and sugar. The enthalpy of this mixture is calculated as the sum of the component enthalpies. (For liquid and solid bodies the enthalpy is practically equal to the internal energy). The enthalpy of sugar is the sum of the physical enthalpy and devaluation enthalpy. The physical enthalpy of sugar is calculated for the temperature range from T_0 to T at the constant specific heat, $c_{su} = 430.227$ kJ/(kmol K), assumed, e.g., for an oak according to data given by Wisniowski (1979).

The values of devaluation enthalpies d_n for the standard parameters (pressure $p_n = 101.325$ kPa and temperature $T_n = 298.15$ K) are tabulated, e.g., by Szargut et al. (1988). Variation of the devaluation enthalpy within the considered temperature range is negligible. Due to a lack of data, the devaluation enthalpy $d_{n,C_6H_{12}O_6} \equiv d_{n,su} = 2,529,590$ kJ/(kmol of sugar) is assumed to be similar to that for α-D-galactose, predicting that the devaluation enthalpy of the real substance generated in the leaf differs insignificantly. Such an assumption is supported by the fact that the devaluation enthalpies tabulated for the substances of the same chemical formula (α-D-galactose and L-sorbose) differ insignificantly.

12.5.2   Entropy of Substance

In a chemical process some substances are created and some others are consumed. Thus, the so-called *absolute entropy* of each substance, calculated as the entropy increase from a temperature of absolute zero, should be considered according to formula (2.38). The absolute entropy values, s_{tab}, for gaseous CO_2, O_2, and H_2O are tabulated as a function of temperature at pressure $p_{tab} = 0.1$ MPa; e.g., by Burghardt (1982). To calculate actual entropy s at any pressure p, the correction $R \cdot \ln(p/p_{tab})$ is added, where R is the universal gas constant, $R = 8.3147$ kJ/(kmol K), and p is the partial pressure of gas in the environment.

Also, according to Burghardt (1982), the entropy of liquid water at temperature T_n is 69.98 kJ/(kmol K). To calculate the entropy s_w of liquid water at temperature T the correction $c_w \cdot \ln(T/T_n)$ is added.

Due to a lack of data about the absolute entropy of sugar, such data are calculated. Again, sugar generated in the leaf is assumed to be the same as α-D-galactose. To calculate the required absolute standard entropy $s_{n,C_6H_{12}O_6}$, formula (2.65) on the *standard entropy of devaluation reaction* σ_n is used as follows:

$$s_{n,C_6H_{12}O_6} \equiv s_{n,su} = 6(s_{H_2O} + s_{CO_2})_n - (s_{O_2})_n - \sigma_{n,su} \qquad (12.2)$$

where s_{H_2O}, s_{CO_2}, and s_{O2} are the absolute standard entropies of the respective gases and the standard entropy of devaluation $\sigma_{n,su} = -979.71$ kJ/(kmol K) is given by Szargut and Petela (1965b). From equation (12.2) the calculated value of the absolute standard entropy is $s_{n,su} = 2164.6$ kJ/(kmol K). To calculate the absolute entropy s_{su} at any temperature T, the correction $c_{su} \cdot \ln(T/T_n)$ is added.

12.5.3   Exergy of Substance

As mentioned in Section 12.4, from all possible exergy components only the physical b_{ph} and chemical b_{ch} exergy are taken into account to calculate the total exergy $b = b_{ch} + b_{ph}$ of a substance. The exergy of each gas (CO_2, H_2O, and O_2) is zero because in the considered case their states are in full equilibrium with the environment.

The total specific exergy b_w of liquid water is approximated in Section A.7 as a function of temperatures T and T_0, based on data from the Szargut and Petela's diagram (1965b).

The exergy of the generated biomass is the sum of the exergy of the components (liquid water and sugar). The specific chemical exergy of sugar is determined based on the standard tabulated value $b_{n,su} = 2{,}942{,}570$ kJ/kmol, to which the correction for the difference of temperatures T_n and T_0 is added according to equation (2.63):

$$b_{ch,su} = b_{n,su} + \frac{T_n - T_0}{T_n}\,(d_{n,su} - b_{n,su}) \qquad (12.3)$$

According to formula (2.64) the specific physical exergy of sugar is:

$$b_{ph,su} = c_{su}\,(T - T_0) - T_0 c_{su}\,\ln\frac{T}{T_0} \qquad (12.4)$$

12.6 Radiation Properties

12.6.1 Energy of Radiation

12.6.1.1 Energy Radiation of the Sun

In general cases, the temperature and properties (e.g., emissivity) of the source, from which the considered radiation arrives, can be unknown. Energy of such radiation is considered to be arbitrary radiation (i.e., radiation of any irregular spectrum not expressible by the ideal black or gray model). This can be determined by radiosity calculated based on the results of spectrum measurements. In the present considerations the radiation arriving at the leaf surface is considered to be arbitrary radiation arriving from the sun. The radiation is recognized as nonpolarized and uniformly propagating within the solid angle under which the sun is seen from the earth. To calculate the solid angle the radius of the sun $R_S = 695{,}500$ km and the mean distance from the sun to the earth $L_S = 149{,}500{,}000$ km have been assumed.

The formulae regarding radiation are derived based on the discussion in Chapter 7. The radiosity j_S of the solar radiation of the real spectrum can be calculated based on equation (7.15) written using wavelength λ instead of frequency ν:

$$j_S = 2\left(\int_\beta \int_\varphi \cos\beta \sin\beta\ d\beta\ d\varphi\right)\int_\lambda i_{0,\lambda}d\lambda \qquad (12.5)$$

The double integral in the brackets of equation (12.5) was calculated in Example 7.4, formula (7.78). However, because the single integral in equation (12.5) can be solved analytically only for a black radiation—thus for any radiation of arbitrary spectrum—a numerical solution is applied. Using (7.78) the following form of equation (12.5) can be applied:

$$j_S = 4.329 \cdot 10^{-5}\ \pi\ \sum_n (i_{0,\lambda}\Delta\lambda)_n \qquad (12.6)$$

where $i_{0,\lambda}$ is the measured monochromatic intensity of radiation depending on the wave length λ, and n is the successive number of the wavelength intervals within the considered range of wavelengths. For example, especially for green plants, a measurement method has been patented by Obynochnyi et al. (2006). The observed spectra of global direct and diffuse radiation were used, e.g., as in the numerical analysis by Jiacong (1995), and calculation of the sky spectral irradiances were developed by Gueymard (2008).

The monochromatic intensity values $i_{0,s\lambda}$ given by Kondratiew (1954) have been used in the calculations. For the 0–∞ wavelengths range, the sum in equation (12.6) has been calculated by Petela (1962) and the respective radiosity is $j_S = 1.3679 \text{ kW}/\text{m}^2$, as was used in Example 7.5. The j_S value so obtained is slightly smaller than the newest value $1.3661 \text{ kW}/\text{m}^2$ determined by Gueymard (2004). For the PAR arriving from the sun in the highest layer of atmosphere within the wavelengths range (400–700 nm): $\sum_n (i_{0,\lambda}\Delta\lambda)_n = 4013.105 \text{ kW}/(\text{m}^2 \text{ sr})$. Thus, the radiosity of the PAR calculated from equation (12.6) is $j_V = 0.5446 \text{ kW}/\text{m}^2$.

12.6.1.2   Energy Radiation of the Leaf Surface

The radiation emitted by the leaf surface is recognized as the radiation of the determined surface properties. The leaf emission propagates in all directions of the hemisphere as well as the radiation from the environment arrives at the leaf surface from all directions of the hemisphere. Therefore, the energy e_L exchanged between the leaf and the environment is assumed to be:

$$e_L = \alpha_{L,a}\sigma \left(T^4 - T_0^4\right) \tag{12.7}$$

where $\alpha_{L,a}$ is the average absorptivity of the leaf surface, σ is the Boltzmann constant for black radiation, and T_0 is the environment temperature. To simplify the consideration the sky temperature is assumed to be equal to the environment temperature. At small temperatures (T and T_0) the energy of PAR is relatively small, e.g., in comparison to the case of radiation at the temperature of the sun. Therefore, the assumption that the average absorptivity $\alpha_{L,a}$ equals the leaf absorptivity α_L for the non-PAR wavelengths range $\alpha_{L,a} \approx \alpha_L$ slightly affects the value of the energy e_L calculated from equation (12.7).

12.6.2   Entropy of Radiation

12.6.2.1   Entropy Radiation of the Sun

As mentioned in Section 12.6.1.1, the temperature and properties of the surface from which the radiation arrives are unknown and the considerations of Chapter 7 are applied.

The radiosity entropy s_S of the solar radiation of the real spectrum can be calculated based on equation (7.36) written using wavelength λ instead of frequency v:

$$s_S = 2 \left(\int_{\beta} \int_{\varphi} \cos\beta \sin\beta \, d\beta \, d\varphi \right) \int_{\lambda} L_{0,\lambda} d\lambda \qquad (12.8)$$

Again, the double integral in the brackets of equation (12.5) was calculated in Example 7.4, formula (7.78). However, because the single integral in equation (12.8) can be analytically solved only for a black radiation, thus for any radiation of arbitrary spectrum, a numerical solution is applied. Using (7.78) the following form of equation (12.8) can be applied:

$$s_S = 4.329 \cdot 10^{-5} \pi \sum_n (L_{0,\lambda} \Delta\lambda)_n \qquad (12.9)$$

where $L_{0,\lambda}$ is the entropy of the monochromatic intensity of radiation depending on the wavelength λ and determined by equations (7.25). It is noteworthy that the universal formula (7.25) can be used for any radiation of an arbitrary spectral composition.

The monochromatic intensity values $i_{0,\lambda}$, given by Kondratiew (1954), have been also used for entropy calculations. For the $0-\infty$ wavelengths range, the sum in equation (12.9) has been calculated by Petela (1962) as $s_S = 0.307 \cdot 10^{-3}$ kW/(K m^2). Using formula (7.25) for the PAR arriving in the highest layer of atmosphere, the value $\sum_n (L_{0,\lambda} \Delta\lambda)_n = 0.83293$ kW/(K m^2 sr) has been calculated. Thus the entropy of radiosity for the PAR calculated from equation (12.9) is $s_V = 0.113 \cdot 10^{-3}$ kW/(K m^2).

12.6.2.2 Entropy Radiation of the Leaf Surface

Analogously to the discussion of radiation energy (Section 12.6.1.2), and using formula (5.24) at emissivity equal to absorptivity, the entropy s_L of radiant energy exchanged between the leaf and the environment is assumed to be:

$$s_L = \alpha_L \frac{4}{3} \sigma \left(T^3 - T_0^3 \right) \qquad (12.10)$$

12.6.3 Exergy of Radiation

12.6.3.1 Exergy Radiation of the Sun

The calculation of exergy b_S for nonpolarized, uniform, and direct solar radiation arriving in the earth's atmosphere is based on the analytical formula (7.45) in a numerical form, such as formula (7.79) in

Example 7.5 with accounting of formula (7.78). Additionally, by application of the wavelength λ, instead of frequency ν, one obtains:

$$b_S = 4.329 \cdot 10^{-5} \, \pi \left(\sum_n (i_{0,\lambda} \Delta \lambda)_n - T_0 \sum_n (L_{0,\lambda} \Delta \lambda)_n + \frac{9.445 \cdot 10^{-12}}{\pi} T_0^4 \right)$$

(12.11)

Using $T_0 = 293$ K in equation (12.11), the exergy of total solar radiation $b_S = 1.2835$ kW/m^2 and the exergy of PAR $b_V = 0.5155$ kW/m^2 are calculated.

12.6.3.2 Exergy Radiation of the Leaf Surface

By definition, the exergy of radiant emission for any surface at the temperature of the local environment T_0, is zero. Therefore, the exergy of environmental radiation arriving at the leaf surface is zero. Assuming the leaf has emissivity equal to absorptivity, the exergy b_L of emission from the leaf surface at temperature T can be determined by the formula (6.10) as follows:

$$b_L = \alpha_L \, \frac{\sigma}{3} (3T^4 + T_0^4 - 4T_0 T^3)$$

(12.12)

12.7 Balances Equations

12.7.1 Mass Conservation Equations

The mass fluxes, kmol/(m^2 s) of CO_2 and O_2 are determined by the stoichiometric factors of equation (12.1): $n_{CO_2} = 6n_{su}$, $n_{O_2} = 6n_{su}$, where n_{su} is the amount (kmol) of sugar produced within a period of 1 s and per 1 m^2 of irradiated leaf surface.

The mass flux n_w of water entering the leaf includes (a) water n_{wL} within the generated biomass, (b) $6 \cdot n_{su}$ of water entering the chemical reaction (12.1), and (c) water n_{H_2O} vaporized into the environment. Thus, $n_w = n_{wL} + 6n_{su} + n_{H_2O}$ where $n_{wL} = n_{su}(1 - z_{su})/z_{su}$ and where z_{su} is the mole fraction of sugar in the biomass composed of sugar and water.

As discussed by Jørgensen and Svirezhev (2004), an important factor in the determination of the effectiveness of photosynthesis is the mole ratio $r = n_{H_2O}/n_{CO_2}$ of the water vapor and carbon dioxide rates. Water vapor diffuses from the internal surface of the leaf, through the stomata and intercellular space, toward the external surface of the leaf, and then diffuses through the boundary layer to the atmosphere. The water rate, n_{H_2O}, is proportional to the generalized coefficient D_{H_2O} of diffusion and to the difference $(z_{H_2O,L} - z_{H_2O,0})$, where $z_{H_2O,L}$ is the initial mole concentration of vapor at the inner surface and $z_{H_2O,0}$ is the final mole concentration in the environment. Diffusion of carbon dioxide occurs in the opposite direction and is also proportional to the generalized CO_2 diffusion coefficient D_{CO_2}, as well as to the respective difference of mole concentrations $z_{CO_2,0}$ and $z_{CO_2,L}$. Therefore the rates ratio is:

$$r = \frac{D_{H_2O}}{D_{CO_2}} \frac{z_{H_2OL} - z_{H_2O,0}}{z_{CO_2,0} - z_{CO_2,L}} \frac{M_{H_2O}}{M_{CO_2}} \qquad (12.13)$$

where M_{H_2O} and M_{CO_2} are the molecular masses of H_2O and CO_2, respectively. The diffusion coefficients ratio was estimated by Jørgensen and Svirezhev (2004) as $D_{H_2O}/D_{CO_2} \approx 1.32$ and according to Budyko (1977):$z_{CO_2,0} - z_{CO_2,L} \approx 0.1\, z_{CO_2,0}$.

It is also assumed that the concentration of water vapor within the leaf corresponds to the saturation pressure $p_{s,T}$ at temperature T, $z_{H_2O,L} = p_{s,T}/p_0$. Thus, equation (12.13) can be written as:

$$r = 5.4 \frac{p_{s,T} - \varphi_0 p_{s,0}}{p_0 z_{CO_2,0}} \qquad (12.14)$$

The ratio r is determined by the diffusion processes that control the rate of reaction (12.1), accordingly to assumption (xi) in Section 12.4.

12.7.2 Energy Equation

The energy conservation equation was formulated for the system shown schematically in Figure 12.1. The energy delivered consists of absorbed solar radiation and the enthalpies of carbon dioxide and liquid water. The energy increase of the system is determined by the rates of the sugar substance and liquid water in the produced biomass. The extracted energy consists of the enthalpies of oxygen and water vapor as well as convective heat and emission exchanged by the leaf surface:

$$\gamma \left[\alpha_V j_V + \alpha_L (j_S - j_V) \right] + n_{CO_2} h_{CO_2} + n_w h_w = \\ n_{su} h_{su} + n_{wL} h_w + n_{O_2} h_{O_2} + n_{H_2O} h_{H_2O} + q_k + e_L \qquad (12.15)$$

where γ is the radiation weakening factor and α_V and α_L are the absorptivities of the leaf within and beyond the PAR wavelength range, respectively.

According to assumption (vii), Section 12.4, the biomass is an ideal solution of sugar and water. The total enthalpy of the biomass is the sum of the respective components, $n_{su} \cdot h_{su} + n_{wL} \cdot h_w$.

The heat (q_k) transferred by convection from the leaf surface to the environment is $q_k = k(T - T_0)$, where k is the heat transfer coefficient. Equation (12.15) is used to calculate the unknown rate n_{su}. The leaf temperature T is higher than the environment temperature T_0 by the difference Δ:

$$T = T_0 + \Delta \qquad (12.16)$$

as discussed in Section 12.4, assumption (ix).

12.7.3 Entropy Equation

The irreversibility of photosynthesis can be evaluated using overall entropy growth Π, which comprises the entropy changes occurring as a consequence of the process. For example, the sun emits radiation regardless of the presence of the considered leaf, and the space is filled with this radiation (e.g., photon gas) with its respective entropy. When the considered leaf becomes exposed, then a portion of the incident solar radiation from space is absorbed by the leaf. Thus, the respective radiation entropy s_S disappears and this needs to be taken into account. Other consequences are the convective and radiative heat transfer from the leaf to the environment of temperature T_0. Added consequences are the appearing and disappearing entropies of substances taking part in the chemical reactions of photosynthesis.

Therefore, entropy growth Π consists of the positive (appearing) entropies of generated sugar (s_{su}), liquid water (s_{wL}) in the generated biomass, released water vapor (s_{H_2O}), oxygen (s_{O_2}), net emission (s_L) of the leaf, and convective heat (s_q). The negative (disappearing) entropies correspond to carbon dioxide (s_{CO_2}), water (s_w), and absorbed radiation (s_S). Therefore:

$$\Pi = n_{su}s_{su} + n_{wL}s_w + n_{H_2O}s_{H_2O} + n_{O_2}s_{O_2} + s_L + s_q$$
$$-n_{CO_2}s_{CO_2} - n_w s_w - \gamma[\alpha_V s_V + \alpha_L(s_S - s_V)] \qquad (12.17)$$

where $s_q = q_k/T_0$.

As with enthalpy, the total entropy of the biomass is the sum of the respective components, $n_{su} \cdot s_{su} + n_{wL} \cdot s_w$.

12.7.4 Exergy Equations

According to the scheme in Figure 12.1, the following exergy balance equation can be written:

$$\gamma[\alpha_V b_V + \alpha_L(b_S - b_V)] + n_{CO_2}b_{CO_2} + n_w b_w =$$
$$n_{su}b_{su} + n_{wL}b_w + n_{O_2}b_{O_2} + n_{H_2O}b_{H_2O} + b_{qk} + b_{eL} + \delta b$$
$$\qquad (12.18)$$

where δb is the total exergy loss due to all irreversible processes occurring within the system. The exergy loss can be determined by the Gouy–Stodola law expressed by formula (2.60):

$$\delta b = \Pi T_0 \qquad (12.19)$$

Again as for enthalpy and entropy, the total exergy of the biomass is the sum of the respective components, $n_{su} \cdot b_{su} + n_{wL} \cdot b_w$.

12.8 Perfection Degrees of Photosynthesis

As was discussed in Section 4.6.2, to measure the thermodynamic perfection of a chemical process, the energy and exergy degrees of perfection, defined analogously for convenient comparison, can be applied. To determine the degree of perfection, all terms of the energy (or exergy) balance equation are categorized either as useful product, process feeding, or loss. The denominator of the degree of perfection represents the feeding terms, whereas the numerator expresses the useful products. The loss is not included in the formula because it is a compensation of the perfection degree to 100%.

In photosynthesis the produced sugar represents the useful product and the feed is determined by radiation, CO_2, and liquid water. Other components of the balance equations are categorized as the waste. Thus, based on equation (12.15), the energy degree of perfection η_E of the considered photosynthesis is:

$$\eta_E = \frac{n_{su} h_{su}}{\gamma \left[\alpha_V j_V + \alpha_L \left(j_S - j_V \right) \right] + n_{CO_2} h_{CO_2} + n_w h_w} \tag{12.20}$$

whereas the exergy degree of perfection η_B of the photosynthesis, based on equation (12.18), is:

$$\eta_B = \frac{n_{su} b_{su}}{\gamma \left[\alpha_V b_V + \alpha_L \left(b_S - b_V \right) \right] + n_{CO_2} b_{CO_2} + n_w b_w} \tag{12.21}$$

Example 12.1 For the simplified analysis of photosynthesis, the following input values have been used in the computations for the system presented in Figure 12.1:

- Environment temperature $T_0 = 293$ K;
- Temperature difference $\Delta = 5$ K;
- Relative humidity of environment air $\varphi_0 = 0.4$;
- Environment pressure p_0 equal to the standard pressure $p_0 = p_n = 101.325$ kPa;
- Weakening radiation factor $\gamma = 0.7$;
- Leaf absorptivity within PAR wavelength range $\alpha_V = 0.88$;
- Leaf absorptivity beyond the PAR range $\alpha_L = 0.05$;
- Convective heat transfer coefficient $k = 0.003$ kW/(m^2 K);
- Mole fraction of sugar in biomass $z_{su} = 0.08$.

The results obtained using the mathematical model of photosynthesis are as follows. From equation (12.16) the leaf temperature $T = 298$ K and from equation (12.15) the sugar production rate $n_{su} = 3.21 \cdot 10^{-9}$ kmol/(m^2 s). The percentage terms of energy, entropy, and exergy equations—(12.15), (12.17), and (12.18), respectively—are shown in Table 12.1 and in Figure 12.2. The 100% reference for

Term	Energy %	Entropy %	Exergy %
Input:			
PAR	1459.8	11.06	87.76
Non-PAR	125.4	1.08	7.43
Liquid water	−1485.2	87.0	4.81
CO_2	0	0.86	0
Subtotal	100	100	100
Output:			
Convection heat	65.3	8.16	0
Radiation of leaf surface	6.4	0.79	0.003
Liquid water in biomass	−7.1	0.41	0.023
Water vapor	0	280.50	0
$C_6H_{12}O_6$	35.4	1.113	2.608
O_2	0	0.670	0
Irreversible loss	—	—	97.366
Subtotal	100	291.643	100
Total:	0	191.643	0

TABLE 12.1 Results of the Energy, Entropy, and Exergy Calculations (from Petela, 2008a)

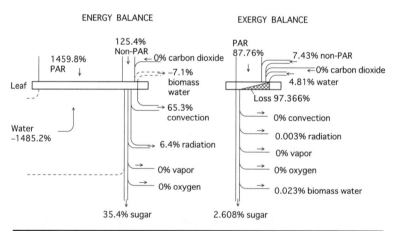

FIGURE 12.2 Bands diagram presenting energy and exergy balances of the considered photosynthesis process shown in Figure 12.1.

the output terms is assumed as the sum (input) of the absorbed radiation and the substances input (CO_2 and liquid water).

Thus. the perfection degree value $\eta_E = 35.4\%$ is larger than $\eta_B = 2.608\%$ (by about $35.4/2.608 \approx 14$ times) mainly because of the denominators in equations (12.20) and (12.21). The exergy of liquid water, in the denominator of equation (12.21), is positive, whereas the energy of this water in the denominator of equation (12.20) is negative, (the water vapor is assumed as the reference phase for enthalpy calculation).

The energy terms (Table 12.1 and Figure 12.2, energy balance) show that the input consists of the positive radiation energy ($1459.8 + 125.4 = 1585.2\%$) and the negative ($-1485.2\%$) liquid water enthalpy. The energy of the consumed carbon dioxide is zero because it enters the system at the reference temperature T_0. The output energy terms show no irreversible loss and the zero energy of both; the produced oxygen and released water vapor leave the system at the reference temperature. Heat transferred by convection and radiation are 65.3% and 6.4%, respectively. The energy of liquid water contained in the produced biomass is negative (-7.1%) because the vapor phase was assumed as the reference substance for water.

The exergy input terms (Table 12.1 and Figure 12.2, exergy balance) are the absorbed radiation ($87.76 + 7.43 = 95.19\%$) and the water's positive value, 4.81%. The exergies of the delivered CO_2, released O_2, and water vapor are zero because these gases, at the external side of the system boundary layer, have parameters equal to the environment parameters. The exergy of convective heat is zero because it is released to the environment. The exergy of the leaf radiation (0.003%) is small due to the relatively low temperature of the leaf. It is significantly smaller than the respective energy (6.4%). The exergy of liquid water contained in the produced biomass is 0.023%, (positive). Unlike the enthalpy and energy analyses, exergy analysis shows the irreversibility loss. In this case it is relatively very large (97.366%).

Table 12.1 presents also the entropy terms of equation (12.17). The terms show the agreement with the Second Law of Thermodynamics and illustrate the role of the particular input or output fluxes in overall entropy growth.

12.9 Some Aspects Inspired by the Example Calculations

12.9.1 Trends Responsive to Varying Input Parameters

The model (Figure 12.1) was also used for computation of the results shown in Table 12.2. These results illustrate the trends of the output data in response to changes in input parameters. The values in column 3 of Table 12.2 are considered as the reference values for studying the influence of the varying input parameters on the output data. Therefore, each of the next columns (4–11) corresponds to the case in which the input is changed only by the value shown in a particular column, whereas the other input parameters remain at the reference level.

Column 4 corresponds to a change in the environment temperature T_0, which increases from 293 to 298 K. The 5-K T_0 increase causes a pressure increase, $\varphi_0 \cdot p_{s0}$, from 0.93 to 1.27 kPa. Note also that the

| Quantity | Units | Ref. value | Mono-variant changes of input parameters and resulting outputs | | | | | | | |
1	2	3	4	5	6	7	8	9	10	11
Input:										
T_O	K	293	298							
Δ	K	5		3						
α_V		0.88			0.85					
α_L		0.05				0.07				
γ		0.7					0.75			
k	W/(m² K)	3						6		
φ_O		0.4							0.9	
z_{CO_2}		0.03								0.06
Output:										
$\varphi_O \cdot p_{s,O}$	kPa	0.93	1.27	0.93	0.93	0.93	0.93	0.93	2.10	0.93
$p_{s,T}$	kPa	3.17	4.25	2.81	3.17	3.17	3.17	3.17	3.17	3.27
r		401	537	336	401	401	401	401	194	200
$n_{CO_2} \cdot 10^8$	kmol/(m² s)	1.93	1.46	2.33	1.87	1.99	2.07	1.85	3.88	3.76
$n_{H_2O} \cdot 10^6$	kmol/(m² s)	7.74	7.82	7.82	7.48	7.98	8.32	7.40	7.35	7.54
$n_{su} \cdot 10^9$	kmol/(m² s)	3.21	2.43	3.88	3.11	3.32	3.46	3.08	6.47	6.27
η_E	%	35.42	28.64	55.36	34.57	35.32	37.27	20.66	55.34	54.39
η_B	%	2.61	1.97	3.14	2.60	2.61	2.62	2.50	5.48	5.09

TABLE 12.2 Responsive Trends of Output to Changes in Some Input Parameters (from Petela, 2008a)

5-K increase in temperature T, from 298 to 303 K, causes the pressure, $p_{s,T}$, to increase from 3.17 to 4.25 kPa. Consequently, the pressure difference, $p_{s,T} - \varphi_0 \cdot p_{s,0}$, increases and thus the ratio r increases from 401 to 537 kmol H_2O/kmol CO_2. The increase of r occurs in conjunction with the increase of n_{H_2O} from $7.74 \cdot 10^{-6}$ to $7.82 \cdot 10^{-6}$ kmol/(m^2 s) and with the decrease of n_{CO_2} from $1.93 \cdot 10^{-8}$ to $1.46 \cdot 10^{-8}$ kmol/(m^2 s). The increased amount of water vaporized from the liquid to the vapor state causes a decrease of the rate of sugar production, n_{su}, from $3.21 \cdot 10^{-9}$ to $2.43 \cdot 10^{-9}$ kmol/(m s). Consequently, the energy degree of perfection, η_E, decreases from 35.42 to 28.64%. The exergy degree of perfection, η_B, decreases from 2.61 to 1.97% due to the increased amount of water in the vapor state.

The results shown in the other columns can be analyzed similarly.

Column 5 shows the effect of reducing the assumed temperature difference Δ from 5 to 3 K. That change causes decreases of the diffusive fluxes for both CO_2 and H_2O, as well as the sugar rate and both degrees of perfection.

Column 6 shows that a 3% drop of α_V (from 0.88 to 0.85) causes a reduction of the sugar rate n_{su} from $3.21 \cdot 10^{-9}$ to $3.11 \cdot 10^{-9}$ kmol/(m^{2}s). Both the energy and exergy degrees of perfection decrease respectively from 35.42 to 34.57% and from 2.61 to 2.60%.

Column 7 shows that a 2% increase of α_L (from 0.05 to 0.07) causes a decrease in the sugar rate from $3.21 \cdot 10^{-9}$ to $3.32 \cdot 10^{-9}$ kmol/(m^2 s). The energy degree of perfection decreases from 35.42 to 35.32%, while the exergy degree of perfection remains practically the same (2.61%).

Column 8 shows that for an increase of the weakening factor γ (e.g., by 5%, from 0.7 to 0.75), the sugar rate and both degrees of perfection respond with appropriate increases.

Column 9 shows that when the heat transfer coefficient k increases by a factor of 2 [from 0.003 to 0.006 kW/(m^2 K), e.g., due to wind], then the sugar rate and both degrees of perfection decrease.

Column 10 shows the effect of a change in the relative humidity, φ_0, of the local environmental air. When φ_0 increases (from 40 to 90%) the partial pressure of water vapor in the environment increases. This causes a decrease in the diffusion flux for the vapor, as well as significant increases in sugar rate and of both degrees of perfection.

As shown in column 11, the twofold increase in the concentration of z_{CO2} causes a significant growth in sugar rate n_{su} (from $3.21 \cdot 10^{-9}$ to $6.27 \cdot 10^{-9}$ kmol/(m^2 s). Both energy and exergy degrees of perfection also grow significantly to 54.39% and 5.09%, respectively. The profiles of the CO_2 concentration in the boundary layers at the leaf surface are discussed in Section 12.9.5.

It was also found that z_{su} has no significant affects on the output parameters considered here.

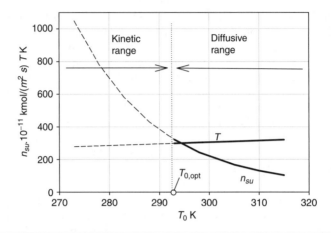

Figure 12.3 Influence of T_0 on leaf temperature T and rate n_{su} of sugar production (from Petela, 2008a).

12.9.2 Relation Between the Environment Temperature, Leaf Temperature, and Rate of Sugar Generation

The photosynthesis model also enables studies of various interrelationships. For example, based on the reference data listed in Example 12.1, Figure 12.3 shows the leaf temperature T and the rate n_{su} of sugar production as functions of the environment temperature T_0. With increasing environment temperature T_0, the leaf temperature T grows, whereas n_{su} decreases. Figure 12.4 shows how both the energy

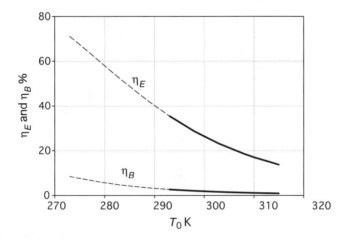

Figure 12.4 Influence of T_0 on the energy (η_E) and exergy (η_B) degrees of perfection (from Petela, 2008a).

degree of perfection η_E and the exergy degree of perfection η_B for photosynthesis decrease with the increasing environment temperature T_0.

Note that the dashed parts of the curves in Figure 12.3 have only a theoretical meaning. This is because it is supposed that for the low values of T_0, the rate n_{su} of sugar production is not controlled by the diffusion of gases but by the reaction kinetics. Such a supposition can be derived from the statement by Jørgensen and Svirezhev (2004) that the optimal temperature of photosynthesis is $T \approx 298$ K. This corresponds to an environment temperature T_0 lower by Δ, i.e., $T_{0,opt} \approx 293$ K.

The statement of the optimum indicates that if on the right-hand side (diffusive range) the rate n_{su} decreases with increasing T_0, then on the left-hand side from the optimum the rate n_{su} also has to decrease (with decreasing T_0). The latter happens due to the critical role played by chemical reaction kinetics: Chemical reaction rates decrease with decreasing temperature; this is not considered in the present analysis. Therefore, the real curve n_{su} in Figure 12.3, instead of following the dashed line, should bend down from the value for a certain optimal environment temperature ($T_{0,opt}$). This is shown subsequently as a thick dashed line in Figure 12.8. The description of the dashed parts of the curves in Figures 12.4–12.6 is the same.

12.9.3 Ratio of Vaporized Water and Assimilated Carbon Dioxide Rates

Another example of the considered relations is the analysis of the important photosynthesis parameter r. For the reference input data

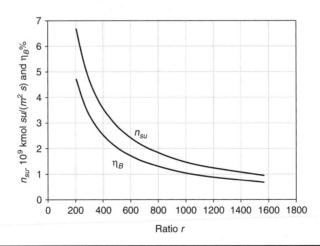

FIGURE 12.5 Influence of ratio r on the sugar rate n_{su} and the exergy degree η_B of photosynthesis perfection (from Petela, 2008a).

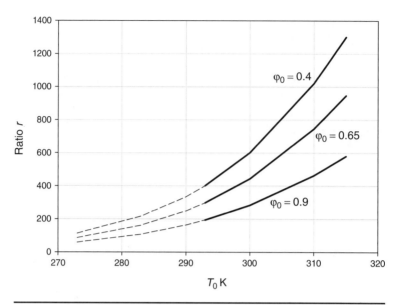

Figure 12.6 Effect of varying T_0 and φ_0 on ratio r (from Petela, 2000a).

listed in Example 12.1, Figure 12.5 shows how the rate n_{su} and degree of perfection η_B increase with decreasing r. Figure 12.6 shows that the greater is the T_0 and the smaller is φ_0, the greater is the r.

12.9.4 Exergy Losses in the Component Processes of Photosynthesis

Categorizing or interpreting exergy losses is, to a certain extent, the subject of agreement. Care should be taken to not overlap the losses of component processes. It is proposed that the degradation of heat from the leaf temperature T to temperature T_0 of the environment, as conducted through the boundary layer, can be considered separately. It can be argued that, in fact, the convective heat leaves the leaf surface at temperature T and then, beyond the leaf, degrades to environment temperature T_0. There is no exergy loss due to radiation through the boundary layer which is transparent to radiation.

In the present simplified approach, the total exergy loss $\delta b = 97\%$ (Table 12.1) is assumed to be the sum of only four components: chemical reaction (12.1), water vaporization, heat exchange by the leaf surface, and convective heat transfer through the boundary layer. The appropriately categorized terms of equation (12.17) can be used in equation (12.19) to calculate the entropy growth in each of the particular component processes.

However, solar radiation absorbed at the leaf surface is expended only on three processes: chemical reaction, vaporization, and heat exchange by the leaf surface. To calculate the three respective exergy losses, the entropy of the absorbed radiation has to be split appropriately in accordance with the split of energy. Based on equation (12.15), each of the respective split fractions is the ratio (each specific energy term divided by total energy of absorbed solar radiation). Therefore, the incident radiation energy is distributed according to the following three split fractions:

$$\mu_{ch} = \frac{n_{su}h_{su} + n_{O_2}h_{O_2} - n_{CO_2}h_{CO_2} - (n_w - n_{H_2O} - n_{wL})h_w}{\gamma\left[\alpha_V j_V + \alpha_L\left(j_S - j_V\right)\right]} \quad (12.22)$$

$$\mu_{vap} = \frac{n_{H_2O}\left(h_{H_2O} - h_w\right)}{\gamma\left[\alpha_V j_V + \alpha_L\left(j_S - j_V\right)\right]} \quad (12.23)$$

$$\mu_{surf} = \frac{q_k + e_L}{\gamma\left[\alpha_V j_V + \alpha_L\left(j_S - j_V\right)\right]} \quad (12.24)$$

where $\mu_{ch} + \mu_{vap} + \mu_{surf} = 1$. It is assumed that the entropy of the absorbed incident solar radiation is split according to the above split fractions.

Each exergy loss is determined according to formula (12.19). Therefore, the exergy loss δb_{ch} for the overall chemical reaction of photosynthesis is equal to the environment temperature multiplied by the respective entropy growth:

$$\delta b_{ch} = T_0\{n_{su}s_{su} + n_{O_2}s_{O_2} - n_{CO_2}s_{CO_2} - (n_w - n_{H_2O} - n_{wL})s_w$$
$$-\mu_{ch}\gamma[\alpha_V s_V + \alpha_L(s_S - s_V)]\} \quad (12.25)$$

Respectively, the exergy loss δb_{vap} for liquid water vaporization is:

$$\delta b_{vap} = T_0\{n_{H_2O}(s_{H_2O} - s_w) - \mu_{vap}\,\gamma[\alpha_V s_V + \alpha_L(s_S - s_V)]\} \quad (12.26)$$

and the exergy loss δb_{surf}, due to radiative and convective heat exchange between the leaf surface and the external environment, is:

$$\delta b_{surf} = T_0\left\{s_L + \frac{q_k}{T} - \mu_{surf}\gamma\left[\alpha_V s_V + \alpha_L\left(s_S - s_V\right)\right]\right\} \quad (12.27)$$

The exergy loss δb_{qk}, due to convection heat q_k transferred through the boundary layer from the leaf surface to the external environment, is:

$$\delta b_{qk} = T_0\,q_k\left(\frac{1}{T_0} - \frac{1}{T}\right) \quad (12.28)$$

Component process of the photosynthesis	Loss %
Overall chemical reaction	0.2
Vaporization of liquid water	93.1
Radiative and convective heat exchange by the leaf surface	4.0
Convection heat transferred through boundary layer	0.1
Total	97.4

Table **12.3** Distribution of the Irreversible Exergy Losses (from Petela, 2008a)

Table 12.3 presents the distribution of exergy losses for the calculation in Example 12.1. The largest loss occurs during vaporization ($\sim$93.1%) and the next largest loss ($\sim$4.0%) is due to heat exchange between the leaf surface and the environment. This latter loss includes heat lost by convection from the leaf surface at temperature T. However, this heat loss causes successive exergy loss ($\sim$0.1%) during the transfer through the boundary layer to the environment at temperature T_0. The exergy loss during chemical reaction is relatively low ($\sim$0.2%), probably because of a relatively very small amount of created biomass.

Future considerations can be developed to explain how, e.g., the exergy loss δb_{ch} is split amongst the exergy losses for chemical reaction assimilating carbon dioxide (δb_{CO_2}) and the exergy loss due to the release of oxygen (δb_{O_2}), $\delta b_{ch} = \delta b_{CO_2} + \delta b_{O_2}$.

12.9.5 Increased Carbon Dioxide Concentration in the Leaf Surroundings

This effect is considered on the basis of the simplified model of layer zones in the vicinity of the leaf surface (Figure 12.7). Adjoining the leaf surface is the boundary layer (Figure 12.1). Between the boundary layer and the environment there exists a zone of mildly blown gas (the blow layer). The blown gas can be, e.g., a cold waste combustion product arising from combustion at a very large excess-air ratio. Thus, the concentration of CO_2 in the combustion product is not significantly larger than the CO_2 concentration in the environment. It is assumed that the concentration of CO_2 is relatively small and still does not affect radiation arriving at the leaf surface. The profiles of CO_2 mole fractions are shown in Figure 12.7 for the case with a blowing gas, $z_{CO_2, B}$ (solid line), and for the case without a blowing gas, $z_{CO_2, 0}$ (dashed line). The blown gas has temperature T_0. The mole fractions $z_{H_2O, B}$ of the water vapor in the blown gas and in the environment are the same, $z_{H_2O, B} = z_{H_2O, 0} = \phi_0 \cdot p_s / p_0$.

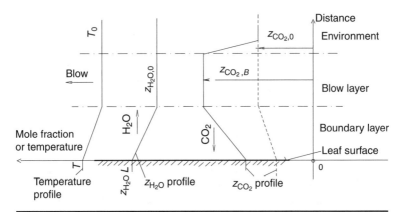

FIGURE 12.7 Simplified theoretical model of the CO_2 fertilization of photosynthesis (from Petela, 2008a).

Due to differences in the respective concentrations, diffusive transportation of species occurs: O_2 out of the surface (not shown in Figure 12.7), H_2O from the leaf to the environment, and CO_2 in the opposite direction into the leaf surface. The length of the diffusion path for H_2O is not significantly changed by the presence of the blown gas zone. The slightly increased concentration of CO_2 insignificantly decreases the relatively large oxygen concentrations $z_{O_2, B}$ in the blow layer, $z_{O_2, B} \approx z_{O_2, 0}$. Thus, the presence of the blown gas zone practically does not change the diffusion flux of O_2. The diffusion rate n_{CO_2} grows due to the increased CO_2 concentration difference, and, contrary to the no-blow case, the exergy of the CO_2 rate entering the system boundary is different than zero. The presence of the blown gas zone impacts formula (12.14) only by the increased CO_2 mole fraction from $z_{CO_2, 0}$ to $z_{CO_2, B}$ (Figure 12.7). The approximate quantitative effects of the increased CO_2 concentration in the leaf surroundings are shown in Table 12.2, column 11.

12.9.6 Remarks on the Photosynthesis Degree of Perfection

The considered degrees of perfection, η_E and η_B, are found to be relatively small. They present the net values because, by neglecting the radiation reflected from the leaf, only the net utilized solar radiation is assumed as the input to the photosynthesis process. If the chemical energy of sugar were related to the total radiation arriving in the leaf surface, as proposed, e.g., by Szargut and Petela (1965a), the degrees of the photosynthesis perfection would become even smaller.

The definition of efficiency is a matter of discussion and agreement. For example, photosynthesis efficiency can be defined simply, based on the ratio of the values for the generated sugar and the absorbed radiation. In such cases the exergy efficiency η_b is always larger than the energy efficiency η_e ($\eta_b/\eta_e > 1$). This is due to the fact that the denominator in energy efficiency is larger than the denominator in exergy efficiency, whereas the numerator in the energy efficiency is smaller than the numerator in exergy efficiency (generally, devaluation enthalpy of the organic substance is smaller than its exergy). For example, neglecting the terms for liquid water and CO_2 as well as non-PAR in equations (12.20) and (12.21), the efficiencies ratio is given by $\eta_b/\eta_e = (b_{su}/b_V)/(h_{su}/j_V)$. For the data assumed in Example 12.1, the ratio $\eta_b/\eta_e = (2.94 \cdot 10^6/0.516)/(2.53 \cdot 10^6/0.545) = 1.22$ at the values $\eta_b = 2.74\%$ and $\eta_e = 2.23\%$.

Unlike photosynthesis, the exergy/energy efficiencies ratio for technical devices, such as those converting solar radiation to heat, is smaller than unity $\eta_b/\eta_e < 1$. For example, for a solar cooker with a cylindrical–parabolic profile, the energy efficiency is always larger than the exergy efficiency. Thus the efficiencies ratio η_b/η_e, as determined experimentally by Ozturk (2004) and theoretically by Petela (2005), is within the approximate range 0.03–0.16.

The degrees of perfection considered in the present analysis are determined for favorable conditions. However, significantly smaller degrees of perfection can be obtained for a vegetation system considered globally during a finite time period. In such a situation the conditions fluctuate beyond favorable values. As mentioned earlier, Szargut and Petela (1965a) obtained a small exergy degree of perfection (~0.033%) from the approximate analysis of the forest vegetation studied for one year in realistic conditions.

The efficiency values given by various authors are similarly small. However, direct comparison of the values is difficult because of differently assumed conditions.

12.10 Concluding Remarks

It is worthwhile commenting on some possible misinterpretations of the entropy and exergy of radiation emitted by the sun's surface. For example, some researchers, instead of the solar radiation entropy s_S, erroneously introduce the smaller value of entropy calculated as heat (exchanged between the sun and the leaf) divided by the sun's temperature (e.g. 6000 K).

In technical calculations of heat exchange the sun's surface is assumed to be at the equilibrium state (the sun's surface receives energy from the sun's interior and emits this energy into the surrounding space). The state of the sun's surface is represented by its effective temperature and the emitted radiation spectrum.

Any surface beyond the sun, exposed to radiation from the sun, can also be at a stable temperature resulting from the energy of radiation both absorbed and emitted. Using the temperatures of the sun and the exposed surface, the effect called *exchanged radiation heat Q* can be calculated from the First Law of Thermodynamics. This effect is real, and the energy efficiency of any device driven by an exposed surface can be related to the exchanged heat.

However, when one examines the situation based on the Second Law of Thermodynamics, the degradation of the exchanged heat Q can be better understood. Note that the entropy of solar radiation incident on the leaf is larger than the entropy of heat at the temperature of the sun's surface. The problem can be also illustrated by the simple example of radiation emission from a surface—not necessarily from the sun, but from any blackbody surface. In these cases the energy $e = \sigma \times T^4$ and entropy of the blackbody emission is $s = (4/3) \times \sigma \times T^3$. In contrast, the entropy s_{err}, determined erroneously as the emission divided by temperature, would be $s_{err} = e/T = \sigma \times T^3 \neq s$.

From the exergy viewpoint, the leaf receives radiation exergy smaller than the exergy of heat Q. Only the exergy of radiation should be used in the fair exergy balance of the leaf surface.

One has to be aware that, as mentioned before, the exergy efficiency can be defined by researchers in different ways and, e.g., the explanation in Section 4.6.3 with respect to heating water by solar radiation can be compared to the case of photosynthesis. The exergy of the sugar produced can be related either to the exergy of heat at the sun's surface, $Q \times (1 - T_0/T_S)$, or to the exergy b_S of the sun's radiation, equation (12.11), or to the exergy of heat absorbed on the leaf surface, $Q \times (1 - T_0/T_L)$. The exergy efficiency increases successively through the above three possibilities due to the decreasing values of the denominators in the efficiency formulas, $Q \times (1 - T_0/T_S) > b_S > Q \times (1 - T_0/T_L)$. The exergy efficiency which relates the process effect to the decrease of the sun's exergy, $Q \times (1 - T_0/T_S)$, is unfair because the exposed surface obtains only the solar radiation exergy, and the leaf's surface is independent of irreversible emissions at the sun's surface. Relating the process effect to the exergy of heat absorbed, $Q \times (1 - T_0/T_L)$, favors the exposed surface by neglecting its imperfectness during the absorption of heat Q. Thus, from these three possibilities, relating the photosynthesis process effect to the exergy b_S of the sun's radiation is the possibility best justified in this analysis.

It is worth noting that the use of the exergy of heat at the sun's surface can be justified only in an unreal theoretical situation where the exposed leaf surface would be entirely in direct contact with the sun's surface and the heat exchange would reversibly occur at a zero temperature gradient.

However, from the comparative viewpoint of entirely different processes, the best justified definition of the efficiency of photosynthesis seems to be according to equation (12.21).

The methodology presented in this chapter for understanding the exergy of photosynthesis outlines a preliminary study of the process based on simultaneous analyses of energy, entropy, and exergy. The study introduces the devaluation enthalpy (for the fair comparison of energy and exergy balances), the formulae for arbitrary radiation (convenient for the use of measurements of any actual radiation spectrum), and formulates the limiting diffusion range of the process. The study determines the effects of the main process input parameters and describes the model of CO_2 photosynthesis. Multi-factored aspects of the problem are presented based on original computation results. However, the developed analyses cannot be directly compared with literature data since the latter are relatively sparse and are based usually on incompatible assumptions.

The interdisciplinary subject of photosynthesis is very complex and involves many areas of knowledge including thermodynamics, theory of exergy, transfer of radiation energy, heat convection, gas diffusion, chemistry, thermochemistry, photochemistry, as well as data dependent on time, day, month, season, weather conditions, geometrical configuration, etc. To obtain even a preliminary understanding of the energy, entropy, and exergy changes occurring during photosynthesis, only a certain model situation, determined with many simplifying assumptions, has been considered. These assumptions can be gradually reduced in the future.

It has been confirmed that plants absorb radiation on their surface, which, due to the endothermic chemical reaction, remains at relatively low temperature, only a little higher than the environment temperature.

In the introduced diffusive model of photosynthesis, the rate n_{su} of global reaction (12.1) is limited by diffusing gases and is schematically presented in Figure 12.8 by the diffusion curve (solid part). The other part (dashed) of the diffusion curve has no practical meaning. The intersection of the diffusion and kinetics curves defines a certain

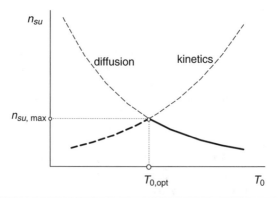

Figure 12.8 Schematic presentation of the optimum of the photosynthesis reaction (from Petela, 2008a).

Output variable M	Input variable N							
	T_0	φ_0	Δ	$z_{CO_2,0}$	γ	α_V	α_L	k
r	−	−	+	−	0	0	0	0
n_{su}	−	+	−	+	+	+	+	−
n_w	+	−	−	−	+	+	+	−
n_{H_2O}	+	−	−	−	+	+	+	−
n_{CO_2}	−	+	−	+	+	+	+	−
n_{O_2}	−	+	−	+	+	+	+	−
η_E	−	+	−	+	+	+	−	−
η_B	−	+	−	+	+	+	+	−
δb	+	−	+	−	−	+	−	+

TABLE **12.4** Algebraic Sign of the Partial Derivative $\partial M/\partial N$ (from Petela, 2008a)

optimal environment temperature $T_{0,opt}$ at which the rate is maximum, $n_{su,max}$. At low leaf temperatures, $T < T_{0,opt}$, the photosynthesis reaction is controlled by its kinetics and the rate n_{su} decreases with decreasing temperature T_0, as shown in Figure 12.8 (dashed thick curve), whereas the dashed thin part of this curve has no practical meaning. The reaction kinetics can be considered in the future.

As a consequence of the many assumptions, the calculated quantitative responses to the varying photosynthesis inputs are of a limited certainty. However, improved certainty can be expected for the direction trends found in response to the varying input parameters. These trends are shown in Table 12.4 by the algebraic signs of the partial derivative $\partial M/\partial N$, where M and N are any output and input variables, respectively. With growing N, M can grow (+), drop (−), or can remain unchanged (0). For example if φ_0 grows, then r decreases (−), n_{su} increases (+), n_w decreases (−), etc.

Some parameters can be controlled (e.g., T_0, k, φ_0, z_{CO_2}) and some cannot. For example, the factor γ (which depends on the weather conditions, time of day, or year), α_V and α_L (which are the properties of the plant, and self-modeling Δ), cannot be controlled. Regarding controllable parameters, the presented diffusion model ($T_0 > T_{0,opt}$) suggests striving for:

- rather low temperature (T_0) of the leaf surroundings;
- low heat transfer coefficient k (e.g., avoid wind);
- high humidity φ_0 to reduce diffusion of vapor; and
- surroundings with increased concentrations of CO_2 to intensify diffusion.

Some of the above effects, which are already known, are confirmed and approximately estimated together with the effects newly disclosed.

The computation example discloses a dramatic difference in the respective values of the terms of energy and exergy balances (Table 12.1 and Figure 12.2). Although both energy and exergy viewpoints are true, the exergy interpretation seems to be more practical.

The observations formulated about the macro interpretation of photosynthesis based on exergy can be confronted with the not considered descriptions of the component's mechanisms of photosynthesis at the micro level. As a result, some new ideas can be found to improve the formulation of the exergy approach. Also, based on exergy, better interpretations of the micro mechanisms can be obtained. For example, the exergy analysis can be applied separately to the light and dark reactions, or different kinds of plants can be examined.

In the future, more detailed exergy investigations on factors such as temperature, light intensity, photochemistry, CO_2 concentration, chlorophyll concentration/property data, and application of fertilizers supplied through air, water, and soil, can be developed for both the diffusive and kinetic ranges of photosynthesis.

Nomenclature for Chapter 12

a	universal constant, $a = 7.561 \cdot 10^{-19}$ kJ/(m^3 K^4)
b	specific exergy of substance, J/kg, or exergy of radiation, W/m^2
c	specific heat, J/(kg K), or speed of light, $c = 2.998 \cdot 10^8$ m/s)
CIE	Commission Internationale del'Eclairage (International Commission on Illumination)
d	devaluation enthalpy, J/kmol
D	generalized coefficient of diffusion
e	energy emission density of surface, W/m^2
h	specific enthalpy, kJ/kg
j	radiosity, W/m^2
k	Boltzmann constant, $k = 138.03 \cdot 10^{-28}$ kJ/K
k	convective heat transfer coefficient, W/(m^2 K)
i_λ	monochromatic intensity of radiation, depending on λ, W/(m^3 sr)
i_ν	monochromatic intensity of radiation, depending on ν, W/(m^2 sr)
L	entropy of the monochromatic intensity of radiation, W/(m^3 K sr)
L_S	mean distance from the sun to the earth, $L_S = 149{,}500{,}000$ km
M	molecular mass, or output variable
n	successive number of the wavelengths interval

n   substance rate, kmol/(m² s)
N   input variable
p   pressure, Pa
q   convective heat, W/m²
Q   heat, J
r   mole ratio of evaporated water to assimilated carbon dioxide
PAR   the photosynthetically active radiation within wavelength
       400–700 nm
R   universal gas constant, $R = 8.3147$ kJ/(kmol K)
R_S radius of the sun, $R_S = 695{,}500$ km
s   specific entropy of substance, J/(kg K), or entropy of
       radiation, W/(m² K)
t   temperature, °C
T   absolute temperature or absolute temperature of the leaf, K
z   mole fraction

Greek
α   surface absorptivity
β   azimuth angle, deg
γ   radiation weakening factor
δ*b*   exergy loss, kW/m²
Δ   difference between temperature of the leaf and environment, K
η_B exergy degree of perfection of photosynthesis
η_b exergy efficiency of photosynthesis
η_E energy degree of perfection of photosynthesis
η_e energy efficiency of photosynthesis
λ   wavelength, m
ν   vibration frequency, 1/s
μ   solar energy split fraction
Π   overall entropy growth, W/(m² K)
σ   entropy of devaluation reaction, J/(kmol K)
σ   Boltzmann constant for black radiation, W/(m² K⁴)
φ   relative humidity or declension angle, deg

Subscripts
a   average
B   blow layer
B   exergetic
b   exergetic
ch   chemical
E   energetic
e   energetic
err   erroneous
k   convective
L   leaf

max	maximal
n	standard (normal), or successive number
opt	optimal
ph	physical
q	heat transferred
s	substance, or saturation
s	sun
su	sugar
surf	surface
tab	tabulated
vap	vaporization
V	PAR wavelengths range
w	liquid water
0	environment or normal (directional radiation)
λ	wavelength
ν	vibration frequency

CHAPTER **13**

Thermodynamic Analysis of the Photovoltaic

13.1 Significance of the Photovoltaic

The present chapter outlines the photovoltaic effect and presents simple energy and exergy analyses of the simultaneous generation of heat and power by photovoltaic (PV) technology. This double conversion of radiation energy categorizes the PV technology to the systems of cogeneration of power and heat. The specificity of the PV effect is that it can generate electricity only as long as continuous light is available. Electrical energy can be stored for later retrieval during a period when there is a lack of radiation. Devices based on the PV effect can serve as power sources in remote terrestrial locations and for different cosmic space applications. The PV effect can also power calculators and other electronic products. In spite of the relatively small power available from the PV effect, plans to utilize solar radiation to power automobiles and aircraft are also being developed.

Usually the term *solar cell* is used for devices that use the PV effect to capture energy from sunlight, whereas the term *PV cell* is used when the light source is unspecified.

PV energy is considered to be the most promising form of solar energy because the energy of light can be converted directly into electric energy without the use of any moving mechanical parts and without the use of fuel. Manufacturing of solar cells and photovoltaic arrays has been noticeably expanding in recent years.

The literature on solar cells has been extensive. For example, many aspects of solar cells including the physics of energy conversion mechanisms and efficiency are presented by Würfel (2005). Badescu (2006) studied the electrical output from the PV array involving latitude, climate, and PV module shape. Recently, e.g., Chow et al. (2009)—based on experimental data and validated numerical models—studied the influence of the glass cover in photothermic and photovoltaic processes and found that energy and exergy viewpoints differ. Based on a short description of the PV effect, this significant difference between energy and exergy evaluation of the PV process is discussed in the following section.

13.2 General Description of the Photovoltaic

Some materials, called *semiconductors*, have the capacity for photoconductivity, which is electrical conductivity affected by exposure to electromagnetic radiation (e.g., light). The capacity of semiconductors for electrical conductivity lies between the abilities of conductors and insulators. Examples of semiconductors applied in PV can be silicon (Si), gallium arsenide (GaAs), copper sulphate (Cu_2S), and different organic substances (e.g., polymers).

The possibility of using sunlight to produce an electric current in solid materials was discovered in 1839 by Becquerel. However, to determine that the conversion of light into electricity occurs at the atomic level took many years.

The photon of the incidental radiation can be absorbed by semiconductors if the energy of the photon is sufficiently high. The absorbed photons knock loose the electrons, negatively (n) charged from their atoms. This allows the electrons to move freely in the semiconductor. The electrons knocked loose leave their positions (so-called *holes*), which behave as complimentary positive (p) charges.

The PV effect is arranged in the PV cell, which generally consists of two regions, like sandwich layers, each as a nonhomogeneous semiconductor with specially added impurities (dopants), such that one region (n) has an excess of electrons (of negative charge) while the other region (p) has an excess of positive holes. The structure of these two regions (a $p–n$ junction) generates an internal electric field. If the photons create free electrons and holes in the vicinity of the $p–n$ junction, then the electric field makes the electrons move toward the side n and the holes move in opposite direction toward the side p. The generated tension between regions p and n is the electromotive force and, using wires, both sides can be connected to any electric energy receiver, e.g., a light bulb through which an electric direct current (DC) runs.

PV cells, of typical size 120 mm × 120 mm, can be assembled to obtain a PV module with an approximately 0.5 m^2 surface area. Several modules can be assembled to obtain a PV system. Connection of modules in series and in parallel allows for high flexibility of the system. The PV cells are connected with silver strips that play the role of an ohmic contact. The PV system can be used stand-alone or can be connected to the power grid or to batteries to store electric energy. The continuous electric current generated by the cells can be converted into an alternating current (AC) with use of an inverter.

Within the semiconductor material, the so-called *recombination* can occur during which the free electron can become bound back to an atom. The recombining electrons do not contribute to the production of electrical current. Therefore, the energy conversion efficiency should take into account only the effective power collected from the solar cell. Photovoltaic efficiency, which is the ratio of electric power generated by a photovoltaic cell at any instant to the power of the sunlight striking the cell, for commercially available cells, does not exceed about 18%.

Different construction of the photovoltaic cell can cover a various range of frequencies of light to produce electricity; however, they cannot cover the whole solar spectrum and, thus, much of incident solar energy is converted to heat or is wasted. The modules can have much higher efficiency if illuminated with monochromatic light. For example, to increase the conversion efficiency the light can be split into different wavelength ranges and the separated beams directed onto appropriately designed PV cells.

An increase in the efficiency of PV cells can be achieved also by a system using lenses or mirrors to concentrate sunlight; however, such high-efficiency solar cells are more expensive than conventional flat-plate photovoltaic cells.

In the future, it would be recommended to develop exergy analysis for the above discussed PV cells irradiated with monochromatic light or for PV cells with concentrated sunlight. However, such analyses can be difficult, and the methodology of exergy analysis will be shown here only for a simple solar cell.

13.3 Simplified Thermodynamic Analysis of a Solar Cell

The principle of a solar cell can be considered for an ideal simple situation in which the sun irradiates the flat surface of the solar cell on earth. The energy streams exchanged by the solar cell are schematically shown in Figure 13.1. As there is no motion of substance in the

FIGURE **13.1**
Scheme of the
energy streams of
a solar cell.

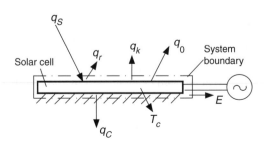

gravitational field, the eZergy consideration has no application. The representative temperature of the solar cell is T_C. Generally, the heat q_S transferred from the sun's surface at temperature T_S to the outer surface of the solar cell on the earth is distributed to the generated electrical energy E, the reflected solar radiation q_r, the useful heat q_C absorbed by the solar cell, and to the convection and radiation heat, q_k and q_0, respectively, both transferred to the environment.

The energy balance equation for the considered system, defined by the system boundary, can be written as follows:

$$q_S = q_r + q_k + q_0 + q_C + E \tag{13.1}$$

For simplicity, the variables in equation (13.1) are related to 1 m^2 area of the solar cell. Calculation of solar energy should account for both direct and diluted radiation. In the ideal case of a clear sky, and assuming the sun as a black surface, heat q_S can be calculated approximately, accounting only for the direct radiation, e.g., as follows:

$$q_S = 2.16 \times 10^{-5} \, \sigma T_S^4 \tag{13.2}$$

where 2.16×10^{-5} is the configuration (sun–earth) factor and $\sigma = 5.6693 \times 10^{-8}$ W/(m^2 K^4) is the Boltzmann constant of black surface.

The reflected solar radiation energy is:

$$q_r = \rho_C q_S \tag{13.3}$$

where ρ_C is the reflectivity of the solar cell surface.

The convection heat is:

$$q_k = k \ (T_C - T_0) \tag{13.4}$$

and the radiation energy is:

$$q_0 = \varepsilon_C \sigma \left(T_C^4 - T_{sky}^4 \right) \tag{13.5}$$

where k is the convective heat transfer coefficient and T_0 is the environment temperature. The sky temperature T_{sky} is assumed to be equal to the environment temperature; $T_{sky} = T_0$. The solar cell surface is assumed to be perfectly gray at emissivity ε_C.

The useful heat q_C can be determined from equation (13.1) if the electrical energy E is known, e.g., from the measurement.

The solar cell can be evaluated by the energy electrical efficiency:

$$\eta_{E,el} = \frac{E}{q_S} \tag{13.6}$$

or by the energy cogeneration efficiency:

$$\eta_{E,\text{cog}} = \frac{E + q_C}{q_S} \qquad (13.7)$$

Interpreting the same process based on the Second Law of Thermodynamics, i.e., including exergy, the degradation of heat is disclosed. The exergy b_S incoming to the considered surface from the sun is split into the exergy b_r of reflected solar radiation, the exergy of heat b_0 radiating to the environment, the exergy of heat b_k transferred to the environment by convection, the exergy of useful heat b_C transferred from the solar cell to its interior, the electric energy E, and the exergy loss δb due to the irreversibility of the considered system. Thus, the exergy balance equation for the system shown in Figure 13.1 is:

$$b_S = b_r + b_k + b_0 + b_C + E + \delta b \qquad (13.8)$$

Analogously to the energy of solar radiation determined by equation (13.2), the exergy of solar radiation for the simple case of a clear sky is:

$$b_S = 2.16 \times 10^{-5} \frac{\sigma}{3} \left(3T_S^4 + T_0^4 - 4T_0 T_S^3\right) \qquad (13.9)$$

The exergy of radiative heat is:

$$b_0 = \varepsilon_C \frac{\sigma}{3} \left(3T_C^4 + T_0^4 - 4T_0 T_C^3\right) \qquad (13.10)$$

The reflected solar radiation exergy is:

$$b_r = \rho_C b_S \qquad (13.11)$$

The exergy of convective heat:

$$b_k = q_k \left(1 - \frac{T_0}{T_C}\right) \qquad (13.12)$$

The exergy of useful heat transferred by conduction or convection is:

$$b_C = q_C \left(1 - \frac{T_0}{T_C}\right) \qquad (13.13)$$

The exergy loss δb can be calculated by completion of equation (13.8). From the exergetic viewpoint the solar cell can be evaluated by the exergy electric efficiency:

$$\eta_{B,el} = \frac{E}{b_S} \qquad (13.14)$$

or by the exergy cogeneration efficiency:

$$\eta_{B,cog} = \frac{E + b_C}{b_S} \qquad (13.15)$$

Because the solar radiation energy is always larger than the solar radiation exergy, $q_S > b_S$, and the electrical exergy and electrical energy are equal, the energy electrical efficiency of the solar cell is always smaller than the exergy electrical efficiency, $\eta_{B,el} > \eta_{E,el}$.

Example 13.1   Consider a 1 m² surface of a polycrystalline silicon photovoltaic cell that generates 152 W of electrical energy. The cell has temperature $T_C = 318$ K, emissivity $\varepsilon_C = 0.95$, and reflectivity $\rho_C = 1 - \varepsilon_C = 0.05$. Environment temperature is $T_0 = 288$ K. The temperature of the sun is assumed as $T_S = 5800$ K. The convective heat transfer coefficient $k = 3$ W/(m² K).

Applying equations (13.1)–(13.14) in the calculation procedure described in Section 13.3, the results presented in Table 13.1 are obtained. The energy of solar radiation $e_S = 1386$ W/m² and the exergy of solar $b_S = 1294$ W/m² were respectively assumed as 100% in the energy and exergy balances.

The instant energy electric efficiency $\eta_{E,el} = 10.48\%$ is smaller from the exergy electric efficiency $\eta_{B,el} = 11.16\%$; however, the energy cogeneration efficiency $\eta_{E,cog} = 10.48 + 65.88 = 76.36\%$ is significantly larger than the exergy

Term	Energy %	Exergy %
Input:		
Solar radiation	100	100
Subtotal	100	100
Output:		
Reflection	5	5
Convection	6.21	0.62
Radiation	12.43	0.67
Useful heat	65.88	6.61
Electricity	10.48	11.16
Loss	—	75.94
Subtotal	100	100
Total	0	0

TABLE **13.1**   Results of the Energy and Exergy Calculations

cogeneration efficiency $\eta_{B,cog} = 11.16 + 6.61 = 17.77\%$. Table 13.1 illustrates also that the low temperature heat (convection, radiation, and useful heat) has small exergy value.

Nomenclature for Chapter 13

AC   alternate current
b exergy of emission, W/m^2
DC   direct current
E electric energy, W/m^2
n negatively charged
p positively charged
PV   photovoltaic
q heat flux, W/m^2
T absolute temperature, K

Greek
ε emissivity
η efficiency
σ Boltzmann constant for black radiation

Subscripts
B exergetic
C solar cell
cog   cogeneration
E energetic
el electric
k convection
r reflection
s sun
sky   sky
0   environment

References

Badescu, V. L'exergie de la radiation solaire directe et diffuse sur la surface de la terre, *Entropie*, 145, 67–72, 1988.

Badescu, V. Simple optimization procedure for silicon-based solar cell interconnection in a series-paralell PV module, *Energy Conversion and Management*, 47, 1146–1158, 2006.

Badescu, V. Exact and approximate statistical approaches for the exergy of black radiation, *Central European Journal of Physics*, 6(2), 344–350, 2008.

Bejan, A. *Advanced Engineering Thermodynamics*, 2nd ed., John Wiley & Sons, New York, 1997.

Bernardes, M. A. dos S., A. Voss, and G. Weinrebe. Thermal and technical analyses of solar chimneys, *Solar Energy*, 75(6), 511–524, 2003.

Bisio, G., and A. Bisio. Some thermodynamic remarks on photosynthetic energy conversion, *Energy Conversion and Management*, 39(8), 741–748, 1998.

Boehm, R. F. Maximum performance of solar heat engines, *Appl. Energy*, 23, 281–296, 1986.

Borror, D. J. *Dictionary of Root Words and Combining Forms*, Mayfield Publishing Co., 1960.

Bosnjakovic, F. *Technische Thermodynamik*, 5th ed., Dresden, Leipzig, 1967 (part I), 1965 (part II).

Brittin, W., and G. Gamow. Negative entropy and photosynthesis, *Proceedings of N.A.S. Physics*, 47, 724–727, 1961.

Budyko, M. I. *Global Ecology*. Mysl', Moscow, 1977.

Burghardt, M. D. *Engineering Thermodynamics with Applications*, 2nd ed., Harper & Row, New York, 1982.

Campbell, N. A., L. G. Mitchell, and J. B. Reece. *Biology: Concepts and Connections*, 5th ed. (and earlier editions), Benjamin/Cummings, Menlo Park, CA, 1999a.

Campbell, N. A., L. G. Mitchell, and J. B. Reece. *Biology: Concepts and Connections*, 3rd ed. (and earlier editions), Benjamin/Cummings, Menlo Park, CA, 1999b.

Candau, Y. On the exergy of radiation, *Solar Energy*, 75(3), 241–247, 2003.

Carnot, S. *Reflections on the Motive Power of Fire, and on Machine Fitted to Develop That Power*, Bachelier, Paris, 1824.

Chow, T. T., G. Pei, K. F. Fong, Z. Lin, A. L. S. Chan, and J. Ji. Energy and exergy analysis of a photovoltaic–thermal collector with and without a glass cover, *Applied Energy*, 86(3), 310–316, 2009.

Chu, S. X., and L. H. Liu. Analysis of terrestrial solar radiation exergy, *Solar Energy* 83(8), 1390–1404, 2009.

CIE Publication. *17.4 International Lighting Vocabulary*, Wien, 1987.

De Groot, S. R., and P. Mazur. *Nonequilibrium Thermodynamics*, North Holland, Amsterdam, 1962; reprinted by Dover, Mineola, NY, 1984.

De Vos, A., and H. Pauwels. Discussion, *Journal of Solar Energy Engineering, Transactions of ASME*, 108, 80–83, 1986.

Duffie, J. A., and W. A. Beckman. *Solar Engineering of Thermal Processes*, 2nd ed., John Wiley and Sons, New York, 1991.

Fraser, R. A., and J. J. Kay. Establishing a role for the thermal remote sensing of eco-systems: exergy analysis. In *Thermal Remote Sensing in Land Surface Processes* (D. Quattrochi and J. Luval, eds.), Ann Arbor Press, Ann Arbor, MI, 2001.

Fujiwara, M. Exergy analysis for the performance of solar collectors, *Journal of Solar Energy Engineering*, 105(2), 163–167, 1983.

Funk, P. Evaluating the international standard procedure for testing solar cookers and reporting performance, *Solar Energy*, 68(1), 1–7, 2000

Gannon, A. J., and T. W. Von Backström. Solar chimney cycle analysis with system loss and solar collector performance, *Journal of Solar Energy Engineering, Transactions of the ASME*, 122(3), 133–137, 2000.

Gates, D. M. *Biophysical Ecology*, Springer-Verlag, New York, 1980.

Gray, W. A., and R. Müller. *Engineering Calculations in Radiative Heat Transfer*, Pergamon Press, Oxford, 1974.

Gribik, J. A., and J. F. Osterle. The second law efficiency of solar energy conversion, *Transactions of ASME, Journal of Solar Energy Engineering*, 106(2), 16–21, 1984.

Gribik, J. A., and J. F. Osterle. The second law efficiency of solar energy conversion, *Transactions of ASME, Journal of Solar Energy Engineering*, 108(2), 78–84, 1986.

Gueymard, C. A. The sun's total and spectral irradiance for solar energy applications and solar radiation models, *Solar Energy*, 76(4), 423-453, 2004.

Gueymard, C. A. Prediction and validation of cloudless shortwave solar spectra incident on horizontal, tilted, or tracking surfaces, *Solar Energy*, 82(3), 260–271, 2008.

Guggenheim, E. A. *Thermodynamics*, North-Holland, Amsterdam, 1957.

Haaf, W., K. Friedrich, G. Mayr, and J. Schlaich. Solar chimneys, *International Journal of Solar Energy*, 2, 3–20, 1983.

Halliday, D., and R. Resnick. In: *Physics*, Parts I and II, John Wiley & Sons, New York, p. 996, Chapter 40, 1967.

Holman, J. P. *Heat Transfer*, 10th ed., McGraw-Hill, New York, 2009.

Iqbal, M. *An Introduction to Solar Radiation*. Academic Press, Toronto, 1983.

Jacob, M. *Heat Transfer*, vol. II, John Wiley & Sons, New York, 1957.

Jeter, S. M. Maximum conversion efficiency for the utilization of direct solar radiation, *Solar Energy*, 26(3), 231–236, 1981.

Jiacong, C. Second law analysis of global irradiance based on observed spectral distributions, *Journal of Dong Hua University* (English edition), 12(2), 22–31, 1995.

Jørgensen, S. E., and Y. M. Svirezhev. *Towards a Thermodynamic Theory for Ecological Systems*, Elsevier, Amsterdam, 2004.

Jurevic, D., and P. Zupanovic. Photosynthetic models with maximum entropy production in irreversible charge transfer steps, *Computational Biology and Chemistry*, 27, 541–553, 2003.

Karlsson, S. The exergy of incoherent electromagnetic radiation, *Physica Scripta*, 26, 329–332, 1982.

Kondratiew, K. Ya. Radiation energy of the Sun. *GIMIS* (in Russian), 1954.

Korn, G. A., and T. M. Korn. *Mathematical Handbook for Scientists and Engineers*, McGraw-Hill, New York, 1968.

Kuiken, G. D. C. *Thermodynamics of Irreversible Processes*, John Wiley & Sons, New York, 1994.

Kundapur, A. Review of solar cooker designs, *TIDE*, 8(1), 1–37, 1988. (Also http://solcooker.tripod.com.)

Landsberg, P. T., and G. Tonge. Thermodynamics of the conversion of diluted radiation, *Journal of Physics (A)*, 12, 551–562, 1979.

Lebedev, P. N. Experimental examination of light pressure, *Annals of Physics (Leipzig)*, 6, 433, (in German), 1901.

Marchuk, W. N. *A Life Science Lexicon*, Wm. C. Brown Publishers, Dubuque, IA, 1992.

McAdams, W. H. *Heat Transmission*, 3rd ed., McGraw-Hill, New York, 1954.

Ming, T., W. Liu, and G. Xu. Analytical and numerical investigation of the solar chimney power plant systems, *International Journal of Energy Research*, 30(11), 861–873, 2006.

Moran, M. J., and H. N. Shapiro. *Fundamentals of Engineering Thermodynamics*, 2nd ed., John Wiley & Sons, New York, 1992.

Moreno, J., J. Canada, and J. Bosca. Statistical and physical analysis of the external factors perturbation on solar radiation exergy, *Entropy*, 5, 452–466, 2003.

Mullet, L. B. The solar chimney overall efficiency, design, and performance, *International Journal of Ambient Energy*, 8(1), 35–40, 1987.

Muneer, T. *Solar Radiation and Daylight Models*, with a chapter on solar spectral radiation by C. Gueymard and H. Kambezidis, 2nd ed., Elsevier Butterworth-Heinemann, Oxford, U.K., and Burlington, MA, 2004.

Nichols, E. F., and G. F. Hull. A preliminary communication on the pressure of heat and light radiation, *Physics Review*, 13, 307–320 (1901).

Obynochnyi, A. N., I. I. Sventitskii, and L. Y. Yuferev. Method and device for determination of exergy of optical radiation in plant growing. Russian Patent Application RU 2005103491, 20050210, 2006.

Ozturk, H. Experimental determination of energy and exergy efficiency of the solar parabolic cooker, *Solar Energy*, 77(1), 67–71, 2004.

Padki, M. M., and S. A. Sherif. A mathematical model for a solar chimney, *Proceedings of 1992 International Renewable Energy Conference*, vol. 1, M.S. Audi (ed.), University of Jordan, Faculty of Engineering and Technology, Amman, Jordan, pp. 289–294, 1992.

Parrott, J. E. Theoretical upper limit to the conversion efficiency of solar energy, *Solar Energy*, 21(3), 227–229, 1978.

Parrott, J. E. Letters to the editor, *Solar Energy*, 22(6), 572–573, 1979.

Pastohr, H., O. Kornadt, and K. Gurlebeck. Numerical and analytical calculations of the temperature and flow field in the upwind power plant, *International Journal of Energy Research*, 28, 495–510, 2004.

Pasumarthi, N., and S. A. Sherif. Experimental and theoretical performance of a demonstration solar chimney model, Part I: mathematical model development, *International Journal of Energy Research*, 22, 277–288, 1998a.

Pasumarthi, N., and S. A. Sherif. Experimental and theoretical performance of a demonstration solar chimney model, Part II: experimental and theoretical results and economic analysis. *International Journal of Energy Research*, 22, 443–461, 1998b.

Perry, R. H., and D. W. Green. *Perry's Chemical Engineers' Handbook*, 7th ed., McGraw-Hill, New York, 1997.

Petela, R. Exergy of heat radiation. Ph.D. thesis, Faculty of Mechanical Energy Technology, Silesian Technical University, Gliwice (in Polish), 1961a.

Petela, R. Exergy of radiation of a perfect gray body, *Zeszyty Naukowe Politechniki Slaskiej, Energetyka* 5, 33–45 (in Polish), 1961b.

Petela R. Exergy of radiation radiosity. *Zeszyty Naukowe Politechniki Slaskiej, Energetyka* 9, 43–70 (in Polish), 1962.

Petela, R. Exergy of heat radiation. *Transactions of ASME, Journal of Heat Transfer*, 86(2), 187–192, 1964.

Petela, R. The problem of derivation of formula for heat radiation exergy, *Zeszyty Naukowe Politechniki Slaskiej, Energetyka* 50, 105–109 (in Polish), 1974.

Petela, R., and A. Piotrowicz. Exergy of plasma, *Archiwum Termodynamiki i Spalania*, 3, 381–391, 1977.

Petela, R. *Heat Transfer*, PWN, Warsaw (in Polish), 1983.

Petela, R. Some exergy properties of heat radiation, *Archiwum Termodynamiki*, 2, 189–193, 1984.

Petela, R. Exergetic analysis of atomization process of liquid, *Fuel*, 3, 419–422, 1984a.

Petela, R. Exergetic efficiency of comminution of solid substances, *Fuel*, 3, 414–418, 1984b.

Petela, R. Exergy analysis of processes occurring spontaneously, CSME Mechanical Engineering Forum, June 3–9, University of Toronto, vol. I, pp. 427–431, 1990.

Petela, R. Exergy of undiluted thermal radiation. *Solar Energy*, 74, 469–488, 2003.

Petela, R. Exergy analysis of the solar cylindrical–parabolic cooker, *Solar Energy*, 79, 221–233, 2005.

Petela, R. Influence of gravity on the exergy of substance, *International Journal of Exergy*, 5(1), 1–17, 2008.

Petela, R. An approach to the exergy analysis of photosynthesis, *Solar Energy*, 82/4, 311–328, 2008a.

Petela, R. Thermodynamic study of a simplified model of the solar chimney power plant, *Solar Energy*, 83, 94–107, 2009.

Petela, R. Gravity influence on the exergy balance, *International Journal of Exergy*, 6(3), 343–356, 2009a.

Petela, R. Thermodynamic analysis of chimney, *International Journal of Exergy*, 6(6), 868–880, 2009b.

Petela, R. Radiation spectra of surface, *International Journal of Exergy*, 7(1), 89–109, 2010.

Planck, M. *The Theory of Heat Radiation*, translation from German by M. Masius, Dover, New York, 1914; 2nd ed., 1959; reprinted in the *History of Modern Physics*, Vol. 11, American Institute of Physics, 1988.

Press, W. H. Theoretical maximum for energy from direct and diffuse sunlight, *Nature*, 264, 734–735, 1976.

Purves, W. K., D. Sadava, G. H. Orians, and H. C. Heller. *Life: The Science of Biology*, 7th ed., Sinauer Associates, Sunderland, MA, 2003.

Reis, A. H., and A. F. Miguel. Analysis of the exergy balance of green leaves, *International Journal of Exergy*, 3(3), 231–238, 2006.

Schlaich, J. *The Solar Chimney: Electricity from the Sun*, Edition Axel Menges, Stuttgart, 1995.

Schmidt, E. *Einführung in die technische Thermodynamik*, Springer-Verlag, Berlin, 1963.

Spanner, D. C. *Introduction to Thermodynamics*, Academic Press, London, 1964.

Swinbank, W. C. Long-wave radiation from clear skies, *Quarterly Journal of the Royal Meteorological Society*, 89, 339, 1963.

Szargut, J., and Z. Kolenda. Theory of co-ordination of the material and energy balances of chemical processes in metallurgy, *Archiwum Hutnictwa*, 2, 153, 1968.

Szargut, J., and R. Petela. *Exergy*, WNT, Warsaw (in Polish), 1965.

Szargut, J., and R. Petela. *Exergy*, Energija, Moscow (in Russian), 1968.

Szargut, J., D. R. Morris, and F.R. Steward. *Exergy Analysis of Thermal, Chemical, and Metallurgical Processes*, Hemisphere, New York, 1988.

Vermaas, W. An introduction to photosynthesis and its applications, *The World & I*, 158–165, 1998.

Von Backström, T. W, and A. J. Gannon. Compressible flow through solar power plant chimney, *Transactions of the ASME, Journal of Solar Energy Engineering*, 122(8), 139–145, 2000.

Von Backström, T. W., and T. P. Fluri. Maximum fluid power condition in solar chimney power plants: An analytical approach, *Solar Energy*, 80(11), 1417–1423, 2006.

Wall, G. Exergy, ecology, and democracy: Concepts of a vital society or a proposal for an exergy tax. In *ENSEC'93, International Conference on Energy Systems and Ecology*, Cracow, 5–9 July 1993, pp. 111–121, J. Szargut et al. (eds.); also 2nd European Congress on Economic and Management of Energy in Industry, Estoril, 5–9 April 1994; also www.exergy.se/goran/eed/.

Wright, S., M. Rosen, D. Scott, and J. Haddow. The exergy flux of radiative heat transfer for the special case of blackbody radiation, *International Journal of Exergy*, 2, 24–33, 2002.

Würfel, P. *Physics of Solar Cells: From Principles to New Concepts*, Wiley-VCH, Weinheim, 2005.

Yourgrau, W., and A. van der Merve. Entropy balance in photosynthesis, *Proceedings of National Academy of Science, Physics*, 58, 734–737, 1968.

APPENDIX

Fractional units, $n < 0$			Multiplied units, $n > 0$		
Prefix	Symbol	10^n	Prefix	Symbol	10^n
deci	d	10^{-1}	deca	da	10
centi	c	10^{-2}	hecto	h	10^2
milli	m	10^{-3}	kilo	k	10^3
micro	μ	10^{-6}	mega	M	10^6
nano	n	10^{-9}	giga	G	10^9
pico	p	10^{-12}	tera	T	10^{12}
femto	f	10^{-15}			
atto	a	10^{-18}			

A.2 Typical Constant Values for Radiation and Substance

Radiation

$a = 7.564 \times 10^{-16}$ J/(m^3 K^4)   universal constant

$h = 6.625 \times 10^{-34}$ J s   Planck's constant

$\sigma = 5.6693 \times 10^{-8}$ W/(m^2 K^4)   Boltzmann constant for black radiation

$C_b = 5.6693$ W/(m^2 K^4)   constant for black radiation

$c_0 = 2.9979 \times 10^8$ m/s   velocity of light in vacuum

$c_1 = 3.743 \times 10^{-16}$ W m^2 first Planck-law constant ($2\pi \times h \times c_0^2$)

$c_2 = 1.4388 \times 10^{-2}$ m K   second Planck-law constant

$c_3 = 2.8976 \times 10^{-3}$ m K   third Planck-law constant (in the Wien displacement law)

$c_4 = 1.2866 \times 10^{-5}$ W/(m^3 K^5)   fourth Planck-law constant

Substance

$k = 1.3805 \times 10^{-23}$ J/K   Boltzmann constant (general)

$R = 8314.3$ J/(kmol K)   universal gas constant

A.3 Application of Mathematics to Some Thermodynamic Relations

In thermodynamics, especially significant is the case in which the considered system contains only the medium. Such a case is called a *closed system*. It is assumed that the surrounding walls do not play a role and that both the potential and kinetic energy of the medium are neglected. In such a case the system energy is directly equal to the internal energy of the considered medium. The First Law of Thermodynamics for a closed system considered on a unitary variable basis (i.e., calorific properties are related to a unit of amount, e.g., unit mass, in case of radiation to unit volume), states differentially that the delivered heat dq is equal to the internal energy growth du and performed work $p \times dv$:

$$dq = du + p\, dv \qquad (A.1)$$

If the considered process is reversible, then heat can be expressed with temperature T and entropy s:

$$ds = \frac{dq}{T} \qquad (A.2)$$

and using both equations:

$$du = T\, ds - p\, dv \qquad (A.3)$$

Enthalpy h is defined as $h = u + pv$ and differentiating $dh = du + pdv + vdp$. Thus:

$$dh = T\, ds + v\, dp \qquad (A.4)$$

Then, considering internal energy as a function of specific volume and entropy, $u = f(s,v)$, from a mathematical viewpoint:

$$du = \left(\frac{\partial u}{\partial s}\right)_v ds + \left(\frac{\partial u}{\partial v}\right)_s dv \qquad (A.5)$$

or analogously for a function $h(s, p)$:

$$dh = \left(\frac{\partial h}{\partial s}\right)_p ds + \left(\frac{\partial h}{\partial p}\right)_s dp \qquad (A.6)$$

By comparison of equations (A.3) and (A.5):

$$\left(\frac{\partial u}{\partial s}\right)_v = T \qquad (A.7a)$$

$$\left(\frac{\partial u}{\partial v}\right)_s = -p \qquad (A.7b)$$

or comparing equations (A.4) and (A.6):

$$\left(\frac{\partial h}{\partial s}\right)_p = T \tag{A.7c}$$

$$\left(\frac{\partial h}{\partial p}\right)_s = v \tag{A.7d}$$

Also another group of relations, called Maxwell's relations, may be developed. For example, differentiating (A.7a) and (A.7b):

$$\left(\frac{\partial T}{\partial v}\right)_s = \frac{\partial^2 u}{\partial v\, \partial s} \tag{A.8}$$

$$-\left(\frac{\partial p}{\partial v}\right)_v = \frac{\partial^2 u}{\partial s\, \partial v} \tag{A.9}$$

As:

$$\frac{\partial^2 u}{\partial v\, \partial s} = \frac{\partial^2 u}{\partial s\, \partial v} \tag{A.10}$$

thus:

$$\left(\frac{\partial T}{\partial v}\right)_s = -\left(\frac{\partial p}{\partial s}\right)_v \tag{A.11a}$$

and in a similar manner other Maxwell's relations may be derived:

$$\left(\frac{\partial T}{\partial p}\right)_s = \left(\frac{\partial v}{\partial s}\right)_p \tag{A.11b}$$

$$\left(\frac{\partial p}{\partial T}\right)_v = \left(\frac{\partial s}{\partial v}\right)_T \tag{A.11c}$$

$$\left(\frac{\partial v}{\partial T}\right)_p = -\left(\frac{\partial s}{\partial p}\right)_T \tag{A.11d}$$

The specific heat c_v at constant volume, defined as $c_v = (\partial u/\partial T)_v$, and after using equations (A.7a), can be differentiated as follows:

$$c_v = \left(\frac{\partial u}{\partial s}\right)_v \left(\frac{\partial s}{\partial T}\right)_v = T\left(\frac{\partial s}{\partial T}\right)_v \tag{A.12}$$

In a similar manner the specific heat c_p at constant pressure, defined as $c_p = (\partial h/\partial T)_p$, can be differentiated as follows:

$$c_p = \left(\frac{\partial h}{\partial s}\right)_p \left(\frac{\partial s}{\partial T}\right)_p = T\left(\frac{\partial s}{\partial T}\right)_p \tag{A.13}$$

Another example of a mathematical relation applied to thermodynamics can be the formulae that use the measurable properties to obtain nonmeasurable properties. For example, consider internal energy as function of temperature and specific volume, $u = f(T, v)$. Differentiating the function one obtains:

$$du = \left(\frac{\partial u}{\partial T}\right)_v dT + \left(\frac{\partial u}{\partial v}\right)_T dv \qquad (A.14)$$

Differentiating equation (A.3) at constant T yields:

$$T\left(\frac{\partial s}{\partial v}\right)_T = \left(\frac{\partial u}{\partial v}\right)_T + p \qquad (A.15)$$

The Maxwell's relation (A.11c) can be used to replace the entropy derivative in (A.15), which after rearranging leads to the following relation:

$$\left(\frac{\partial u}{\partial v}\right)_T = T\left(\frac{\partial p}{\partial T}\right)_v - p \qquad (A.16)$$

Equation (A.14), after taking into account $(\partial u / \partial T)_v = c_v$, and equation (A.16), becomes finally:

$$du = c_v\, dT + \left[T\left(\frac{\partial p}{\partial T}\right)_v - p\right] dv \qquad (A.17)$$

In a similar manner the equation for calculation of enthalpy (only for substance) can be derived as:

$$dh = c_p\, dT + \left[v - T\left(\frac{\partial v}{\partial T}\right)_p\right] dp \qquad (A.18)$$

A.4 Review of Some Radiation Energy Variables

#	Variable	Energy Symbol	Energy Units	Energy Formula
1	Black radiation energy in a volume V, e.g., formula (5.13)	U	J	$Va T^4$
2	Density of black radiation energy in a volume V, e.g., formula (5.12)	u	J/m^3	$a T^4$
3	Density of monochromatic radiation in a volume V, depending on λ	u_λ	J/m^4	(3.12)

#	Variable	Symbol	Energy Units	Energy Formula
4	Energy emission of a surface area A, which is radiation of temperature T emitted into the forward hemisphere from the surface at temperature T and emissivity ε	E	W	$A\varepsilon\sigma T^4$
5	Emission density, which is the emission E related to the gray surface area, e.g., formula (3.22)	e	W/m^2	$\varepsilon\sigma T^4$
6	Emission density of a black surface, e.g., formula (3.21)	e_b	W/m^2	σT^4
7	Density of monochromatic emission of a black surface into the front hemisphere (within solid angle 2π) depending on λ, e.g., formula (3.13)	$e_{b,\lambda}$	W/m^3	$\dfrac{c_1}{\lambda^5\left(\exp\frac{c_2}{\lambda T} - 1\right)}$
8	Density of monochromatic emission of any surface (e.g., gray) into the front hemisphere (within solid angle 2π), e.g., formula (3.10)	e_λ	W/m^3	$e_\lambda = \dfrac{de}{d\lambda}$
9	Radiosity, which is the total surface radiation composed of emission and reflected radiation of different temperatures	J	J	(7.1)
10	Radiosity, density which is the radiosity related to the surface area, e.g., formula (3.8)	j	W/m^2	$j = \dfrac{J}{A}$
11	Directional radiation intensity, which expresses the total radiation propagating within a solid angle $d\omega$ and along a direction determined by the flat angle β with the normal to the surface, e.g., formula (3.27), (3.29)	i_β	$W/(m^2\ sr)$	$\dfrac{j}{\pi}\cos\beta$
12	Directional normal radiation intensity at $\beta = 0$, e.g., formula (3.28)	i_0	$W/(m^2\ sr)$	$\dfrac{j}{\pi}$
13	Directional normal ($\beta = 0$) black radiation intensity	$i_{b,0}$	$W/(m^2\ sr)$	(3.28)
14	Directional black radiation intensity	$i_{b,\beta}$	$W/(m^2\ sr)$	(3.29)
15	Directional normal monochromatic radiation intensity of nonpolarized (linearly polarized) radiation, depending on ν	$i_{0,\nu}$	$J/(m^2\ sr)$	(7.3)

#	Variable	Energy		
		Symbol	Units	Formula
16	Directional normal monochromatic radiation intensity of nonpolarized (linearly polarized) radiation, depending on λ	$i_{0,\lambda}$	W/(m³ sr)	(7.4)
17	Principal (smallest and largest) directional normal monochromatic components of radiation intensity of nonpolarized (linearly polarized) radiation	$i_{0,\nu,\min}$ $i_{0,\nu,\max}$	J/(m² sr)	(7.2)
18	Directional normal monochromatic intensity of black radiation linearly polarized propagating within unit solid angle, dependent on wavelength ν	$i_{b,0,\nu}$	J/(m² sr)	(7.9)
19	Directional normal monochromatic intensity of black radiation linearly polarized propagating within unit solid angle, dependent on wavelength λ	$i_{b,0,\lambda}$	W/(m³ sr)	(7.8)
20	Any radiation energy arriving from a certain surface A' in the considered surface A, introduced for general considerations	$j_{A'}$	W/m²	(7.10)

A.5 Review of Some Radiation Entropy Variables

#	Variable	Entropy		
		Symbol	Units	Formula
1	Radiosity entropy	S	W/K	(7.20)
2	Entropy density of a photon gas in the equilibrium state, residing in a system	s_S	J/(K m³)	(5.23)
3	Entropy density of radiation emitted by unit surface area of a body in all the directions of the front hemisphere in unit time	s	W/(m² K)	(5.24) (7.32)
4	Entropy of directional normal radiation intensity, which expresses the entropy passing within a unitary solid angle, in unit time and through a unitary surface area perpendicular to propagation direction	L_0	W/(K m² sr)	(7.21) (7.26)

#	Variable	Entropy		
		Symbol	**Units**	**Formula**
5	Principal (smallest and largest) mutually independent (incoherent), polarized at right angles to each other, values of the monochromatic component of the entropy of radiation intensity	$L_{0,\nu,min}$ $L_{0,\nu,max}$	$J/(K\ m^2\ sr)$	(7.21)
6	Entropy of monochromatic intensity of linearly polarized radiation dependent on frequency	$L_{0,\nu}$	$J/(m^2\ K\ sr)$	(7.22) (7.24)
7	Entropy of monochromatic intensity of linearly polarized radiation dependent on wavelength	$L_{0,\lambda}$	$W/(m^3\ K\ sr)$	(7.23) (7.25)
8	Entropy of emission emitted within solid angle $\omega \leq 2\pi$ in which L_0 is constant	S_ω	$W/(m^2\ K\ sr)$	(7.28)
9	Entropy of monochromatic directional normal intensity of linearly polarized black radiation propagating within unit solid angle and dependent on frequency	$L_{b,0,\nu}$	$J/(m^2\ K\ sr)$	(7.24) (7.30)
10	Entropy of monochromatic directional normal intensity for linearly polarized black radiation propagating within unit solid angle and dependent on wavelength	$L_{b,0,\lambda}$	$W/(m^3\ K)$	(7.25)
11	Entropy density of radiation emitted by the unit black surface area of a body in all the directions of the front hemisphere in unit time	s_b	$W/(m^2\ K)$	(7.31)
12	Entropy of radiosity density	s_j	$W/(m^2\ K)$	(7.37)
13	Entropy of radiosity density propagating within solid angle ω	$s_{j,\omega}$	$W/(m^2\ K\ sr)$	(7.38)
14	Entropy of radiosity density passing the unit control surface area A' in a space and in the unit time and falling upon the element dA of the considered surface A, introduced for general considerations	$s_{j,A'}$	$W/(m^2\ K)$	(7.33)

A.6　Review of Some Radiation Exergy Variables

#	Variable	Exergy		
		Symbol	Units	Formula
1	Exergy of photon gas, i.e., exergy of black radiation enclosed within a system	$b_{b,S}$	J/m^3	(5.29)
2	Exergy density of emission from a black surface	b_b	W/m^2	(6.8) (7.49)
3	Exergy density of emission from a gray surface of emissivity ε	b	W/m^2	(6.10) (6.13)
4	Exergy of radiosity density	$b \equiv b_{A'}$	W/m^2	(7.41)
5	Exergy of radiosity	$B \equiv B_{A' \to A}$	W	(7.42)
6	Exergy of nonpolarized, uniform, black radiation propagating within a solid angle ω	$b_{b,\omega}$	$W/(m^2 \, sr)$	(7.50)
7	Exergy of monochromatic black radiation propagating within an elemental solid angle $d\omega$ and within wavelength range $d\lambda$	$b_{b,\omega,\lambda}$	$W/(m^3 \, sr)$	(8.18)
8	Exergy of monochromatic radiation propagating within an elemental solid angle $d\omega$ and within wavelength range $d\lambda$	$b_{\omega,\lambda}$	$W/(m^3 \, sr)$	(8.25)
9	Incorrect exergy of enclosed black radiation according to Jeter, e.g., formula (9.11)	b_J	J/m^3	$b_J = a\left(T^4 - T_0^4\right)\left(1 - \dfrac{T_0}{T}\right)$
10	Incorrect exergy of enclosed black radiation according to Spanner, e.g., formula (9.9)	b_S	J/m^3	$b_S = \dfrac{a}{3}\left(3T^4 - 4T_0T^3\right)$
11	Exergy of enclosed black radiation according to Petela, e.g., formula (9.8)	b_P	J/m^3	$b_P = \dfrac{a}{3}\left(3T^4 + T_0^4 - 4T_0T^3\right)$
12	Exergy/energy ratio according to Petela	ψ		(9.7)
13	Exergy/energy ratio according to Jeter (incorrect)	η_J		(9.11)
14	Exergy/energy ratio according to Spanner (incorrect)	η_S		(9.5)

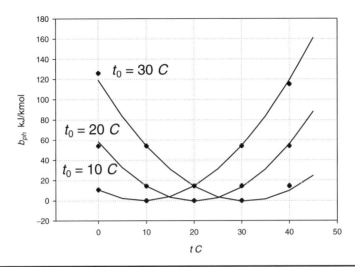

Figure A.1 Approximated physical exergy of water (from Petela, 2008a).

A.7 Exergy of Liquid Water

Exergy of liquid water b_w, kJ/kmol, is the sum $b_w = b_{ph} + b_{ch}$ of the physical part b_{ph} and chemical part b_{ch}, where $b_{ch} = R \cdot T_0 \cdot \ln{(1/\varphi_0)}$. Using the Szargut and Petela (1965b) diagrams the approximation formula for calculation of the physical exergy b_{ph} of liquid water is $b_{ph} = a + bt + ct^2$, where $a = -23.22 + 2.718 \cdot t_0 + 0.0675 \cdot t_0^2$, $b = 2.689 - 0.5787 \cdot t_0 + 0.00767 \cdot t_0^2$, and $c = 0.117 - 1.05 \cdot 10^{-3} \cdot t_0 + 2.7 \cdot 10^{-4} \cdot t_0^2 - 7.5 \cdot 10^{-6} \cdot t_0^3$ and where $t_0 = T_0 - 273$.

Acceptable accuracy of approximation, as shown in Figure A.1, is obtained within the ranges of the water temperature $t = 10 - 30°C$ and environmental temperature $t_0 = 10 - 30°C$. If $t = t_0$, then precisely $b_{ph} = 0$. Any negative values of a calculated b_{ph} result from the imperfectness of the approximation and should be rounded up to zero. Imperfectness is illustrated, e.g., by the values of the most inconvenient discrepancy, which occurs for the minimum of the curves. Instead of the required zero, the interpolation formula gives –0.6 kJ/kmol. However, the interpolation formula, even with such imperfectness, is useful because it allows for significant simplification of the computations.

Index